MOLECULAR-GENETIC MECHANISMS OF DEVELOPMENT

MOLECULAR-GENETIC MECHANISMS OF DEVELOPMENT

ZHORES A. MEDVEDEV
Laboratory of Molecular Radiobiology
Institute of Medical Radiology
Obninsk, USSR

Translated from Russian by Basil Haigh
Cambridge, England

℗ PLENUM PRESS • NEW YORK–LONDON • 1970

The original Russian text, published by Meditsina Press in Moscow in 1968 for the Academy of Medical Sciences of the USSR, has been corrected by the author for this edition. The translation is published under an agreement with Mezhdunarodnaya Kniga, the Soviet book export agency.

Медведев Жорес Александрович

МОЛЕКУЛЯРНО-ГЕНЕТИЧЕСКИЕ МЕХАНИЗМЫ РАЗВИТИЯ

MOLEKULYARNO-GENETICHESKIE MEKHANIZMY RAZVITIYA

MOLECULAR-GENETIC MECHANISMS OF DEVELOPMENT

Library of Congress Catalog Card Number 71-80754
SBN 306-30403-1

© 1970 Plenum Press, New York
A Division of Plenum Publishing Corporation
227 West 17th Street, New York, N.Y. 10011

United Kingdom edition published by Plenum Press, London
A Division of Plenum Publishing Company, Ltd.
Donington House, 30 Norfolk Street, London W.C.2, England

All rights reserved

No part of this publication may be reproduced in any form
without written permission from the publisher

Printed in the United States of America

Preface

Although as part of my general plan, this book is a continuation of my earlier monograph "Protein Biosynthesis and Problems of Ontogenesis,"* published in 1963, in all other respects it is an independent work.

The earlier monograph was devoted to the analysis of many of the aspects of the problem of protein biosynthesis, and problems of inheritance and development were discussed only insofar as they are derivatives of the problems of biosynthesis.

The complex act of protein biosynthesis, comprising autoreproduction of the genetic material (DNA), formation of the templates of protein synthesis (messenger RNA), synthesis of amino acid carriers (transfer RNA), formation of ribosomes and polysomes, activation of amino acids, and so on, was examined in the previous monograph not merely from the standpoint of interaction between the components of this system, but also from that of their manifestation in actual biological systems during morphogenesis and aging of the organism. However, both morphogenesis and aging were investigated very generally, simply as models, without any detailed analysis of their specific features and complexity. The present book is therefore a logical continuation of its predecessor. It rests largely on a comprehensive analysis of the molecular-genetic and biochemical aspects of development and differentiation of living organisms, and questions of protein biosynthesis are discussed briefly and generally, and only so far as is necessary for fulfillment of the primary task.

---
* Zh. A. Medvedev, 1966, Protein Biosynthesis and Problems of Heredity, Developement, and Ageing, Oliver and Boyd, Ltd. Edinburgh and Plenum Press, N. Y.

Later I intend to prepare another monograph to examine the molecular-genetic basis of aging and this will be to some extent a continuation of the present book.

Although I outlined the order of my studies of ontogenesis long ago, it has proved to be absolutely in agreement with modern tendencies in the development of biochemistry and genetics. Several years ago, when I was starting to prepare my previous book, the study of the vanguard of biochemistry, and its progress had led to significant developments in the whole of biological science. At the present time many problems which then appeared highly complex have been solved in principle, and it is now mainly a question of filling in the details in the general picture of protein biosynthesis.

Meanwhile, the molecular mechanisms of development and differentiation have now become the dominant problem, and investigations of morphogenesis from this standpoint have increased in number by leaps and bounds. It is perfectly logical to suppose, once the question of why a complete and perfect biological system develops has been solved in principle, that we naturally will want to know how a particular level of development is preserved and maintained for a definite period of time, characteristic of each species, which we call the "life span," and why the organism gradually accumulates the group of changes which form the basis of aging.

The author is grateful to Doctor of Biological Sciences G. V. Lopashov (Institute of Biology of Development, Academy of Sciences of the USSR) for his great help in the preparation of the manuscript for publication and for his valuable advice and criticisms. Gratitude is also due to Doctors of Biological Sciences N. V. Timofeev-Resovskii, V. I. Korogodin, and N. V. Luchnik, and to O. A. Khoperskaya, of the staff of the Institute of Biology of Development, Academy of Sciences of the USSR, for their valuable remarks which were taken into account during preparation of the manuscript for the press. A special word of thanks is due to M. N. Medvedeva for technical help in preparation of the manuscript.

Zh. A. Medvedev

Contents

CHAPTER 1. The Elementary Acts of Morphogenesis: Processes of Biosynthesis of Informational Macromolecules. The Biochemical Mechanisms of Transfer of Genetic Information 1
Introduction. 1
§ 1. DNA Synthesis as a Process of Reproduction of Genetic (Programming) Information. 3
§ 2. RNA Synthesis as a Process of Transfer of Genetic Information (Transcription) from DNA to Systems of Protein Synthesis 5
§ 3. Protein Synthesis as a Process of Translation of Genetic Information into a Functionally Active State. . 10
  a) Activation of Amino Acids and Their Fixation by Adaptor Molecules of sRNA ...., 11
  b) Ribosomes and Polyribosomes (Polysomes), Their Functions in the Structural Organization of Protein Biosynthesis 14
  c) Biosynthesis of Proteins and the Problem of the Genetic Code 18
Literature Cited. 20

CHAPTER 2. Molecular-Genetic Mechanisms of Reproduction and Morphogenesis of Viruses 23
Introduction. 23
§ 1. Mechanisms of Morphogenesis of the Simplest Ribonucleoprotein (Two-Component) Viruses of Plants 24
§ 2. Mechanisms of Morphogenesis of RNA-Containing Viruses of Animal Cells 30

§ 3. Morphogenesis of RNA-Phages (Viruses of Bacteria) . 36
§ 4. Mechanisms of Morphogenesis of Small DNA-Phages Containing Single-Stranded DNA 41
§ 5. The Molecular Mechanisms of Morphogenesis of Animal DNA-Viruses 46
§ 6. Molecular Mechanisms of Morphogenesis of Large DNA-Phages 49
   a) Regulation of Temporal Sequence of Synthesis of Early and Late Proteins Coded by Different Segments of the Same Polynucleotide (Polycistron) of Phage DNA 53
   b) Genetic Function of the Phage DNA Macromolecule during Phage Replication and Reproduction 61
   c) Study of the Nature of Assembling of the Phage Particle from Its Various Components 71
Conclusion 72
Literature Cited 73

CHAPTER 3. Molecular-Genetic Mechanisms of Morphogenesis and Intracellular Biochemical Differentiation of Monochromosomal Bacterial Cells 79
Introduction 79
§ 1. Types of Morphogenetic Reactions in Bacteria 80
§ 2. Substrate-Induced Synthesis of Bacterial Enzymes as a Morphogenetic Process 82
   a) The Principle of Enzymic Induction 82
   b) Concept of the Operon, a Group of Linearly and Functionally Connected Genes, Exemplified by Synthesis of β-Galactosidase. Regulator Genes and Repressor Genes. Scheme of the Lactose Operon 83
§ 3. The Principle of Combination of Successively Working Genes Responsible for Synthesis of Successively Working Enzymes Concerned Jointly with the Formation of Various Metabolites into Linear Structures or Operons 88
   a) The Histidine Operon 88
   b) The Tryptophan Operon 91
   c) Other Integrated Single-Operon Structural Combinations of Genes 92

CONTENTS

§ 4. Aspects of the One Operon—One Molecule of Messenger RNA Theory. Modulators of the Velocity of Synthesis. 93

§ 5. Genetic Determination of Functionally Interconnected and Successive Reactions by Structurally Separate Genes and by Groups of Genes Located in Different Parts of the Bacterial Chromosome. Arginine Synthesis by *E. coli* and Cysteine Synthesis by Salmonella 96

§ 6. Possible Nature of Genetic Repressors and Mechanisms of Repression of Operons and Cistrons of the Bacterial Chromosome. 99

§ 7. Functioning of the Bacterial Chromosome as Carrier of the Entire Program of Morphogenesis and Metabolism of Bacterial Cells 104
  a) Genetic Map of the Bacterial Chromosome 104
  b) Biochemical Structure of the Chromosome of *E. coli* and Its Replication 105
    1) Onset of Replication of *E. coli* Chromosome and its Connection with the Cell Cycle 106
    2) Rate of Replication of DNA of the Bacterial Chromosome and Its Regulation 107
    3) Molecular Mechanism and Linear Sequence of the Process of Replication of the *E. coli* Chromosome from the Starting Point 108
    4) Models of Regulation of the Replication Process of the *E. coli* Chromosome 111
  c) Replication of the Chromosome of *B. subtilis* 114
  d) Transcription of Genetic Information from DNA of the Bacterial Chromosome. 115

§ 8. Differences in Development of Genetic Information of the Bacterial Chromosome Connected with the Cell Cycle and Morphogenetic Phenomena. 117

Literature Cited 121

CHAPTER 4. Molecular-Gentic Systems Controlling Morphogenesis and Differentiation in Multicellular Organisms. Molecular Structure of Chromosomes of the Cell Nucleus and Structural and Biochemical Features of Genetic and Functional Differentiation of Chromosomes 127

Introduction.................................... 127
§ 1. Structural Organization of Genetic Elements of the
     Nucleus and Chromosomes....................... 129
§ 2. Reproduction (Replication) of Genetic Structures of
     the Cell Nucleus.............................. 139
§ 3. Asynchrony of DNA Replication in Different
     Parts of Chromosomes and in Different
     Chromosomes and an Indication of Linear Genetic
     and Functional Heterogeneity of Chromosomal DNA ... 142
  a) Asynchrony of Linear Replication of DNA at
     Different Loci of the Individual Chromosome......... 144
  b) Asynchrony of DNA Replication in Different
     Chromosomes of Haploid and Diploid Set 148
§ 4. Functional (RNA and Protein Synthesis),
     Morphological, and Genetic Differentiation
     of Chromosomes. The Study of Puffs on Polytene
     Chromosomes and Dynamics of Distribution
     in Connection with Morphogensis and Specialization
     of Zones of Active RNA Synthesis in Chromosomes ... 153
  a) Dynamics of Formation and Disappearance of
     Chromosomal Puffs in Connection with
     Morphogenetic Processes 158
  b) Experimental Changes in the Pattern and
     Activity of Puffs Caused by Hormones,
     Chemical Agents, and Environmental Factors 164
  c) Biochemical Specificity of Chromosomal Puffs
     and Characteristics of Forms of RNA Synthesized
     by Puffs..................................... 167
§ 5. Study of Differentiation of RNA Synthesis by Loci
     and in Time in Lampbrush Chromosomes of
     Vertebrate Oocytes............................ 169
§ 6. Differentiation of the Lampbrush Type Y
     Chromosome of the Spermatocytes of Some
     Drosophila Species............................ 178
Literature Cited.................................. 178

CHAPTER 5. Biochemical Realization of the Morphogenetic
     Program. Changes in Proteins and Nucleic Acids
     During Processes of Differentiation and Development.. 183
Introduction..................................... 183

§ 1. Morphogenetic Changes in the Protein Spectrum
of Cells and Tissues 184
 a) Genetic Control of Synthesis of Individual Proteins
in Cells and Tissues 185
  1) Genetic Control of the Structure and Synthesis of
  Hemoglobins 185
  2) Genetics of Lactate Dehydrogenase and Some
  Enzymes 189
 b) Quantitative and Qualitative Changes in the
 Pattern of the Structural Proteins of Cells
 during Morphogenesis and Differentiation 192
 c) Morphogenesis of the Plasma Proteins 197
 d) Biochemical Morphogenesis of the Hemoglobins.
 Genetic Mechanisms of the Change from
 Embryonic Hemoglobin to Adult Forms 200
  1) Structure and Functions of Fetal Hemoglobin 201
  2) Morphogenetic Switchover from the System
  of Synthesis of Fetal Hemoglobin to Adult Hemoglobin. . 202
  3) Phylogenetic Distribution of the Change in
  Hemoglobin Forms with Morphogenesis 206
  4) Genetics of the Control System of Morphogenetic
  Switchover from HbF Synthesis to HbA Synthesis
  and its Congenital Anomalies 207
 e) Morphogenetic Changes in the Pattern of
 Enzymes and Isozymes in Tissues and Organs 218
  1) Morphogenetic Changes in Patterns of Different
  Enzymes 218
  2) Morphogenetic Changes in Isozyme Groups and
  Their Significance in Differentiation 226
  3) Regulation of Enzyme Activity in Ontogenesis 230
 f) The Immunology of Development 232
§ 2. Morphogenetic Changes in the Pattern of
Ribonucleic Acids Synthesized in the Nucleus 233
 a) Changes in the RNA Pattern during Oogenesis
 and Early Embryogenesis 233
 b) Pattern of Ribonucleic Acids during Differentiation
 Processes. Appearance of Tissue Differences in
 RNA Composition 241
§ 3. Biochemical Composition of DNA during
Morphogenesis and Embryonic Differentiation 245

Literature Cited... 249

CHAPTER 6. Chromosomal Proteins and Their Role in
    Regulation of Selective Gene Activity During
    Differentiation and in Specialized Cells 259
Introduction... 259
§1. Chemical Characteristics of Histones and
    Protamines and the Nature of Their Complexes
    with DNA in the Chromosomes 261
§2. Heterogeneity and Specificity of Histones in
    Connection with Their Possible Genetic Function..... 266
    a) Heterogeneity of Histones in Relation to
       Amino Acid Composition, Molecular Weight, and
       Other Biochemical Properties 267
    b) Tissue and Morphogenetic Differences between
       Patterns of Heterogeneity of Histones (Tissue
       Specificity of Histones) 270
    c) Effect of Histones on the Transcription of Genetic
       Information of DNA (on Synthesis of RNA and
       Proteins by Chromosomal Structure). The Histones
       as Repressors................................ 273
    d) Biosynthesis of Histones as a Genetic Problem 285
       1) Theoretical Aspect of the Problem of
          Morphogenetic Repressors..................... 285
       2) Biosynthesis of Histones
    e) Experimental Approaches to Determination of the
       Role of Histones in Morphogenetic Processes 288
    Conclusion. Discussion of the Genetic Function of
       Histones...................................... 288
Literature Cited... 290

CHAPTER 7. Induction of Differentiated Activity of the
    Genes by Cytoplasmic and Chromosomal Factors
    and its Role in the General Organization of
    Morphogenesis................................. 295
Introduction... 295
§1. Effect of the Cytoplasm on the Character of Functional
    Differentiation of Nuclear Structures............... 296
    a) Differentiation of the Cytoplasm During Cleavage
       of the Oocyte and its Morphogenetic Importance
       (Ooplasmic Segregation) 298

b) Nuclear Transplanatation Experiments in the Study of
   Primary Mechanisms of Differentiation 303
§ 2. Molecular-Genetic Aspects of the Phenomena of
   Embryonic Induction 309
§ 3. Hormonal Control of Differentiation and the
   Influence of Hormones on Functional Activity of
   Genetic Systems 320
§ 4. Mechanism of Rapid Induction at the Cytoplasmic
   Level. A Possible Model 327
§ 5. Intrachromosomal Factors Determining the Level of
   Activity and Pattern of Function of Genes and Gene
   Systems 327
§ 6. Periodicity and Rhythm of Cell Biochemical
   Activity and its Genetic Programming.
   The Genetic Clock 334
Literature Cited 337

CHAPTER 8. Molecular Mechanisms Programming
   Morphogenesis and Differentiation.
   A Theoretical Analysis 343
Introduction 343
§ 1. Outlines of the Problem of Programmed
   Morphogenesis. Questions to be Answered 344
§ 2. Viruses and Bacteria; What Is Present and What Is
   Not in Their Genetic Systems Regulating and
   Programming Development 351
   a) Viruses 352
   b) Bacteria 356
§ 3. Molecular-Genetic Mechanisms of Morphogenesis
   and Differentiation in Multicellular Organisms 360
   a) Formulation of the General Oulines of a Basic
      Hypothesis 360
   b) Hypotheses of Mechanisms of Programmed
      Differentiation 365
   c) Theoretical Simulation of the Molecular Mechanisms
      of Morphogenesis and Differentiation. Concluding
      Remarks 379
Literature Cited 389

CHAPTER 9. Recent Advances in the Study of Molecular-
Genetic Mechanisms of Development 391

§ 1. New Material on the Mechanism of Transfer
of Genetic Information................................ 391
§ 2. New Data Concerning Mechanisms of
Morphogenesis of Viruses 392
§ 3. New Data on the Molecular Mechanisms of
Regulation of the Morphogenesis of Bacteria........ 394
§ 4. Functional Activity of the Chromosomes in
Morphogenesis and Differentiation.................. 396
§ 5. New Facts Concerning Changes in Proteins
and Nucleic Acids during Morphogenetic Processes... 400
§ 6. Chromosomal Proteins as Genetic Regulators 402
§ 7. Induction and Repression of Genetic Loci
During Differentiation............................... 404
Literature Cited... 406
Index... 411

Chapter 1

The Elementary Acts of Morphogenesis: Processes of Biosynthesis of Informational Macromolecules. The Biochemical Mechanisms of Transfer of Genetic Information

Introduction

    Every complex process consists of simpler, elementary acts organized in a certain fashion in space and time. Before considering the nature of mechanisms controlling the proper spatial and temporal organization of the elementary acts and processes which together form the complex process of development and differentiation, we naturally must learn something about the nature of these elementary acts of development.

    These elementary processes of morphogenesis at all levels of evolution are acts of biosynthesis of macromolecules: proteins, ribonucleic acids (RNA), and desoxyribonucleic acids (DNA), because all biological systems consist basically of these substances in their various forms and combinations.

    Morphogenesis, the principal "creative" phase of individual development of the organism, is the most interesting, the most important, and the most complex manifestation of life.

    Each living organism, fantastically complex yet composed of simple elements, is created rapidly and almost perfectly in the course of morphogenesis from a single cell under the influence of genetic information contained in the original cell, under favorable conditions for its nutrition.

The possibility of such a process was pure speculation until the material, molecular nature of this information was established. But the problem of morphogenesis is not solved even by an understanding of how a vast quantity of information, of encyclopedic proportions, can be concentrated in the form of nucleotide combinations in the small mass of DNA in the chromosomes. The translation of this information in the required volume, at the essential moment of development, and at the essential place is one of the chief problems in morphogenesis. The molecular genetic "memory" of the cell is not simply a memory, not simply an organizer of proper functioning of cells, but a controlling system of development, and this is its principal feature.

If we are to study the problem of development logically and effectively from this standpoint, it is very important that we adhere as far as possible to the correct and interconnected sequence of events in our examination. In my opinion it is most convenient to commence this examination with the study of forms of organization of molecular genetic material into systems responsible for the programming of purposive temporal and spatial interaction between tens, hundreds, and even thousands of genes in organisms at various stages of evolutionary development of life—in viruses, bacteria, and multicellular organisms. It is during the study of multicellular organisms that we encounter the typical form of morphogenesis accompanied by cell differentiation.

However, in all forms of living beings, with the exception of the simplest ribonucleoprotein viruses, the elementary unit of genetic programming consists of a system of processes embodied in the classical scheme DNA → RNA → protein, with all its attributes in the form of ribosomes, polyribosomes, transfer RNA, activating enzymes, and so on. A short, general examination of these initial acts of morphogenesis is naturally essential before problems of development and differentiation can be analyzed directly.

Because of the subordinate character of this material on the synthesis of proteins and nucleic acids, I do not propose to examine the problems and difficulties, or to paint a complete picture of all the relevant issues. I shall simply discuss briefly, in the manner of a terminological dictionary, the more important aspects of the present situation in this field on the assumption that a reader, wishing to obtain a deeper understanding of the problem of morphogenesis, must be familiar with the fundamental problems of synthesis of proteins and nucleic acids, in which case this chapter will provide him with a general introduction to the remainder of the book.

§ 1. DNA Synthesis as a Process of Reproduction of Genetic (Programming) Information

The desoxyribonucleic acids (DNA) are the principal carriers of specific genetic information, the material from which genes, discrete units of this information, are built. The discrete character of the genes located within the limits of DNA as a chemical compound is one of its special functional properties, because one DNA macromolecule can be used to record information of many genes in essentially the same way as one magnetic tape can record several different texts simultaneously.

DNA synthesis takes place by a mechanism of complementary autoreplication. DNA is present in nearly all biological systems (with the exception of certain phages) as a double helix, formed by poly-desoxyribonucleotides, linked together by complementary hydrogen bonds between base pairs (A-T) (G-C), twisted around a common axis. Under the influence of a special enzyme, polymerase, it can unwind into single polynucleotides, each of which becomes a template for formation of the complementary chain. This process is shown diagrammatically in Fig. 1.

This scheme of semiconservative replication, based on the classical DNA model of Watson and Crick (1953), with subsequent refinements (Pauling and Corey, 1956), corresponds to recent experimental findings, and its chemical and biochemical principles are well known and have been incorporated in all the textbooks of organic chemistry, biochemistry, and genetics. There is no need, therefore, to describe here the nature of the internucleotide (intermonomer) bonds, models of the tertiary structure of DNA, the results of its physical and physico-chemical study, and other material relevant to DNA structure.

The upper limit of the molecular weight of native DNA mole-

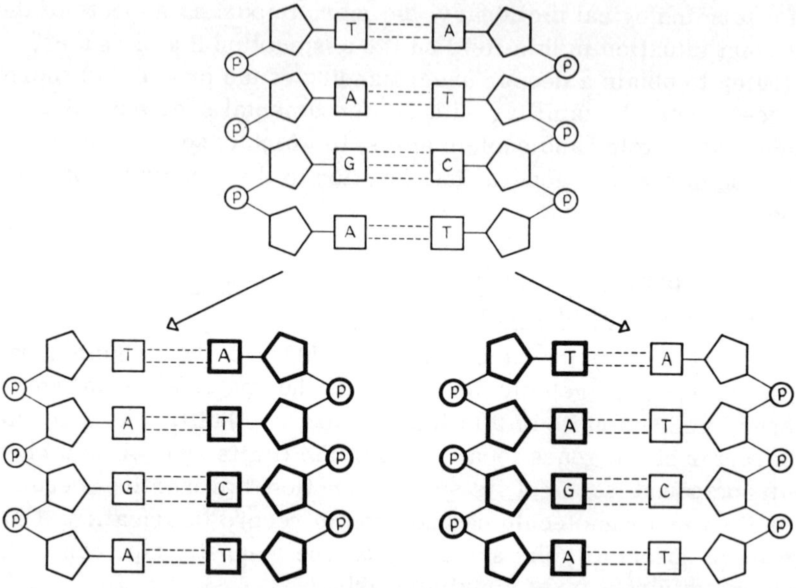

Fig. 1. Scheme showing replication of DNA based on the Watson and Crick model. Thick lines denote polynucleotide chains consisting of two newly synthesized daughter chains. Complementary base-pairs (adenine-thymine and guanine-cytosine) are indicated by dotted lines.

cules has not yet been determined exactly. The DNA macromolecule in the head of T4 bacteriophage has a molecular weight of about 160×10^6. If, however, the bacterial chromosome is based on a continuous DNA helix, as is sometimes assumed, the molecular weight of such a DNA must be 10 or 100 times greater than this.

Replication of the DNA molecule by the complementary duplication of polynucleotide chains satisfies at least the fundamental condition for existence of living matter in all its forms: reproduction and preservation of genetic information, recording nature by means of a nucleotide alphabet or code. The elementary units of this code are known as codons, groups of three nucleotides, each of which codes the position of one amino acid residue in the polypeptide chain.

The functional organization of all biological processes is associated with a number of different proteins, each possessing a unique structure which determines a particular function or reaction in the complex system of metabolism. Reproduction of the structure of protein chains is known to be dependent on information obtained from successive short polynucleotide segments of DNA in which groups of three nucleotides predetermine the composition and position of each amino acid residue.

In the double helical DNA molecule only one of the chains carries "protein" information and, if functioning normally, participates in protein synthesis (through the stage of messenger RNA). The second complementary chain has a purely replicative function by reproducing the messenger chain during DNA synthesis. Concrete forms of this subdivision of function between the two DNA chains in organisms at different levels of development will be examined in later chapters.

These functions of DNA are associated primarily with the control of RNA and protein synthesis. DNA synthesis is correlated as a rule with cell proliferation and precedes cell division.

The immediate substrate for DNA synthesis by DNA-polymerases consists of nucleoside triphosphates, and an essential condition for this synthesis is the presence of primer DNA, along whose unwound polynucleotides new polynucleotide chains are built.

The whole problem of DNA biosynthesis *in vivo* and *in vitro* is one of considerable extent and complexity and we have dealt with it here only in a very general manner. We shall not examine many aspects of this problem both because they have no direct bearing on the purpose of this book and also because the excellent surveys and monographs published every year (Tikhonenko, 1965; Vol'-kenshtein, 1970; Ingram, 1965; Zbarskii and Debov, 1967) make a more comprehensive study of the problem of DNA synthesis superfluous in this book.

Those aspects of DNA synthesis which have a direct bearing on our analysis of mechanisms of morphogenesis (replication of DNA during reproduction of viruses and phages, during division of the bacterial chromosome, during mitosis, and so on) will be examined in the appropriate chapters.

§ 2. RNA Synthesis as a Process of Transfer of Genetic Information (Transcription) from DNA to Systems of Protein Synthesis

The cell DNA, as the main storehouse of specific genetic information, utilizes it for controlling biochemical processes and processes of morphogenesis and regulation through a system of discrete information carriers, namely molecules of ribonucleic acids (RNA) which translate this information into different types of proteins performing specific functions.

Because of the infinite variety of their biochemical, physicochemical, and steric (configuration) properties and potentialities, proteins play the main role in the direct building operations of morphogenesis and cell specialization and also in their day-to-day biochemical, physiological, and functional activity.

There are three principal types of RNA with widely different, but related functions: messenger RNA (abbreviation mRNA); ribosomal RNA (abbreviation rRNA); and soluble or transfer RNA (abbreviations sRNA, tRNA). The functions of these three types of RNA are described below.

Messenger RNA, formed by structural genes (segments of DNA) is the active template for synthesis of protein molecules (polypeptide chains). This RNA is complementary in the sequence of its nucleotides to particular genetic segments of one of the polynucleotides of DNA, and because of this it receives (transcribes) its information. This information is later translated in a special biochemical system of decoding into the form of a definite amino acid sequence.

Ribosomal RNA, also formed in the nucleus (mainly in the nucleolus) under the control of special genetic loci, consists of two types of polymer molecules (molecular weight $0.6-0.7 \times 10^6$ and $1.3-1.7 \times 10^6$) and participates mainly in the formation of the specific morphological structure of ribosomes and is not used as template for protein synthesis. This RNA does not possess tissue specificity and has very weak species-specificity.

Transfer (acceptor) RNA is found in the cytoplasm and nuclear juice as fractions of comparatively small molecules (molecular weight about 25,000), heterogeneous in their nucleotide composition. The functions of these ribonucleic acids are concerned with the transfer of activated amino acids to suitably oriented messenger RNA molecules (templates), thereby permitting interaction between amino acids in the specific order in which they must exist in a given polypeptide chain. This is achieved by having a special group of three nucleotides (an anticodon) in each sRNA molecule, the composition of its nucleotides being complementary to the three-nucleotide coding groups (a codon) of the messenger RNA. To correspond to this function, the number of functional sRNA fractions is equal to the number of amino acids participating in protein synthesis, and in each fraction there are subfractions in connection with what is termed degeneracy of the code (the existence of several codons for one amino acid).

We shall examine the character of interaction between all these amino acid fractions during protein synthesis a little later and here we shall consider the question of RNA synthesis under the control of DNA (DNA-dependent RNA synthesis). Our examination of this problem, which is now one of considerable extent, will also be very general because many aspects of RNA synthesis associated with morphogenesis of viruses, bacteria, and multicellular organisms will be examined in later chapters, and this section of the book is only a general introduction to this material. General aspects of the structure and synthesis of RNA, and especially of messenger RNA, are examined along modern lines in a series of surveys and monographs (Spirin, 1963; Bresler, 1966; Georgiev, 1964, 1965; Osterman, 1965; Tongur, 1965; Hurwitz and August, 1963; Ingram, 1965; Cohen, 1966).

Synthesis of RNA, especially messenger RNA, by the chromosomes is a process of transcription of all genetic information contained in the genome into discrete, functionally active messenger molecules, which subsequently use it to operate the specific metabolism of various cells.

Ribosomal RNA (both fractions) does not possess tissue specificity, and a small series of identical genetic loci can therefore maintain synthesis of ribosomal RNA for all cells of the body. The mass of DNA complementary to ribosomal RNA in this case is very small (about 0.2%; Yankofsky and Spiegelman, 1962, 1963); this DNA fraction is mainly concentrated in the nucleolus, where the ribosomes are formed (McConkey and Hopkins, 1964).

Transfer RNA (sRNA) likewise does not possess tissue specificity, and identical forms of sRNA can be used equally successfully in different cells. However, the total number of fractions and of corresponding genetic loci is much greater for sRNA than for rRNA, because each triple nucleotide codon (and more than 50 have been discovered) has an anticodon in the form of a particular locus in one of the sRNA fractions. The total number of codons may be 64, but some do not carry information (nonsense triplets), marking boundary zones in the templates. However, because of the low molecular weight of the polynucleotides of sRNA, a comparatively small part of the cell genome can provide as many as 50 different types of these molecules in the cells.

A large part of the genome (DNA of the chromosome) is specialized for synthesis of messenger RNA, although the absolute content of this type of RNA usually does not exceed 2-4% of the total stock of RNA in the cell. The pattern of messenger RNA fractions is tissue-specific and corresponds to the spectrum of proteins synthesized in the cells. Thousands of types of RNA, differing in nucleotide sequence, are therefore possible and corresponding complementary loci exist in the composition of the DNA. It has now been precisely established that the manifestation of genetic activity of DNA at all levels of development of living matter can be reduced ultimately to the synthesis of molecules of RNA, principally messenger RNA, by genetic loci in the DNA molecule, and these RNA molecules subsequently participate in protein (polypeptide) synthesis.

Under normal conditions, *in vivo*, only one of the two polynucleotides of the DNA double helix is concerned in this synthesis (*in vitro*, when partially denatured DNA serves as template, both polynucleotides of DNA are concerned in it).

The complementarity of messenger RNA synthesized *in vivo* relative to one of the DNA polynucleotides has been demonstrated in many investigations by different methods (comparative study of nucleotide composition, sequence of nucleotides, and molecular hybridization).

In this case RNA is synthesized only in the presence of all four ribonucleoside triphosphates, magnesium and manganese ions, and the enzyme RNA polymerase (sometimes called DNA-dependent RNA polymerase, to distinguish it from certain forms of virus-specific RNA polymerase responsible for RNA synthesis in the presence of primer RNA).

During formation of complementary RNA the first stage of

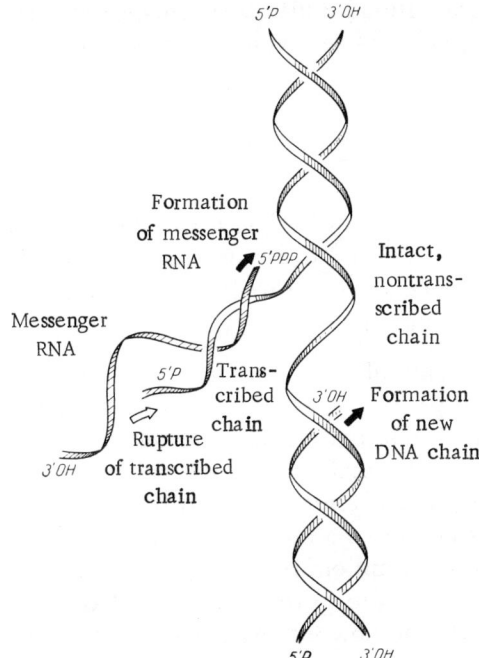

Fig. 2. Model of DNA-dependent RNA synthesis. Synthesis takes place on only one DNA chain (Jones and Truman, 1964).

synthesis is uncovering of part of a single polynucleotide of the DNA double helix with the formation of a mixed DNA/RNA hybrid. Free RNA does not appear until the ratio between DNA and RNA in the hybrid reaches a certain value (Chamberlin and Berg, 1964; Sinsheimer and Lawrence, 1964; Warner et al., 1963). Biological activity of the complementary RNA synthesized in this system, as a specific template for protein synthesis, was clearly demonstrated by the work of Bonner and co-workers (1963). By bringing together DNA-dependent RNA synthesis (using chromatin of pea seedlings as source of DNA) and ribosomal fraction of *Escherichia coli* together into one system, these workers observed synthesis of specific proteins typical of pea seedlings. DNA-dependent RNA synthesis also lies at the basis of reproduction processes in mitochondria and chloroplasts (Schweiger and Berger, 1964; Gibor and Granik, 1964). These cell organelles possess their own genetic system and have the power of intracellular multiplication. A scheme of DNA-dependent RNA synthesis is given in Fig. 2.

§ 3. Protein Synthesis as a Process of Translation of Genetic Information into a Functionally Active State

Intranuclear synthesis of the three forms of RNA (mRNA, rRNA, and sRNA) which we examined above, is essentially the synthesis of components which interact in the cytoplasm during synthesis of protein molecules.

All these forms of RNA are components of a complex protein-synthesizing system producing activation of amino acids, transferring them to templates, and bringing about interaction between amino acids in a strict order programmed by the corresponding gene. Besides different forms of RNA, this system also includes enzymes activating amino acids and forming acyladenylates of these compounds. These same enzymes attach the amino acid residue of the aminoacyladenylate to transfer RNA (tRNA or sRNA), and so, in accordance with the new international classification of enzymes, they have been named according to the type of amino acid which they activate (seryl-sRNA-synthetase, alanyl-sRNA-synthetase, valyl-sRNA-synthetase, tyrosyl-sRNA-synthetase, and so on). The protein-synthesizing system also includes enzymes which transfer complexes of sRNA with amino acid to mRNA templates, and ribosomes, consisting of a group of ribosomal proteins and rRNA. For such a system to function, it naturally must also have substrates and generators of energy (ATP and GTP), Mg^{++} and enzymes, and substrates for renewal of the universal terminal CCA group of different types of sRNA molecules.

I do not propose to examine further those fundamental aspects of the general problem of protein synthesis, either because they are not relevant to the present issue or because the present state of this problem has been fully analyzed in recent monographs and surveys (Medvedev, 1963, 1965, 1966; Bass and Gvozdev, 1965; Campbell, 1965; Sisakyan and Gladilin, 1965).

We shall examine current concepts and basic facts concerning the principles of interaction between components of the protein-synthesizing system only in the most general manner to avoid the necessity of returning to this question in our later discussion of special morphogenetic problems in cases when, in order to explain some of the molecular-genetic mechanisms of morphogenesis, facts and concepts dealing with mechanisms of protein biosynthesis are mentioned.

a) Activation of Amino Acids and Their Fixation by Adaptor Molecules of sRNA. As a rule, amino acids must be activated when they enter the cell, because the formation of peptide bonds during synthesis of polypeptide chains raises a definite energy barrier requiring expenditure of energy to surmount it.

This activation is brought about by special enzymes which we mentioned above, the synthetases of sRNA-amino acid complexes* (seryl-sRNA-synthetase, valyl-sRNA-synthetase, and so on). Usually not less than 20 such enzymes must be present, because each amino acid is activated by its own particular enzyme. Some amino acids have been shown to require more than one activating enzyme. The presence of more than one sRNA for one amino acid, in connection with degeneracy of the genetic code, may also influence the character of enzyme heterogeneity. The enzyme initially activates the amino acid by forming an aminoacyladenylate, bringing the amino acid to react with adenosine triphosphate (ATP) in the manner illustrated in Fig. 3, after which it transfers the amino acid to a molecule of sRNA, attaching the amino acid residue to the specific terminal group of this RNA, to its terminal adenine nucleotide. The enzyme molecules thus have several special reactive groups (allosteric groups). One of the groups in all the enzymes reacts with ATP and is the same in the whole series of enzymes. Another active group reacts selectively with a particular amino acid: this group must be specific in the sense of identifying the structure of the amino acids and choosing the correct molecule. Finally, the third active group of the enzyme is that part of its molecule which reacts with sRNA molecules, and this is distinguished by individual specificity, because from the complement of 50-55 different sRNA

---

*According to the International Classification and Nomenclature of enzymes adopted by the Commission of the International Biochemical Union, with the participation of Soviet biochemists, in the abbreviation of the English name of these enzymes and its Russian translation, transfer RNA is given with the symbol (sRNA) from soluble RNA, and not with the symbol t, frequently used in scientific papers. In future, therefore, we shall use the abbreviation sRNA.

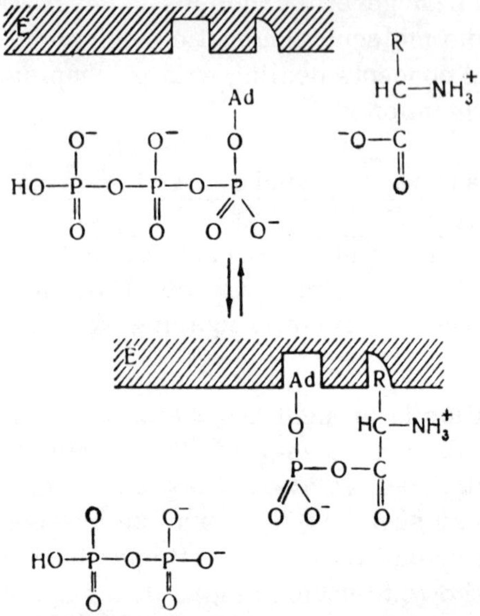

Fig. 3. Scheme of activation of amino acids. e) Enzyme molecule; ad) adenosine; r) side chain of amino acid.

molecules it has to choose those few molecules possessing anticodons related to the triplet codons for the given amino acid in the composition of the template, i.e., the messenger RNA molecule. (Findings have recently started to be published indicating that the number of activating enzymes is measured not by the number of amino acids, but by the number of sRNA fractions). The most important stage of protein biosynthesis is that involving interaction between aminoacyl-sRNA complexes and messenger RNA molecules, because it is this interaction which embodies reading the genetic code and translating it into a definite sequence of amino acid residues.

Transfer RNA is present in cells as a population of polynucleotide molecules of about equal molecular weight (about 24,000) and equal lengths (about 70 nucleotides). Each of these molecules has an identical terminal acceptor three-nucleotide group (cytosyl-cytosyl-adenyl nucleotide, CCA), taking part independently in metabolism, and functioning as the point of attachment of the amino acid residue. The remainder of the polynucleotide portion of each sRNA molecule has many oligonucleotide groups which are complementary to one another, and it thus forms a series of interacting zones, as a result of which a secondary structure appears in sRNA molecules enabling the anticodon to be exposed in such a way that these molecules can interact with the mRNA template by precisely this group.

The study of all aspects of the biochemistry of this RNA fraction has advanced in recent years by leaps and bounds and has been exceptionally fruitful. After discovery of transfer RNA and identification of its principal functions in 1957, within a short period methods were developed for dividing this RNA into fractions based on the accepted amino acid.

Results of an experiment to fractionate total sRNA from *E. coli* by the method of countercurrent distribution (Goldstein et al., 1964) are given in Fig. 4. Clear separation into fractions with different amino acid groups can be seen.

Because of the low molecular weight of the polynucleotides of this RNA, a number of laboratories have recently begun to investigate the complete sequence of nucleotides in individual sRNA mole-

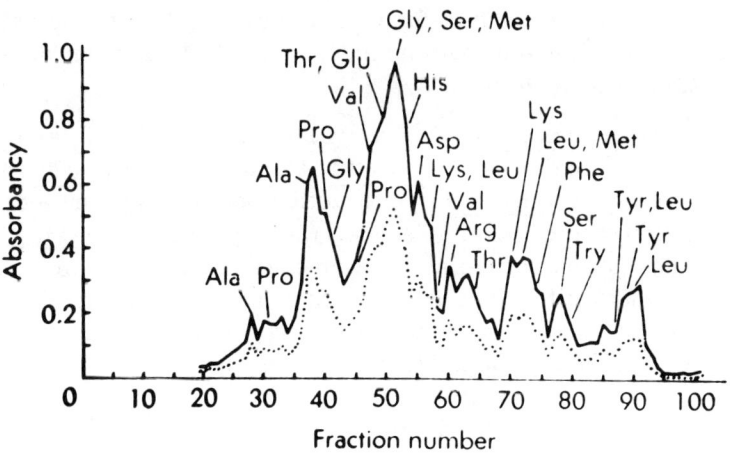

Fig. 4. The distribution pattern obtained by countercurrent distribution for 970 transfers of *E. coli* B sRNA ———, absorbency at 260 mμ; ...., absorbency at 280 mμ. Peak fractions containing the various amino acid acceptor activities are indicated by the arrows (Goldstein et al., 1964).

cules. This vast and laborious task has recently been completed for alanine sRNA of yeast (Holley, Apgar et al., 1965). Several other fractions of sRNA are being studied in turn. In the USSR most progress has been made with the deciphering of the sequence of valine sRNA (Baev, 1965).

During determination of the sequence of alanine sRNA in the investigation cited above, Holley and co-workers found that during interaction between complementary bases, the molecules of alanine sRNA may acquire several configurations, models of which are represented in Fig. 5. The several unpaired areas evidently determine selective contact of the sRNA molecules, first, with the mRNA template, second, with alanyl-sRNA-synthetase, and third, with the enzyme for transferring sRNA to the template. The functions and biochemistry of transfer RNA are discussed in more detail at the present level of knowledge in surveys by Baev (1965), Kiselev (1964), and Chapeville (1964).

As we shall see later, transfer RNA plays an important role in a number of regulatory morphogenetic processes, because on its function ultimately depends the rate of synthesis of individual proteins. The character of the fractional composition of sRNA and quantitative proportions of the different codons and anticodons determine the relative speed of reading the information contained in the genes, and we must therefore direct our attention to the nature of interaction between sRNA and messenger RNA, this fundamental act in protein synthesis.

b) Ribosomes and Polyribosomes (Polysomes), Their Functions in the Structural Organization of Protein Biosynthesis. Interaction between RNA molecules carrying amino acids and messenger RNA molecules carrying genetic information regarding the sequence in which the amino acid must be connected together in order to synthesize a particular protein is carried out by means of special organelles, known as ribosomes, formed in the nucleus (in the nucleolus), but present in all parts of the cell capable of synthesizing protein (cytoplasm, nucleus, plastids, mitochondria).

The ribosomes are ribonucleoprotein particles consisting of high-polymer RNA and a number of structural proteins.

In all cells the functional ribosomes are particles of two

types with sedimentation constants of about 30S and 50S for bacteria and about 40S and 60S for animals, and forming a dimer with sedimentation constant of 70-80S, capable of reversible dissociation under certain conditions.

The approximate shapes of the monomer and dimer are shown in Fig. 6. RNA is responsible for about half the molecular weight of the ribosomes, so that more of the high-polymer component of ribosomal RNA is present in the 50-60S monomer than in the 30-40S monomer. Each monomer contains one molecule of RNA.

It was thought originally that messenger RNA is combined with one dimer ribosome, forming an active protein-synthesizing complex as a result of this combination. However, it was very soon (in 1963) discovered in several laboratories that messenger RNA (the length of the polynucleotide of this RNA for hemoglobin is about 1500-2000 Å) is combined, not with one ribosome, but with a group of several ribosomes (230 Å), hanging from this RNA like

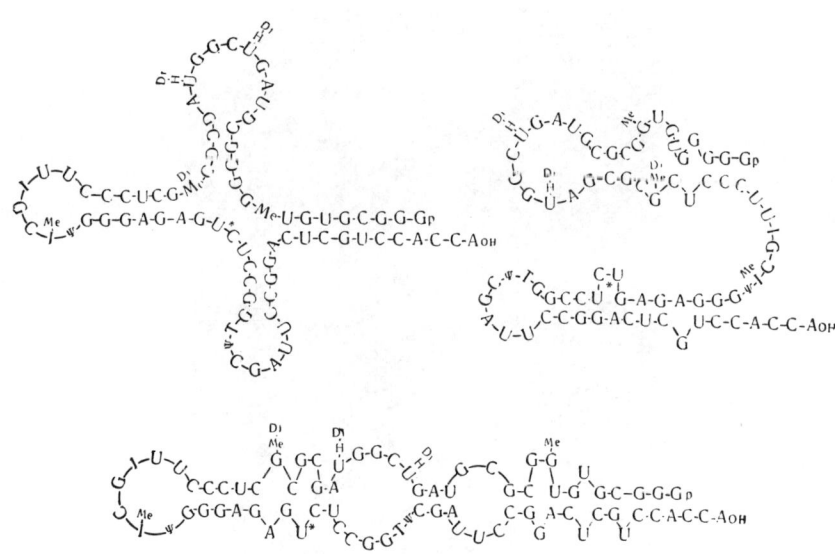

Fig. 5. Scheme illustrating possible methods of formation of the secondary structure (conformation) of alanine-sRNA on the basis of the determined sequence of its nucleotides (Holley et al., 1965).

Fig. 6. Approximate shape of monomers, dimer, and tetramer of *Escherichia coli* ribosomes.

clusters of berries (Fig. 7) (Warner et al., 1963). The longer the messenger RNA molecule, the larger the cluster it forms, and the larger the number of ribosomes connected. These formations have been called polyribosomes, or polysomes, and they have been shown to be a special structure for the synthesis of polypeptide chains.

The existence of a series of excellent surveys dealing specially with the functions of ribosomes and polysomes (Rich, 1963;

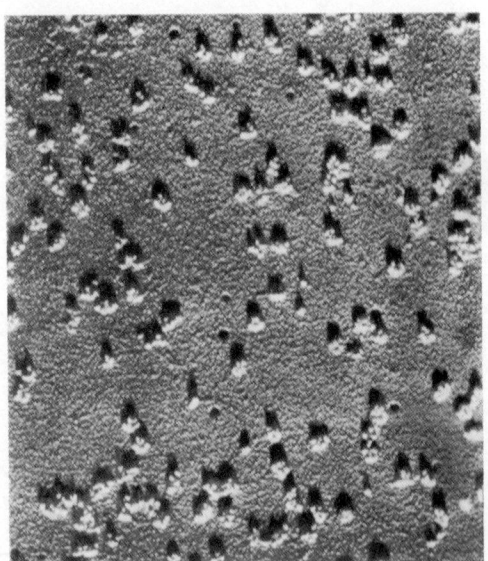

Fig. 7. Electronphotomicrograph of shadowed polysomes from rabbit reticulocytes. Aggregates of 5 or 6 ribosomal particles can be seen (Rich, 1963).

Campbell, 1965; Bogdanov and Shakulov, 1965; Spirin, 1965) makes it unnecessary for us to dwell on this subject in detail as would be necessary in a work specifically devoted to the biochemistry of protein synthesis. All we need do here is to give an up-to-date, soundly based interpretation (or model) of the dynamics of interaction between ribosomes, sRNA, and mRNA during synthesis of the polypeptide chain by the polysome.

The model of the mechanism of this interaction assumes that messenger RNA is a special conveyor along which ribosomes move, and each ribosome, because of periodic attachment of sRNA molecules, forms a finished polypeptide chain (Fig. 8) (Zubay, 1963). The extreme right ribosome in this scheme has just become attached to the sRNA and has started to synthesize a polypeptide chain. The ribosomes move along the mRNA chain by means of a mechanism not permitting them to go backward. At each stop made by the ribosome the corresponding amino acid, supplied by transfer RNA, is selected from the cell medium and attached to the growing polypeptide chain. At the completion of synthesis the ribosome liberates the polypeptide chain and itself leaves the chain of messenger RNA. Almost simultaneously, another ribosome attaches itself to the opposite end of the chain.

During synthesis of each polypeptide subunit of hemoglobin, one ribosome was found to travel along the mRNA chain and to form one polypeptide chain per minute. Polysomes of the largest size were discovered during the study of cells infected with poliomyeli-

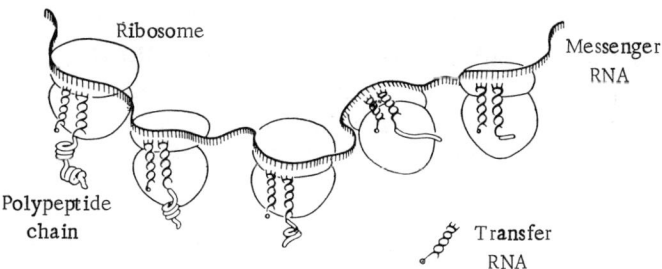

Fig. 8. Model of protein synthesis by a polysome. Ribosomes move along the messenger RNA from right to left (Zubay, 1963).

tis virus. The polycistron RNA of this virus forms polysomes consisting of 50-70 ribosomes (Rich, 1963).

c) Biosynthesis of Proteins and the Problem of the Genetic Code. The whole synthetic machinery described above, forming protein molecules by the conveyor principle, is constructed mainly so that it can take account of the genetic code, incorporating information of the genes represented in molecules of messenger RNA. The essence of this process of translation of information (the text) recorded as triplet nucleotide combinations of four nucleotides in the "text" of polypeptide chains composed of 20 amino acids is very simple. Every three nucleotides in the polynucleotide chain contain information or code the position of one amino acid residue. Since the number of possible triplet nucleotide combinations is 64, and there are 20 amino acids, for each amino acid there are several nucleotide combinations (this phenomenon is known as degeneracy of the code).

The pattern of this relationship between genetic information structures and the processes of protein synthesis which they control have been known a long time, but the experimental determination of the concrete nature of the code, beginning in 1961 with the brilliant work of Nirenberg, first described at the Fifth International Biochemical Congress in Moscow, required several years of very intensive work in many laboratories and in different experimental directions (the use of synthetic polyribonucleotides of known composition in protein synthesis, the study of changes in RNA and protein during mutations of tobacco mosaic virus and mutation of hemoglobins, the study of the mechanism of direct and suppressor mutations of tryptophan synthetase, statistical comparisons, and so on).

As a result of all this work, by 1967 the following system of relationships (dictionary) between amino acids and the nucleotide triplets (codons) coding them had been established and was universal for all living systems which had been studied.

The code as given in this table is not yet final. Not all the possible triplets have been established for all amino acids; not all sequences in the codons themselves have been determined. Considerable work must still be done to identify anticodons in sRNA,

The Present State of Deciphering of the Genetic Code
(Caskey et al., 1968)

Amino acids	Codons	Amino acids	Codons
Alanine	GCA, GCG, GCC, GCU	Lysine	AAA, AAG
		Methionine	AUG
Arginine	CGG, CGA, CGC, AGA, AGG, CGU	Phenylalanine	UUU, UUC
		Proline	CCA, CCC, CCU, CCG
Aspartic acid	GAC, GAU	Serine	AGC, AGU, UCA, UCG, UCC, UCU
Asparagine	AAC, AAU		
Cystein	UGU, UGC	Threonine	ACA, ACG, ACC, ACU
Glutamic acid	GAA, GAG		
Glutamine	CAA, CAG	Tryptophan	UGG
Glycine	GGA, GGC, GGU, GGG	Tyrosine	UAC, UAU
		Valine	GUA, GUG, GUC, GUU
Histidine	CAC, CAU		
Isoleucine	AUA, AUC, AUU	Nonsense triplets (terminal)	UAA, UAG, UGA
Leucine	CUG, CUC, CUU, UUA, UUG, CUA		

Note. Sequence is shown in the direction from 3' to 5' hydroxyl groups counting from right to left in protein synthesis.

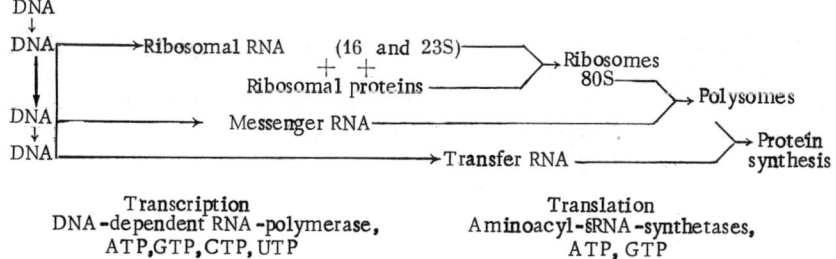

Fig. 9. General scheme of protein synthesis.

and the first steps in this direction have shown that success is unlikely to be rapid. However, there is no need in this book to describe all the details, the history, and the approaches to the solution of the vast problem of genetic coding, which is the subject of special investigations at the present time.

In conclusion it will be useful to give a general model of protein biosynthesis demonstrating molecular interaction between all the principal components of the genetic and cytoplasmic system concerned in this synthesis (Fig. 9).

It can thus be concluded that the chief problems concerned with the mechanism of biosynthesis of proteins and nucleic acids have now been solved. Analysis of this problem has now shifted to the level of filling in all the minor details (discovery of the nucleotide sequence of individual sRNA molecules, the terminal codons, heterogeneity of the aminoacyl-sRNA-synthetases, species specificity of these enzymes for sRNA and for other substrates). The solution of each major problem always brings forth a tremendous number of new problems, and creates a new branch of science. At the same time, however, it provides a decisive stimulus to the analysis of allied problems, sometimes even more important.

From this point of view the discovery of the mechanisms of biosynthesis of macromolecules was a particularly outstanding event, and one connected with a chain of fundamental discoveries in adjacent fields of biology. Unquestionably it acted as the greatest stimulus to research into two cardinal and closely interconnected biological problems, heredity and individual development.

Literature Cited

Baev, A. A., 1965, In: Progress in Biological Chemistry, Yearbook, Vol. 7, ed. B. N. Stepanenko et al., Izd. Nauka, Moscow, p. 67.
Bass, I. A., and Gvozdev, V. A., 1965, In: Biosynthesis of Protein and Nucleic Acids, ed. A. S. Spirin, Izd. Nauka, Moscow, p. 50.
Bogdanov, A. A., and Shakulov, R. S., 1965, In: Biosynthesis of Protein and Nucleic Acids, ed. A. S. Spirin, Izd. Nauka, Moscow, p. 86.
Bonner, J., Huang, R. C., and Gilden, R. V., 1963, Proc. Nat. Acad. Sci., USA, 50:893.
Bresler, S. E., 1966, Introduction to Molecular Biology, Gordon and Breach, New York.
Campbell, P. N., 1965, in: Progress in Biophysics and Molecular Biology, Vol. 15, Pergamon Press, Oxford—London, p. 3.

LITERATURE CITED

Caskey, T., Wilcox, M., Anderson, F., Scolnick, E., Tompkins, R., and Nirenberg, M., 1968, Proc. XII Intern. Congress of Genetics, Vol. II, Tokyo, p. 37.
Chamberlin, M., and Berg, P., 1964, J. Mol. Biol., 8:297.
Chapeville, F., 1964, In: Molecular Biology, Problems and Perspectives, ed. A. E. Braunshtein, Izd. Nauka, Moscow, p. 117.
Cohen, N. P., 1966, Biol. Rev., 41:503.
Georgiev, G. P., 1964, In: Molecular Biology, Problems and Perspectives, ed. A. E. Braunshtein, Izd. Nauka, Moscow, p. 109.
Georgiev, G. P., 1965, In: Biosynthesis of Protein and Nucleic Acids, ed. A. S. Spirin, Izd. Nauka, Moscow, p. 312.
Gibor, A., and Granik, S., 1964, Science, 145:890.
Goldstein, J., Bennett, T. P., and Craig, L. C., 1964, Proc. Nat. Acad. Sci., USA, 51:120.
Goodman, H. M., and Rich, A., 1962, Proc. Nat. Acad. Sci., USA, 48:2101.
Goodman, H. M., and Rich, A., 1963, Nature, 199:318.
Holley, R. W., Apgar, J., Everett, G., Madison, J. T., Marquisee, M., Merrill, S. H., Penswick, J. R., and Zamir, A., 1965, Science, 147:1462.
Hurwitz, J., and August, J. T., 1963, In: Progress in Nucleic Acid Research, Vol. 1, eds. J. N. Davidson and W. E. Cohen, Academic Press, New York.
Ingram, V. M., 1965, The Biosynthesis of Macromolecules, W. A., Benjamin, Inc., New York—Amsterdam.
Jones, K. W., and Truman, D. E. S., 1964, Nature, 202:1264.
Kiselev, L. L., 1964, Usp. Sovr. Biol., 58:177.
McConkey, E. N., and Hopkins, J. W., 1964, Proc. Nat. Acad. Sci., USA, 51:1197.
Medvedev, Zh. A., 1963, Protein Biosynthesis and Problems of Ontogenesis, Medgiz, Moscow.
Medvedev, Zh. A., 1965, Usp. Sovr. Biol., 59:333.
Medvedev, Zh. A., 1966, Protein Biosynthesis and Problems of Heredity, Development and Aging, Oliver and Boyd Ltd., Edinburgh.
Osterman, L. A., 1965, In: Progress in Biological Chemistry, Yearbook, Vol. 7, ed. B. N. Stepanenko et al., Nauka, Moscow, p. 116.
Pauling, L., and Corey, B., 1956, Arch. Biochem. Biophys., 65:164.
Rich, A., 1963, Scient. Am., 209:44.
Schweiger, H. G., and Berger, S., 1964, Biochim. Biophys. Acta, 87:533.
Sinsheimer, R., and Lawrence, M., 1964, J. Mol. Biol., 8:289.
Sisakyan, N. M., and Gladilin, K. L., 1965, In: Progress in Biological Chemistry, Yearbook, Vol. 6, ed. B. N. Stepanenko et al., Izd. Nauka, Moscow, p. 3.
Spirin, A. S., 1963, Some Problems in the Macromolecular Structure of Ribonucleic Acids, Izd. AN SSSR, Moscow.
Spirin, A. S., 1965, In: Nucleic Acids, Izd. Mir, Moscow, p. 341.
Tikhonenko, T. I., 1965, In: Biosynthesis of Protein and Nucleic Acids, ed. A. S. Spirin, Izd. Nauka, Moscow, p. 193.
Tongur, V. S., 1965, In: Biosynthesis of Protein and Nucleic Acids, ed. A. S. Spirin, Izd. Nauka, Moscow, p. 277.
Vol'kenshtein, M. V., 1970, Molecules and Life, Plenum Press, New York.
Warner, J. R., Knopf, P. M., and Rich, A., 1963, Proc. Nat. Acad. Sci., USA, 49:132.

Warner, J., Rich, A., and Hall, C., 1962, Science, 138:1399.
Watson, J. D., and Crick, F. H. C., 1953, Nature, 171:737.
Yankofsky, S. A., and Spiegelman, S., 1962, Proc. Nat. Acad. Sci., USA, 48:1069, 1466; 1963, 49:638.
Zbarskii, I. B., and Debov, S. S. (eds.), 1967, Chemistry and Biochemistry of Nucleic Acids, Izd. Meditzina, Moscow.
Zubay, G., 1963, Science, 140:1092.

Chapter 2

Molecular-Genetic Mechanisms of Reproduction and Morphogenesis of Viruses

Introduction

Viruses are the simplest biological structures possessing an autonomous genetic system, usually consisting of one macromolecule of RNA or DNA, which controls the development and reproduction of the virion inside the host cell. In the simplest cases, virus RNA undergoes the same fate in the host's cells as the messenger RNA which is synthesized in cell nuclei. For some period of time this virus RNA is fully adapted to conditions in the cell, using all the residual metabolic system of the cytoplasm or nucleus for synthesis of virus proteins. In some cases virus DNA becomes an element built into the host's genetic system. Nevertheless, viruses can be regarded as autonomous genetic systems actively programming their development and reproduction.

Within the confines of the virus form of life we find a very great variety of morphological forms and genetic structures, consisting in the simplest viruses of one or two genes, while in the most complex bacterial viruses they consist of between 60 and 100 genes, producing particles with comparatively complex morphology. Examination of the evolution of methods of controlling their purposively interconnected programmed activity is of very great importance also for an understanding of the purposive control of the activity of more complex genetic systems, which in bacteria consist of thousands, and in higher animals of hundreds of thousands, of genes.

In the course of our examination of the morphogenesis and reproduction of viruses we shall follow the chemical principle of classification suggested by Cooper (1961), based on chemical structure of the genetic macromolecule of the virus (RNA, single-stranded DNA, double-helical DNA) and their classification in accordance with the principal evolutionary groups (viruses of plants, of animals, and of bacteria). Viruses of bacteria are usually called bacteriophages.

§ 1. Mechanisms of Morphogenesis of the Simplest Ribonucleoprotein (Two-Component) Viruses of Plants

Virus particles have an inner core which consists in the mature state of one RNA molecule, while their outer coat consists of practically identical molecules of homogeneous protein. They are the simplest of all known forms of life possessing biological individuality and the property of inheritance, i.e., of reproducing under favorable conditions identical forms as a result of the active functioning of a molecular controlling system carrying information in the form of a genetic program, and consisting in the present case of RNA molecules. Many viruses of plants have a structure which is in accordance with this simplest biomolecular principle. However, the simplicity of structure of these viruses in the mature state does not conceal the existence of a process of development, of morphogenesis of these virus particles in the period of their intracellular propagation. Two-component (protein + RNA) viruses are extremely varied in their morphology.

A model of the structure of one of the simplest viruses of this type, tobacco mosaic virus, is shown in Fig. 10.

The ontogenesis of these viruses, after they have penetrated into the host cell, consists in the simplest case of the following processes: 1) release of an RNA-virus particle from the protein coat; 2) autoreplication of RNA and formation of types of RNA capable of playing the role of messenger RNA, interacting with ribosomes of infected cells and forming a system for the synthesis of virus-specific protein or proteins; 3) synthesis of virus-specific proteins using enzyme systems of activation and transfer of amino acids belonging to the host cell; 4) aggregation of new virus RNA and virus protein with the formation of virus particles.

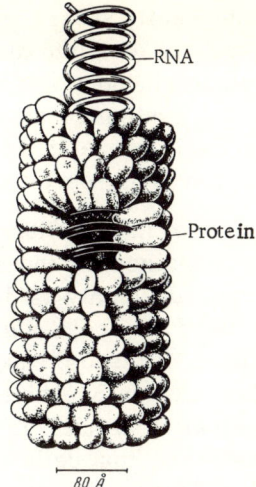

Fig. 10. Model of the molecular structure of tobacco mosaic virus (fragment). Overall size of particle 150 × 3000 Å. Inner black helix represents RNA; structures surrounding it are protein molecules, each of which consists of 158 amino acid residues.

The molecular aspects of reproduction of viruses are surveyed by an extensive literature (Vainshtein and Kiselev, 1964; Gendon, 1964; Bykovskii, 1964; Cohen, 1963; Wildy and Horne, 1963; Tsugita and Fraenkel-Conrat, 1963; Erikson and Franklin, 1966). Only one theme in this field concerns us at this moment: the molecular mechanism of control of the simplest morphogenetic processes of reproduction of the virus particle, and to discuss this theme theoretically we need pick out only those facts which have a direct bearing on it.

Of all the viruses at present known and which have been studied biochemically, the simplest is a special virus described by Reichman (1964) as "satellite of the tobacco necrosis virus." This virus is a ribonucleoprotein with the smallest RNA molecule of any virus (mol. wt. 395,000). This RNA is composed of 1200 nucleotides. Given the triplet nucleotide character of the genetic code, such an RNA cannot code more than 400 amino acid residues. Since each protein molecule of the virus coat contains about 370-380 amino acid residues, clearly the RNA of this virus can code only one protein.

In Spiegelman's laboratory (Clark, Chang et al., 1965) RNA of this virus has been used as template for protein synthesis in an artificial system containing ribosomes and cell extract from *E. coli* Protein synthesized in this system consisted of 400 amino acid residues, and in its immunologic properties and the results of peptide analysis, it corresponded to virus-specific protein.

However, the volume of information inscribed in 1200 nucleotides is insufficient for spontaneous propagation of the virus in plant cells, and it can reproduce only as the satellite of another virus—virus of tobacco necrosis, possessing RNA with a molecular weight of 2×10^6. Proteins of these viruses are serologically different, but the basic virus evidently enables certain factors to be synthesized which are equally essential for the formation of both viruses. RNA synthesis in the host cells takes place on DNA templates and is catalyzed by DNA-dependent RNA-polymerase. It may be assumed that a special factor—the enzyme RNA-replicase—which is not used for reproduction of the satellite virus is essential for synthesis of virus RNA on RNA templates (autoreplication of RNA).

It is thus considered that the simplest virus morphogenesis which actually exists requires at least two genes localized in virus RNA.

The virus RNA entering the cell is thus a very simple biological control system containing and putting into effect a particular purposive program of synthetic and morphogenetic processes resulting in the more extensive autoreproduction of the virus.

The RNA molecule of one of the most thoroughly studied viruses, tobacco mosaic virus (TMV), has the structure of a single-stranded polynucleotide consisting of approximately 6000 nucleotides. The fact that this RNA contains the information determining amino acid sequence in the protein of the virus coat (158 amino acid residues in each protein molecule) is shown not only by the widely known and extensive data on infectivity of pure tobacco mosaic virus RNA and its ability to reproduce whole virus, but also, in a more direct form, by investigations of inherited changes in the amino acid sequence of this protein as a result of experimental modifications of the nucleotide composition of the virus RNA by the action of nitric acid and other substances under carefully controlled conditions. Such treatment causes deamination of the three bases containing an amino group: adenine is converted into hypoxanthine, guanine into xanthine, and cytosine into uracil.

Changes in nucleotide composition are accompanied to a greater or lesser degree by strictly determined changes in the amino acid sequence of virus protein, and some very important work in this direction has been done in recent years to study the character of the amino acid code, i.e., which combinations of nucleotides of virus RNA code the position of a particular amino acid residue in the molecule of virus protein (Wittman, 1963; Wittman and Wittman-Liebold, 1963; Tsugita and Fraenkel-Conrat, 1960, 1962).

At the same time, these investigations showed that virus RNA does not contain information programming synthesis of virus-specific coat protein throughout its length, because only some of the changes in nucleotide composition are revealed as changes in virus protein.

Wittman (1963) studied 133 mutants of tobacco mosaic virus (117 obtained by the action of nitric acid and 16 spontaneous mutants) in this respect. Of the 117 nitric acid mutants, 82 had no change in the protein particle of the coat, and only 35 had various substitutions in the amino acid chain. These facts are regarded as proof that only about one-third of the polynucleotide particle of tobacco mosaic virus RNA codes the protein subunits of the virus coat

After penetrating into the cell, virus RNA at once starts to put into action its "program," by organizing the synthesis of complementary forms of RNA, attachment of virus-specific template RNA to ribosomes, and switching over the protein-synthesizing system of the cell to synthesis of virus-specific protein, in the simplest case, only one protein throughout the active zone of the template.

The suggestion has been made (Doi and Spiegelman, 1963) that virus RNA, when it has penetrated into the host cell, programs synthesis of a special RNA-polymerase for use in synthesis of virus RNA just as in the case of RNA-dependent complementary RNA synthesis. This hypothesis has so far been confirmed for TMV only by the work of Karasek and Schramm (1962), who isolated a special polymerase, catalyzing incorporation of C^{14}-adenine into polyribonucleotide, from cell-free extracts of tobacco leaves infected 3-14 days before the experiment.

In experiments *in vitro*, TMV-RNA can also stimulate synthesis of polypeptides, even if a synthetic system of ribosomes and supernatant fraction of homogenates of *E. coli* is used (Tsugita, Fraenkel-Conrat et al., 1962). Some of the products of this synthesis (about 10%) were identical in antigenic and biochemical properties with TMV protein. The rest of the products of synthesis were similar to TMV protein only in certain respects.

TMV is the virus whose reproduction has been most thoroughly studied, but nevertheless it is not the simplest virus of this type. The molecular weight of TMV-RNA is 2×10^6. However, broad bean mottle virus has a molecular weight of the order of $5.2-5.6 \times 10^6$ and contains an RNA particle with a molecular weight of about 1.1×10^6 (Yamazaki, Bancroft, and Kaesberg, 1961), while brome-grass mosaic virus has a molecular weight of 4.6×10^6 and an RNA with a molecular weight of 1×10^6 (Bockstahler and Kaesberg, 1962). The molecular weight of the RNA of barley striped mosaic virus is also 1×10^6 (Atabekov and Novikov, 1966), i.e., almost half that of TMV-RNA. Examples of the opposite character are also found: the existence of unusually large RNA molecules in plant viruses. The RNA of wound tumor virus, for example, has a molecular weight five times greater than that of TMV-RNA (Gomatos and Tamm, 1963a).

The "ontogenesis" of the simplest virus, as a process of definite development, of increasing the complexity of its organization, its elementary morphogenesis, thus consists of two processes: one structural component of the virus, namely its RNA molecule (parent RNA) carries out the synthesis of two heterogeneous structures in the host cell, both different from the parent RNA itself: a) complementary RNA (replica RNA) and b) virus-specific protein, or two-three proteins; complementary RNA later is used to synthesize virus-specific RNA.

Autoreplication of TMV-RNA takes place in accordance with a principle common to the synthesis of nucleic acids, i.e., by the preliminary formation of complementary forms of RNA (Shipp and Haselkorn, 1964; Burdon, Billeter, et al., 1964; Ralph et al., 1965; Erikson and Franklin, 1966). The intermediate form of this replication is the formation of a special virus-specific double-helical RNA, a distinctive DNA analog. This special form of virus RNA has physical properties distinguishing it from single-helical RNA, greater resistance to ribonuclease, a sudden thermal transition from resistance to ribonuclease to sensitivity to that enzyme, a higher sedimentation coefficient (S) and certain other special features. Shipp and Haselkorn (1964) give in their paper preliminary calculations of the ratio between the number of double-helical virus RNA molecules and the number of single-helical. In one plant cell on the 16th day after infection of the leaf about 10^2-10^3 double-helical RNA molecules are present, compared with 10^5-10^6 mature virus particles. Another group of authors (Burdon, Billeter, et al., 1964) calculated that on the 10th day of infection about 0.5% of virus RNA in tobacco leaves is present in the form of double-helical molecules.

It is not yet clear what regulates the "division of labor" in virus RNA in the period of intracellular reproduction of virus particles, i.e., the relationship between the process of autoreplication of virus RNA (RNA → RNA → RNA → and so on) and the process of its template activity in the synthesis of virus protein (RNA → protein). Evidently the localization of synthesis plays an important role here, in the sense that virus RNA after attachment to ribosomes begins to synthesize protein, i.e., starts to perform the functions characteristic of messenger RNA; in the nuclei, however, or in the unattached form, it carries out autoreplication. Interaction between virus RNA and ribosomes, as an essential act for induction of the synthesis of virus protein, has been demonstrated in several investigations (Haselkorn, Fried, and Dahlberg, 1963; Haselkorn and Fried, 1964; Van Kammen, 1963). In experiments *in vitro*, virus RNA stimulates incorporation of amino acids into proteins also after formation of a complex with ribosomes. Meanwhile, autoreplication of TMV-RNA, as shown by the results of a series of investigations, takes place in cell nuclei (Zech and Vogt-Kohne, 1955; Cochran, Dhaliwal, et al., 1962; Cornuet and Manifacier, 1962; Kim and Wildman, 1962). According to Reddi (1966), RNA from TMV particles (P^{32}-labeled maternal RNA) entering a plant cell as a result of infection is localized in an undegraded form in the cell nucleus. It is evidently in the nucleus that the conditions for its replication are created. On the other hand, its complementary copies enter the cytoplasm for protein synthesis.

Although the RNA molecule of most plant viruses contains no fewer than two or three structural cistrons, on entering the cell it becomes the template for synthesis of two or three virus-specific proteins (just as in the case of messenger RNA) without fragmentation into monocistronic segments. In this case, one template links itself to the ribosomes and carries out the synthesis of several proteins simultaneously. Indirect evidence obtained during the study of polysomes formed by RNA or turnip yellow mosaic virus (mol. wt. of RNA 2.3×10^6) show that each cistron of the template can work as an independent template, rather than in turn after its neighbor (Voorma, Gout, et al., 1964). This type of operation of the template is necessary because different proteins from different parts of the template are synthesized in different quantities.

§2. Mechanisms of Morphogenesis of RNA-Containing Viruses of Animal Cells

The composition of many animal viruses is much more complex than that of the simplest plant viruses, and their morphogenesis consists of a large number of stages. Despite their complexity, the RNA of these viruses, like that of plant viruses, possesses infectivity, i.e., the whole program of intracellular development is contained in the virus RNA molecule.

In this case one RNA molecule can program the synthesis not only of one, but of two or more proteins, i.e., it contains two or more cistrons (genes) (it is a polygene or polycistron). In addition, during intracellular proliferation of a number of species of these viruses (large species), protein or proteins are formed which are not found in the mature virus, but which perform certain functions during intracellular replication of virus particles. They are called "early" proteins in contrast to the "late" proteins which compose mature virus particles.

It has been shown, for example, that in cells in which the proliferation of virus RNA is beginning, synthesis of host cell RNA or DNA templates in the nucleus comes to an end (Bukrinskaya et al., 1964; Schafer, 1963). Inhibition of RNA synthesis in the nuclei after virus infection was also demonstrated by earlier studies (Luria, 1958; Franklin, Wecker, and Henry, 1959). In connection with this, synthesis of cell proteins also is inhibited. This inhibition also appears when cells are infected with virus RNA only. It has therefore been suggested that virus RNA is the initial template for synthesis of a special protein, a repressor of host cell DNA-dependent RNA synthesis (Reich, Franklin, et al., 1961). It is interesting to note that among the group of animal RNA-viruses one has been found (the virus of mouse leukemia) which contains RNA with the highest molecular weight (13×10^6 of all forms of RNA so far known [Mora et al., 1966]).

Just as with plant viruses, RNA synthesis by certain animal viruses takes place in the cytoplasm, and by others in the nucleus; synthesis of virus proteins, however, takes place only in the cytoplasm, on the ribosomes (polysomes).

Another "early" protein of RNA-containing animal viruses (mengovirus and poliovirus) is virus-specific RNA-polymerase (Baltimore and Franklin, 1962, 1963; Baltimore, 1964; Baltimore, Eggers, et al., 1964). This polymerase is localized in cytoplasmic structures and is not inhibited by actinomycin, which brings to a halt all other forms of intracellular RNA synthesis. It is this RNA-polymerase which is active with respect to virus RNA synthesis. Among the reaction products of this polymerase, double helices of RNA were found. Data indicating that "early" proteins can be synthesized during infection by poliovirus have also been obtained experimentally (Watanabe, Watanabe, and Hinume, 1962), although without precise demonstration of the functions of these proteins in virus reproduction.

RNA-dependent polymerase (RNA-synthetase) brings about RNA replication through a stage of formation of double-helical RNA molecules, consisting of complementary chains. The existence of double-helical forms of RNA as a particular phase in virus reproduction has been demonstrated even more clearly for animal viruses than in the investigations with tobacco mosaic virus we have just described. Double-helical RNA is found during intracellular propagation of reovirus (Gomatos and Tamm, 1963a, b; Langridge and Gomatos, 1963), virus of encephalomyocarditis (Montagnier and Sanders, 1963), poliovirus (Baltimore, Becker, and Darnell, 1964; Baltimore, 1966), and certain others.

Until recently it was not known whether virus primer RNA, identical with the RNA which is a component of the virus particle, or a temporary replicative form of RNA is the active template of protein synthesis, or whether each of them participates in the synthesis of different "early" and "late" proteins. Zhdanov and coworkers (1964) have postulated that messenger and infective properties of RNA can be separated by using the principle of associating them with complementary polynucleotides. In this case the virus particle contains RNA which plays no direct part in protein synthesis. Its complementary copies are the templates for protein synthesis.

This hypothesis has recently been confirmed by investigations showing that complementary RNA plays the dominant role in interaction with polysomes, and it is also supported by the observed formation of double-helical replicative virus RNA molecules and the production of complementary RNA in the cytoplasm (Zhdanov, 1966; Kingsbury, 1966; Shapiro and August, 1966; Baltimore, 1966; Baltimore, Girard, and Darnell, 1966; Plageman and Swim, 1966).

The template properties of virus RNA are manifested, as in the case of messenger RNA of the host cell, during its interaction with ribosomes. Just as in the case of normal intracellular synthesis, polysome structures of the type illustrated in Fig. 8 are formed under these circumstances. The formation of active polysomes during interaction between virus RNA and cell ribosomes have been studied in particular detail in the case of poliovirus infections of HeLa cell cultures (Penman et al., 1963; Scharff et al., 1963; Rich et al., 1963). In this condition virus-specific polysomes are formed which are much bigger than the polysomes of uninfected HeLa cells, as a result of the polycistron character of virus RNA. Some of them contained as many as 60 ribosomes.

Since virus RNA of poliovirus, despite its polycistron character, does not undergo fragmentation when supplying information for synthesis of several proteins (early and late) or when functioning in the cell, but preserves its stability and activity for a long time, we can fully understand the desire to see this information in the linear order in which it is read and the reason underlying the temporal distribution of proteins synthesized. In this case it can be postulated that if, for example, the cistrons A, B, C, D, E, F, G, and H are located successively along the polynucleotide chain of the virus RNA, the polysome structure will first "read" information from segment A, then from segment B, C, and so on, or it will start from the area AB, followed by CD, and so on.

Messenger RNA in reticulocytes coding hemoglobin subunits have about 450 nucleotides and form polysomes consisting of from 4 to 6 ribosomes (Warner et al., 1963). The RNA of polioviruses contains about 6000 nucleotides. This suggests that this RNA codes from 7 to 10 cistrons (genes). Some of these proteins, as already mentioned, belong to the early group (RNA-synthetase, repressor of DNA-dependent RNA synthesis in the nuclei of host cells), while others (proteins of the virus envelope) belong to the late group. Proteins immunologically similar to proteins of the poliovirus envelope have also been found in the polysome fraction of infected cells (Scharff et al., 1963). From a consideration of all these facts, Rich and co-workers (1963) have postulated the following scheme of synthesis of virus proteins on a polycistron template, which serves at the same time as a scheme of morphogenesis of viruses of this type (Fig. 11).

The scheme of protein synthesis on simple polysome structures (Fig. 8) assumes that the ribosomes are attached (statistically) to one end of the polysome, move along the messenger RNA molecules as along a conveyor belt, and leave the other end together with liberation of the newly formed protein. The hypothetical scheme of polysome function for an 8-cistron template illustrated in Fig. 11 envisages two alternative mechanisms of synthesis of 8 proteins. The lower variant (b) illustrates a possible case in which polycistron messenger RNA is one point for terminal attachment of ribosomes. The attached ribosome begins to move along the template as the polypeptide chain grows, and at the end of the first cistron (A) it liberates the polypeptide chain and moves further along, starting to form the next polypeptide. In this way the successive production of a series of proteins takes place as the result of successive reading of the linear information of the template. In this case polysomes for which a larger or smaller part of the template is not occupied by ribosomes can be found at the beginning of synthesis.

The other variant (Fig. 11a) envisages that each separate cistron has its own specific point for attachment of ribosomes, so that the cistrons can carry out synthesis of individual proteins independently and at different speeds. In this case the partially loaded templates have the structure shown in Fig. 11a (top). Regulation and switching individual cistrons "on" and "off" in this case likewise can be independent. In the opinion of Rich and co-workers, intermediate cases are also possible.

In their opinion, the study of electron micrographs of polysomes of HeLa cells infected by poliovirus corresponds to the latter scheme (Fig. 11a), because templates can be found which are incompletely laden with ribosomes, in which the individual ribosomes are separated by empty areas of the templates about 500-1000 Å in length. These workers admit, however, that structures of this type could be artefacts and they do not regard these findings as definite evidence in support of either model. It has also been shown that the number of molecules, for example, of virus-specific RNA-polymerase, that are synthesized may be less than the number of virus envelope protein molecules synthesized. Approximately 50% of C^{14}-amino acids in HeLa cells infected with poliomyelitis virus (DNA-dependent RNA synthesis being inhibited by actinomycin D) is found in the envelope proteins. Meanwhile, protein synthesis in accordance with type B (in which the ribosomes pass along the whole template) provides for the synthesis of all types of virus-specific protein in equal quantities, if one polycistronic protein were formed which was then separated into eight proteins. Alternatively, the quantities of individual proteins could be var-

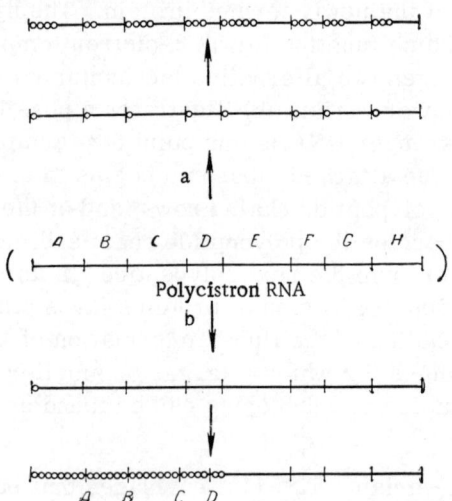

Fig. 11. Scheme of two possible methods of reading a polycistron messenger RNA. The messenger RNA in this case codes eight proteins (A-H) and the small circles represent ribosomes. a) Ribosomes are attached to each cistron and initiate independent synthesis of several proteins; b) there is only one point of initial attachment for the whole RNA molecule and proteins are synthesized sequentially (Rich et al., 1963).

ied by premature termination of translation which would produce polycistronic proteins that contained fewer protein species from the mRNA cistrons to the right in Fig. 11b. However, scheme B would give a simpler explanation for the temporal sequence of protein formation (early, middle, late) without the need for postulating the existence of a regulator, external to the template, for switching the process on.

The scheme (Fig. 11a) thus provides a possible explanation of the synthesis of different virus proteins at different times and in different amounts (at different speeds), but it does require the existence of an as yet unknown mechanism of independent operation of individual cistrons in accordance with the program for virus development. In other words, it requires a programming mechanism, controlling protein synthesis by the polysome in the same way as a tune is produced from notes obtained by striking the keys of a piano. The scheme (Fig. 11b) resembles the simple repetition of a scale of successive sounds. In this case no notes, i.e., programmed regulation of activity, are required; it is assumed that control is by simple linear succession of the cistron-keys.

However, the complete program of development of RNA-viruses, as mentioned above, is contained in their RNA. It acts as template for protein synthesis and predetermines the spectrum of this synthesis in time and space. The existence of some additional system, purely for programming, for use in reproduction of viruses of this type is therefore improbable.

It must therefore be postulated that during protein synthesis in the manner illustrated in Fig. 11a certain meaningful boundary groups of nucleotides must exist between individual cistrons, on the one hand enabling each cistron to work independently, and on the other hand determining the speed at which the sRNA molecules transferring amino acids interact with a given cistron of virus RNA. The same problem arises also in connection with polycistron messenger RNA in bacteria, containing information about a complete operon. In this case it has also been shown that each cistron (segment) of this RNA works at a different speed. To explain this a very ingeneous idea was put forward, namely the existence of speed modulators — nucleotide triplets whose anticodons are represented by a larger or smaller number of specific sRNA molecules (Ames and Hartman, 1963). In the latter case, a deficiency of sRNA of a particular type (containing anticodon to the modulator triplet) delays the process of reading information.

An interesting model of morphogenesis (representing the dynamics of different types of synthesis in time) of RNA-viruses of animals was proposed by Zhdanov (1966) in his concluding remarks at the symposium on reproduction of viruses at the Ninth International Congress on Microbiology in Moscow. Zhdanov based his model mainly on data obtained in his laboratory on the morphogenesis and localization of syntheses in Sendai virus. RNA synthesis by this virus starts in the nucleolus, where synthesis of an early protein (virus S antigen) and formation of ribonucleoprotein take place, whereas another, late protein (V antigen) is formed in the cytoplasm.

On the basis of his study of the quantitative dynamics of these syntheses, Zhdanov postulates that these processes take place in the following order. He accepts that virus RNA is linked to an enzyme which is either present already in the cell or is introduced by the virus, which synthesizes an additional strand, as a result of which replicative double-stranded RNA is formed. Cell RNA-polymerase, working on the additional strand, synthesizes mRNA for the virus RNA-polymerase, and this, attaching itself to the ribosomes, codes the synthesis of this enzyme. Virus RNA-polymerase, working on the whole length of the additional strand, synthesizes a large number of molecules of virus (infective) RNA. On these newly formed strands (on the corresponding cistrons), cell RNA-polymerase synthesizes mRNA for structural proteins (S and V antigens). As S antigen accumulates, free strands of virus RNA join with it to form ribonucleoprotein, which is carried to the places where V antigen is synthesized. There the formation of virus particles, which gradually fill the cell, is completed.

In this case each of the two complementary virus RNA chains at a particular moment of replication forms a particular protein and, consequently, in each RNA molecule there are segments of information and purely replicative segments.

§ 3. Morphogenesis of RNA-Phages
(Viruses of Bacteria)

Bacterial viruses or phages are mostly complex structures with a well-marked differentiation, and in the case of large phages, with motility. Nearly all have DNA as their genetic material, and RNA is formed by them only as an intermediate carrier of information. However, in this group of living systems there are also structurally simplified forms, known as small phages, which resemble in their structure the simple forms of RNA-viruses of animals and plants. Depending on the type of their genetic material, small phages can be divided into three groups: a) phages containing DNA of low molecular weight, but with the normal double-stranded DNA structure (consisting of two complementary polynucleotides); b) phages containing an unusual single-stranded DNA, with a molecule of similar size to that of RNA of plant viruses; c) phages whose genetic material is RNA. This last group of phages was discovered only comparatively recently, and only a few members of it are yet known: phage f2, reproducing only on male strains of *E. coli* (F^+ strains), discovered by Loeb (Loeb, 1960; Loeb and Zinder, 1961), phage MS2 which is similar to it (Davis et al., 1961; Doi and Spiegelman, 1962; Weissman and Borst, 1963), phage R17 (Paranchych and Graham, 1962), phage fr (Kaerner and Hoffman-Berling, 1964), phage β (Nomoyama and Ikeda, 1964), and several of their mutants. The coefficient of reproduction of phage f2 is very high: the yield of infective particles from one *E. coli* cell varies from 100 to 20,000. Purified RNA of these phages possesses infectivity (Davis et al., 1961; Paranchych, 1963), although after purification of the RNA the yield of infective particles is very small.

It is obvious that in our approach to the discussion of possible molecular mechanisms of morphogenèsis of this group of phages we shall encounter the same problems as those arising during the investigation of RNA of plant and animal viruses. The volume of information contained in RNA of the small phages is about the same as is present in the RNA of animal viruses and, consequently, it can provide for the synthesis of only a few proteins. The RNA of these phages is a polycistron, but it cannot contain more than four or five cistrons. RNA of phage f2 has a molecular weight of about 700,000 (Leob and Zinder, 1961). RNA of phage MS2 has a molecular weight of about 1,000,000 (Ohtaka and Spiegelman, 1963). The earliest of the proteins synthesized during reproduction of this virus must be virus-specific RNA-synthetase (polymerase), because synthesis of the new RNA in this case takes place on RNA-templates with the intermediate formation of a double (replicative) RNA helix (Langridge et al., 1964; Weissman, 1964; Kelly and Sinsheimer, 1964; Amman et al., 1964). It has in fact been shown that the RNA of phage MS2 and of the similar phage f2 acts as template for synthesis of RNA-synthetase (Weissman and Borst, 1963; Weissman, Simon, and Ochoa, 1963), and perhaps, of two different RNA-synthetases (Delius and Hofschneider, 1964), which catalyze the formation of double-stranded RNA and the subsequent synthesis of phage RNA by means of complementary RNA-dependent synthesis. Mitomycin C, an inhibitor of DNA synthesis, had no effect on replication of RNA of phage f2, while blocking synthesis of DNA in cells of infected bacteria (Cooper and Zinger, 1962). Actinomycin D, inhibiting DNA-dependent RNA synthesis, likewise had no appreciable action on phage RNA-synthetase (Weissman, Simon, and Ochoa, 1963; Kelly et al., 1965).

Ribonucleoside triphosphates are the substrate for RNA-synthetase, and an assortment of all four triphosphates is necessary for its action. This RNA-synthetase differs in many of its properties from other RNA- and DNA-polymerases (August et al., 1965). According to preliminary findings obtained by these workers, phage RNA-synthetase has the character of a nucleoprotein in which the enzyme protein is bound to "primer" RNA.

RNA-synthetase (or RNA-replicase) formed by phage MS2 has been isolated in a purified form (Haruno et al., 1963). This enzyme possessed definite substrate specificity relative to phage RNA and was inactive in systems containing sRNA or ribosomal RNA as primer.

Doi and Spiegelman (1963) showed that RNA molecules of phage MS2, when entering cells of $E.\,coli$ (labeled with P^{32} and N^{15}), despite active autoreplication (through a phase of complementary replication) and template activity, do not fragment to begin with. They behave as stable templates, but they cannot be numbered among the RNA molecules participating in the formation of particles of the abundant progeny of the initial virus particles. This was discovered somewhat earlier by Davis and Sinsheimer (1963), who could not find parent RNA in the progeny of this phage and postulated that its molecules had fragmented during reproduction. However, as the work of Doi and Spiegelman showed, parent RNA is not present as such in the composition of the phage progeny, but is bound with the ribosome fraction. Having become attached to ribosomes to perform their template functions, molecules of parent RNA remain in this fraction until total lysis of the cell. These workers consider that it is this primary form of RNA which contains the program of synthesis of RNA-synthetase. This is a logical view because synthesis of this enzyme must precede replication of RNA itself. So far as the complementary form of RNA, formed by the action of RNA-synthetase, is concerned, according to these workers it has no "meaning" for protein synthesis and serves simply for subsequent replication of active template.

Conversions of parent RNA of MS2 phage in the period of intracellular development have been studied in particular detail by Kelly and co-workers (1965). They used an interesting method for this purpose. They determined the fate of phage infective RNA (sedimentation coefficient 27S), previously labeled with P^{32} by studying localization of the label at short intervals after s y n - c h r o n i z e d infection. In this case the processes of replication of RNA in the *E. coli* culture were synchronized and by determining the fate of the label at five-second intervals after the beginning of reproduction, it was possible to observe the interesting process of p e r i o d i c i t y of synthesis of new amounts of phage-specific RNA polymer.

DNA-dependent RNA synthesis in *E. coli* cells was inhibited under experimental conditions by actinomycin D. After 50-90 sec, almost 70% of the P^{32} of the parent RNA was found in one fraction of RNA with a sedimentation coefficient of about 20S, which these workers called the replicative form of RNA, because it was resistant to ribonuclease (double-helical form). The quantity of P^{32} of ribonuclease-resistant RNA then fell sharply, and a new peak of this RNA appeared after 3.5 min (Fig. 12). In one case this wave of change of the parent RNA into the double-helical RNAase-resistant form was observed on four successive times. In addition, slight but definite incorporation of parent P^{32} material was observed into a special low-molecular weight 6S fraction (mol. wt. from 1 to 30×10^5) was observed, which, however, contained much newly synthesized RNA (in this fraction the ratio of parent RNA/new RNA was much lower than in the 20S fraction of RNA).

The role of the 6S fraction of RNA in virus reproduction has not yet been explained. Kelly and co-workers consider that this component is perhaps an adaptation for replication of a specific part of the virus genome.

Be that as it may, we can see that phages of the RNA-group do not differ in principle from RNA-viruses, and the elucidation

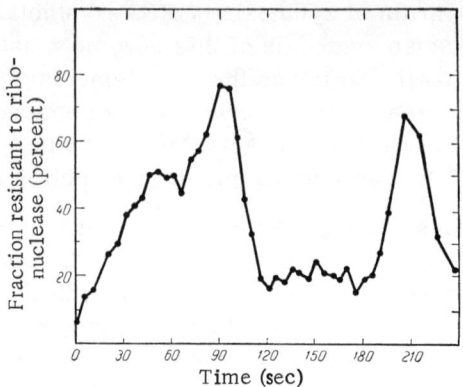

Fig. 12. Percentage of the input P^{32} from labeled MS2 phage into a form resistant to ribonuclease in a lysate of cells as a function of time after removal of actinomycin (Kelly et al., 1965).

of mechanisms of their morphogenesis does not present evolution with any new tasks compared with those solved in the case we have just examined. This comparison can, of course, be made in the other direction, if the RNA-phages are an earlier form in evolution than the RNA-viruses.

These morphogenetic problems consist essentially of regulation of the temporal sequence of synthesis of phage-specific proteins and formation of each protein in the amounts necessary for phage reproduction. In this case coat proteins or protein are formed in much larger amounts than enzyme proteins, both early (RNA-synthetase) and late (lysozyme).

Definite experimental proof that these problems are solved during reproduction of these phages was obtained by Ohtaka and Spiegelman (1963), who studied synthesis of phage-specific proteins after addition of purified RNA of phage MS2 (mol. wt. about 1×10^6) to cell-free extracts of *E. coli*. Under favorable conditions, three phage-specific proteins were synthesized in succession in this system *in vitro*. Initially and throughout the period of syntheses, large quantities of histidine-free phage coat protein were formed, however, 6 minutes of reaction time elapsed before a histidine-containing phage protein was detected in lesser quantity.

The time of appearance of the third component, evidently RNA-synthetase, which was produced in very small quantities, could not be established and these workers do not rule out the possibility that it was formed because synthesis of the coat protein began.

Ohtaka and Spiegelman suggest that the temporal sequence of synthesis of different proteins from one stable polycistronic template is determined by the linear sequence of the cistrons. However, according to their hypothesis, the fact that different neighboring cistrons work at different speeds and produce very different quantities of protein molecules may be associated with the existence of a special molecular system for controlling the frequency of protein synthesis (translation) on a segment of each cistron by means either of inhibition by the product of synthesis or of interaction between a component of the system of synthesis (translation) and a special segment built into the RNA-template. This segment, consisting of a definite sequence of nucleotides, may be connected in some way or other with a system of boundary zones between cistrons (intercistron punctuation).

It is interesting to note that RNA of phage f2 can synthesize phage-specific proteins of the phage coat by interacting not only with ribosomes from *E. coli*, but also with a ribosomal system from *Euglena gracilis*, an organism very far distant from *E. coli* in the scale of evolution (Schwartz et al., 1965).

§4. Mechanisms of Morphogenesis of Small DNA-Phages Containing Single-Stranded DNA

Biological systems whose mechanisms of reproduction were examined in the preceding sections are—structurally speaking—the simplest living organisms. They are all obligate parasites, and they cannot therefore be regarded as primary from the standpoint of evolution, and their simplicity must be the result of reduction. We do not yet know definitely what were the primary evolutionary forms of life, but the arrangement of forms existing at the present time in accordance with the principle of increasing biochemical complexity of their organization and the complexity of their form of individual morphogenesis can be imitated to some degree by the natural evolution of living matter. From this point of view the small phages containing single-stranded DNA, of which phage φX174 has been most thoroughly studied from the biochemical aspect, is of special interest.

In the group of viruses just considered, succession of generations was due to ribonucleic acid, molecules of which, possessing the ability of autoreplication, were at the same time the actual templates of synthesis of specific proteins. However, at the stage of evolution of life when the genetic system of the individual form of life was represented by a single macromolecule, combination of the template and genetic functions in the same polynucleotide chain has definite limitations. If the molecule is too long it can no longer be a convenient template, more especially if the template utilizes ready-made systems of protein synthesis adapted to comparatively short molecules of messenger RNA, corresponding to the length of the polypeptide (or group of polypeptides coded by a single operon) multiplied by the length of the elementary codon. However, the type of the simplest virus structure and of virus reproduction does not provide for a polymolecular (chromosomal) structure of the genetic system. From this point of view, the structural dissociation of the genetic function and the direct function of protein synthesis, which took place with the appearance in evolution of desoxyribonucleic acid, was of fundamental significance in the development of living forms.

In this case the genetic macromolecules became a form of preservation and reproduction of genetic information in a series of generations and did not carry the functions of direct participation in protein synthesis. An intermediate temporary functional unit, a carrier of genetic information nowadays called messenger RNA, appeared between the genetic macromolecule and protein synthesis. This carrier of information brought about the transcription and translation of particulate genetic information separately from each structural cistron of DNA. This type of differentiation of genetic and protein-synthesizing functions made it possible to increase the capacity of the macromolecule for genetic information, and as we now know, the molecular weight of DNA in a group of phages has risen in some forms to an extremely high value (160×10^6). However, in small phages of the φX174 type, the DNA molecules still do not possess a large store of information and their molecular weight is about 1.6×10^6, i.e., they lie within the limits of molecular weight of virus RNA. This is in fact the smallest weight known for the various types of genetic DNA. Phage φX174 was discovered in 1935, but it was not until 1959 that Sinsheimer showed that the DNA of this phage has an unusual single-stranded structure and he studied several of its physical and chemical properties (Sinsheimer, 1959). Since low-molecular-weight and single-stranded forms of DNA had not previously been known in any living organisms, it was natural that phage ϕX174 should have received intensive study in the subsequent years.

In their structure, mature particles of phage ϕX174 are just as simple as the particles of simple viruses of plants and animals. The capsule of this phage consists of 12 morphologically identical protein structures (capsids) located at the apices of an icosahedron (Tromans and Horne, 1961; Maclean and Hall, 1962).

A model of this phage based on examination of electron micrographs is shown in Fig. 13. Each of the twelve morphological units of phage ϕX174 consists, according to several observations, of several subunits composed of individual protein molecules. Inside this complex lies a DNA molecule, comprising 25% of the weight of the phage particles.

It has been clearly shown that it is this DNA molecule which contains the full program of development and reproduction of the phage. If small quantities of pure DNA of phage ϕX174 are added

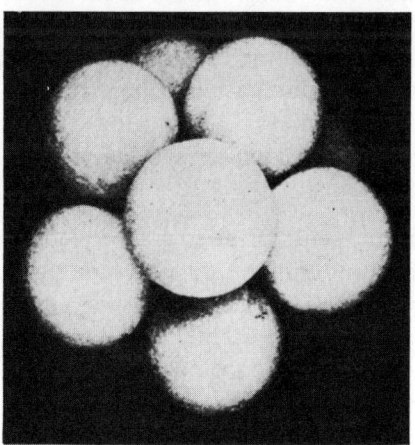

Fig. 13. Model of phage ϕX174. 2,000,000× (Maclean and Hall, 1962).

to bacterial cells partly deprived of their cell wall (protoplasts), the DNA which penetrates into them is infective. Soon after its penetration into the bacterial protoplasts, zones of phage reproduction are formed in them, and within a short time the cells are filled with hundreds of mature phage ϕX174 particles (Guthrie and Sinsheimer, 1960; Hofschneider, 1960; Sekiguchi et al., 1960; Il'yachenko, 1963).

After penetrating into bacterial cells the DNA of this phage undergoes the typical fate which we saw for RNA in cases of virus reproduction considered above: it is replicated by complementary synthesis with the preliminary formation of double-stranded DNA. This process was discovered in later experiments (Sinsheimer, 1961; Sinsheimer et al., 1962) by fractionation of nitrogen- and phosphorus-labeled DNA of infected cells in a density gradient of cesium chloride at various times after infection. This double-stranded form of phage DNA was called the replicative form (RF).

The formation of the replicative double-stranded form of DNA can therefore be regarded as the first act in molecular morphogenesis of this virus. The replicative form also has infective properties, i.e., it preserves the complete program of phage development.

It is interesting to note that molecules of single-stranded DNA of mature phage possess a special form of ring structure closed by a covalent bond (Sinsheimer, 1962; Fiers and Sinsheimer, 1962). An open polynucleotide chain is formed in this case only as a result of procedures breaking the ring. The original parent molecule is not preserved during subsequent reproduction, but gives rise to complementary and true copies, until the end of the intracellular process whereupon it undergoes fragmentation or dispersion (Kozinskii, 1961).

The circular molecule of single-stranded DNA of phage ϕX174 does not lose its circular structure during cell autoreplication when synthesizing its own complementary copy and forming the double-helical replicative form. The newly formed replicative form also has the structure of a closed ring, as has been shown not only by indirect biochemical methods (Burton and Sinsheimer, 1963), but also by direct electronmicroscopic observation (Kleinschmidt et al., 1963).

In the course of RNA synthesis *in vitro* and *in vivo* by phage φX174, DNA/RNA hybrids are formed initially, and these also are circular in shape (Bassel et al., 1964). It is believed that the low-molecular-weight polycistron DNA of this phage, when it undergoes intracellular reproduction, also forms one polycistron RNA template for protein synthesis. In large phages, the genetic DNA macromolecule is known to determine the formation of a series of functionally different molecules of messenger RNA.

As the results of molecular hybridization and determination of base sequence have shown, RNA formed during phage reproduction *in vivo* on the basis of the circular double-helical replicative form of DNA (Hayashi et al., 1964) is complementary to only one of the two DNA chains, i.e., to that which is complementary, in turn, to the single-stranded DNA of the mature phage. *In vitro*, when the circular structure of DNA is disturbed, both strands of replicative DNA can act as substrate for RNA-polymerase. The single-stranded DNA of this phage can also act as template for RNA synthesis *in vitro*, with the preliminary formation of DNA/RNA hybrids (Chamberlin and Berg, 1964).

The sheath of bacteriophage φX174 consists of a single type of protein (Carusi and Sinsheimer, 1961). Synthesis of this protein (identified by its antigenic properties) begins about 8 min after infection of *E. coli* cells and reaches a maximum after 15 min. Mature phage particles begin to appear in the cells after 12 min, and continue to do so parallel to synthesis of virus protein (Rueckert and Zillig, 1962). Homogenized *E. coli* cells also have the property of maintaining synthesis of this type of virus protein (Rueckert, Zillig and Doerfler [1962]).

Synthesis of phage-specific RNA in the case of phage φX174 evidently takes place under the influence of DNA-dependent RNA-polymerase. No repressor of synthesis of host RNA and DNA appears at the beginning of infection, for this synthesis continues parallel to phage formation, almost until the phase of lysis (Rueckert and Zillig, 1962).

The work of Hayashi (1965) showed that RNA formation on the circular DNA of this phage takes place in a certain order. The hybrid of double-helical replicative DNA with RNA is changed in the course of synthesis while its polynucleotide chain also grows.

Ultimately a polycistronic RNA is formed. However, the molecular regulation of early and late proteins for this virus has not been investigated. It has been postulated that a special lytic enzyme producing lysis of bacteria appears in the late phases of infection, although no precise data are yet available concerning the nature of this protein.

The "program of development" in DNA of phage ϕX174 is thus extremely limited, and its execution involves no more complex problems than those occurring during reproduction of the simple viruses of plants and animals.

Other small phages with single-stranded DNA also are known (S13; ϕR; 1ϕ7), but the study of synthesis of their elements is still in its early stages.

§5. The Molecular Mechanisms of Morphogenesis of Animal DNA-Viruses

The group of DNA-containing viruses of animals is exceptionally varied in its special composition, morphology, and cycles of development. It includes comparatively simple viruses from the biochemical point of view, such as the polyhedrosis viruses of insects, and extremely complex forms, such as the virus of smallpox. The morphological structure of one of the largest viruses of this group, the virus of human smallpox, is shown in Fig. 14.

The morphogenesis of the different viruses of this group has been studied very patchily. I do not propose in this section to examine in any great detail the external morphological pattern of morphogenesis of particular viruses of this group. That would be too extensive a task and would be at variance with the main purpose of this book.

A detailed description of the morphological pattern of development of several viruses of this group was given in a survey by Bykovskii (1964).

However, although external morphological aspects of the development of some of these viruses have been studied comparatively thoroughly, research into the molecular genetics and molecular mechanisms of morphogenesis is still very limited. As regards the level of complexity and, consequently, the volume of genetic

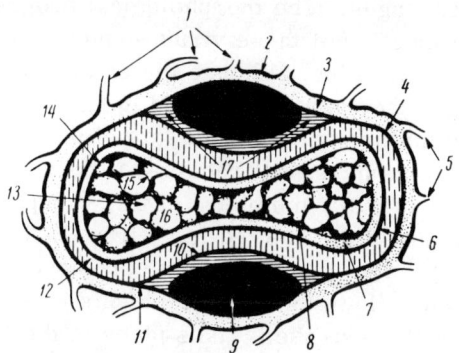

Fig. 14. Structure of mature human smallpox virus (Bykovskii, 1964). 1) Virovilli; 2) outer osmiophilic layer of the virus membrane; 3) middle osmiophobic layer of virus membrane; 4) inner osmiophilic layer of virus membrane; 5) outer opening of canal of virovillus; 6) outer osmiophilic layer of nucleoid membrane; 7) middle (osmiophobic) layer of nucleoid membrane; 8) inner osmiophilic layer of nucleoid membrane; 9) lateral body; 10) membrane of lateral body; 11) area where inner osmiophilic layer of virus membrane splits into two sheets, one of which is the membrane of the lateral body; 12) viroplasm; 13) nucleoidoplasm; 14) site of topographic connection between inner osmiophilic layer of nucleoid membrane and fibrillary component; 15) osmiophilic fibrils filling cavity of nucleoid; 16) hypothetical helices formed by osmiophilic fibrils; 17) osmiophobic spaces, triangular on section.

information contained in their DNA, these viruses resemble the DNA-phages containing double-helical DNA. The small DNA-viruses of animals have a genetic molecule of the same size as the small DNA-phages. The DNA of pseudorabies virus has a molecular weight of about 4.5×10^6 (Ben-Porat and Kaplan, 1962), while the DNA of adenovirus, when it consists of a single molecule, has a molecular weight of 10×10^6, accounting for about 13% of the weight of the mature virus particle (Green, 1962; Green and Pina, 1962). The DNA molecule of one of the largest animal viruses, vaccinia virus, has a molecular weight of about 80×10^6, about half the total weight of DNA in the virus particle (Joklik, 1962, a, c).

Since the possibility has not been ruled out that this DNA may fragment during its isolation, it can be assumed that if the vaccinia virus particles contain not two, but one genetic molecule, its molecular weight must be close to 160×10^6.

Intracellular reproduction of these viruses is a process which continues for several hours, and in its external morphology consists of many stages. The morphological program of DNA, in the case of reproduction of these viruses, must provide for the synthesis of many proteins, although the nature of these proteins, the order in which they are synthesized in time, the mechanism of formation of the mature particle, the fate of the parent DNA, and many other problems have still received little study. Autoreplication of virus DNA in many members of this group of viruses takes place in the nucleus of the host cells. Virus DNA in this case is apparently incorporated into the host's chromosomes and parasitizes its genome. However, DNA synthesis in vaccinia virus, in contrast to other DNA-viruses, takes place in the cytoplasm. The functions of the virus DNA are essentially to synthesize messenger RNA and proteins. Studies of adenovirus reproduction have shown that intracellular synthesis of virus antigens takes place in a definite order (Allison et al., 1960; Wilcox and Ginsberg, 1961), and that it is preceded by synthesis of DNA and RNA (Flanagan and Ginsberg, 1962; Wilcox and Ginsberg, 1963).

The clearest demonstration of the temporal sequence of synthesis of different forms of virus-specific messenger RNA has been given by the work of Becker and Joklik (1964), who studied synthesis of fast-labeled RNA in HeLa cells infected with vaccinia virus.

These workers showed that at the beginning of infection (30 min-1 h after infection) forms of virus-specific RNA of low molecular weight (sedimentation constant 8-12S) are synthesized. After 1 h, forms of virus-specific RNA with a greater molecular weight appear, and the size of the RNA continues to increase with time (14S forms appeared after 2 h, 16S after 3 h, 18S after 4 h, and 20S after 5 h). The nucleotide composition of this RNA corresponded to that of virus DNA, and this correspondence was further confirmed by experiments on DNA/RNA hybridization.

The virus-specific RNA had all the characteristics of messenger RNA: it formed polysomes and competed actively in this respect with messenger RNA of the HeLa cells and completely expelled it from the sites of protein synthesis approximately 3 h after infection.

The heterogeneity of the polyribosomes formed by RNA of vaccinia virus was also demonstrated by other experiments (Scharff et al., 1963; Salzman et al., 1964), which at the same time demonstrated the heterogeneity of the virus protein antigens and heterogeneity of the messenger RNA as regards its molecular weight.

Considering that the temporal sequence of synthesis has also been demonstrated for the proteins of this virus (Loh and Riggs, 1961), it must be supposed that the polygenome structure of double-helical DNA possesses some form of mechanism to provide for temporal differentiation of utilization of its information. It would be easiest to assume that this mechanism coincides with its linear structure, i.e., with the successive arrangement of the cistrons, which are "switched on" one after the other, for example during the successive untwisting of the DNA double helix. It was recently shown that after pseudorabies virus enters rabbit kidney cells, the endogenous DNA in these cells is inhibited by the formation of some form of early virus protein in the cells (Ben-Porat and Kaplan, 1965). The temporal sequence of synthesis of proteins and DNA (between 6 and 28 h after infection) has also been discovered for intracellular reproduction of adenovirus (Palasa and Green, 1965).

A study of the finer details of the DNA of one of the animal viruses, polyoma virus, showed that, like the DNA of phage $\varphi X174$, it is circular (Dulbecco and Vogt, 1963).

The ring is formed by double-helical DNA, each polynucleotide of which is closed. The importance of this phenomenon will be discussed in the next section.

In a recent investigation by Vinograd and co-workers (1965) an electronmicroscopic study was made of the circular double-helical DNA of this virus in various phases of its cycle and after the use of biochemical procedures associated with the formation of breaks in one of the polynucleotides, with twisting, and so on. As a result, these workers showed that the DNA of polyoma virus can exist in several different states (Fig. 15), while maintaining its infectivity.

§ 6. Molecular Mechanisms of Morphogenesis of Large DNA-Phages

The most complete survey of all aspects of molecular biology and genetics of the large bacteriophages is to be found in

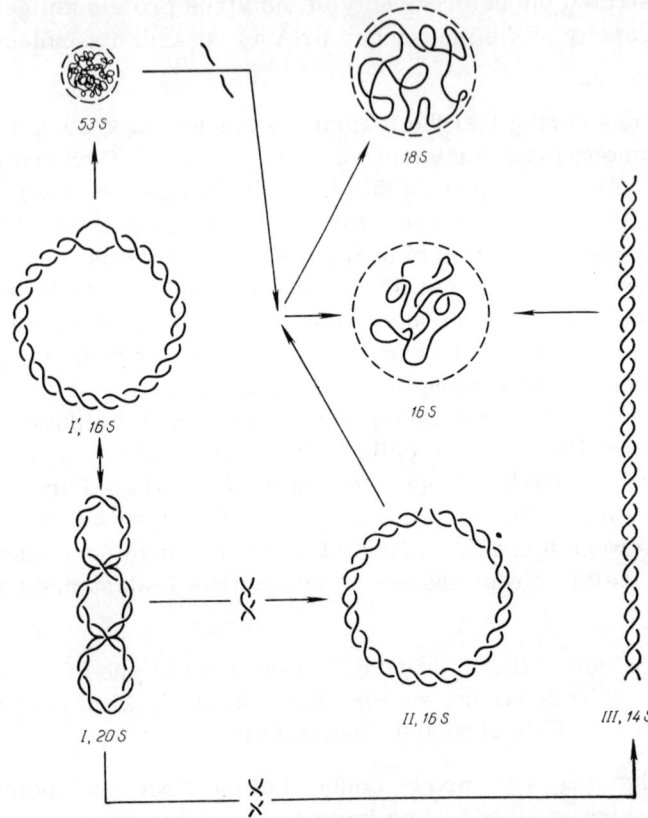

Fig. 15. Diagrammatic representation of the several forms of polyoma DNA. The duplex segment shown contains 12 turns, about one-fortieth of the total number. The twisted circular duplex shown contains one left-hand tertiary turn. 8% of the right-hand duplex turns in the model are unwound to form I'. The dashed circles around the denatured forms indicate the relative hydrodynamic diameters. Sedimentation coefficients (S) are given under each form (Vinograd et al., 1965).

the important monograph by Stent (1963, Russian translation 1965) and Hayes (1964, Russian translation 1965). A detailed analysis of the biochemical aspects of phage reproduction has been given by Cohen (1963) and Thomas (1963). From the wide variety of topics embraced by this problem we shall choose only one for theoretical analysis: the mechanism of action of the system controlling morphogenesis of the phage particle, i.e., the way in which the purposive sequence of individual acts of phage reproduction in time and space is regulated.

The phages of this group are comparatively complex morphological systems capable of motion and of active infection of the bacterial cell (Fig. 16). Reproduction of the most complex phages is associated with the synthesis of more than 30 different phage-specific proteins, from 8 to 10 of which go to compose the mature phage particle and about 20 are enzymes or repressors, participating in intermediate reactions of phage reproduction. To carry out these syntheses in their correct order, the corresponding number of types of messenger RNA is formed. For regulation of all these processes to conform to the time schedule and to yield the correct amount of the different proteins, a self-reproducing genetic control

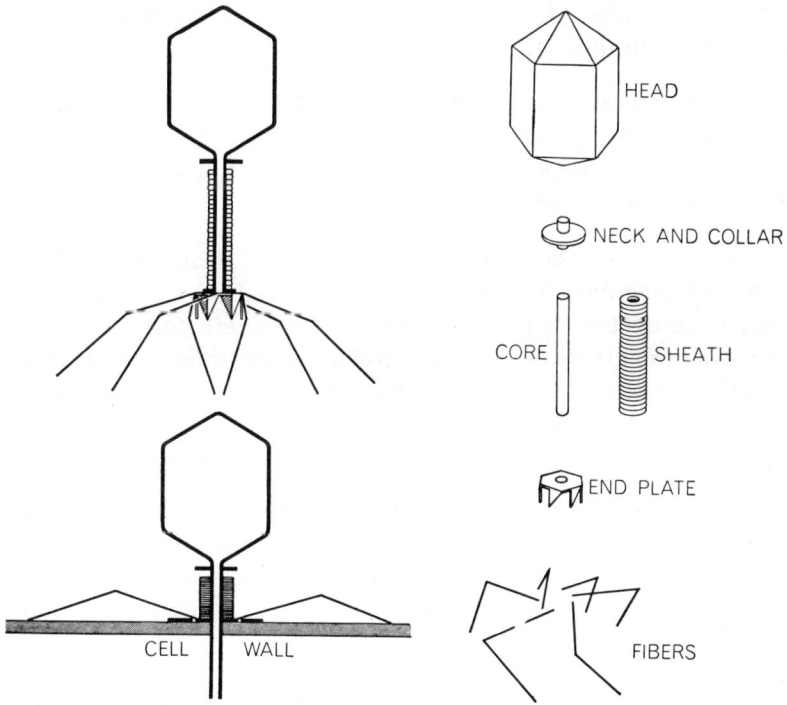

Fig. 16. Diagram showing components of bacteriophage T4. Model of penetration of the bacterial cell wall by the virus shown below on the left (Edgar and Epstein, 1965).

system must exist. The functions of such a system are performed by one giant DNA molecule, compactly arranged in the phage head. The molecular weight of this DNA (in phage T2) is between 130×10^6 and 160×10^6 (Rubenstein et al., 1961). Considering the double-helical character of this DNA and the fact that only one helix contains information of real concern to protein synthesis, it can be calculated that this "messenger" polynucleotide consists of about 250,000 nucleotides. With a nucleotide triplet code this is sufficient for synthesis of a peptide consisting of about 80,000 amino acid residues. Since the simplest proteins are built on the average from 100-200 amino acid residues, it is clear that the volume of information contained by one molecule of phage DNA is far more than is needed to provide for coding the synthesis of 25-30 known phage-specific proteins. Despite their exceptional length, the phage molecules are stable, and there are no weak points for rupture in one of the two chains (Thomas, 1963). With slight variations the DNA molecules of the other phages of this group are built in accordance with the same principle.

Each of the DNA molecules, controlling and programming the system of morphogenesis and reproduction of the phage, when entering a cell in a more or less pure form must provide for the following principal features of the process.

1. The necessary temporal sequence of syntheses of messenger RNA molecules, forming templates for initial synthesis of receptors and systems of synthesis in the host cell, followed by synthesis of early enzymes, organizing the synthesis of certain unusual nucleotides present in the composition of phage DNA and certain other enzymes, followed by synthesis of late proteins of the phage body and proteins responsible for lysis of the host.

2. The necessary quantitative relationships between synthesis of all these proteins, in the sense that synthesis of one molecule of protein No. 1 may correspond to synthesis of three molecules of protein No. 2, five molecules of protein No. 3, eight molecules of protein No. 12, and so on.

3. Autoreplication of the DNA molecule itself, in accordance with the principle of complementary synthesis.

4. Assembly of all details of the mature phage particle.

a) Regulation of Temporal Sequence of Synthesis of Early and Late Proteins Coded by Different Segments of the Same Polynucleotide (Polycistron) of Phage DNA. The DNA molecule of the large T-even phages (T2, T4, etc.) differs from the usual DNA of the host cell in a number of chemical features. Instead of desoxycytosine, the DNA of these phages contains desoxy-5-hydroxymethylcytidine-5-phosphate. After it has been synthesized this DNA is glucosylated. The thymine for this DNA is synthesized rather differently than in the host cells. In order, therefore, to enable synthesis of the DNA of these phages to take place in an infected cell, a series of new enzymes must be provided which are absent from the host cells and are not programmed in its genome. Information for synthesis of these enzymes is contained in the phage DNA itself. A large group of these phage-specific enzymes has now been identified (desoxycytidylate-hydroxymethylase, a special DNA-polymerase, glucosyltransferase, desoxycytidine-pyrophosphatase, desoxycytidyl-deaminase, thymidylate-synthetase, and so on). In addition, phage DNA contains information for synthesis of other enzymes of the usual type (desoxyribonuclease, dehydrofolate-reductase, kinases, etc.). The identification of these enzymes has been undertaken over a number of years in several laboratories (Stent, 1963; Cohen, 1963; Edgar and Epstein, 1965). From our point of view it is important to note that these enzymes belong to the group of "early" proteins, i.e., proteins formed before intensive synthesis of phage DNA itself and also proteins used as building material for the phage particle (mostly proteins of the phage head). These proteins belong to the group of "late" proteins. Early proteins begin to be synthesized soon after infection, after which their synthesis is inhibited and at the 7th or 8th minute the synthesis of late proteins begins.

The group of early proteins also evidently include certain repressors formed by the phage to inhibit RNA and DNA synthesis in the cells of the infected bacteria themselves. If synthesis of only the early proteins is inhibited without inhibition of synthesis of phage RNA (by chloramphenicol), synthesis of ribosomal and soluble RNA in the host cells continues even after infection, but synthesis of bacterial DNA and of messenger RNA is inhibited in this case also (Okamoto et al., 1962; Nomura, Okamoto, and Asano, 1962; Nomura, Matsubara, et al., 1962). Different forms of RNA and DNA synthesis in the host cells may perhaps, therefore, be inhibited by different types of repressors formed by the phage. The existence of phage-induced inhibitors of DNA-dependent RNA-polymerase has also been demonstrated by more direct methods (Khesin, Shemyakin, et al., 1962; Skold and Buchanan, 1964). In phage λ which contains a somewhat smaller DNA than the phages of the T group, differences between the nucleotide composition of phage-specific messenger RNA molecules formed in the early and late periods of phage reproduction have been clearly demonstrated (Skalka, 1966).

Genetic information carried by the DNA molecule is thus utilized at different times, first one group of cistrons and then another being active, and it is evident that taken as a whole more of the late proteins are formed, and also that among these proteins, different types are synthesized in different amounts. Among the early proteins, moreover, not all the enzymes reach the maximum of their synthesis at the same time after infection, but some are earlier than others. The order of events is shown schematically in Fig. 17 (Edgar and Epstein, 1965).

Since synthesis of every protein is preceded by the formation of a specific template on DNA, in the form of molecules of messenger RNA reproducing the nucleotide code of the structural cistron of the particular protein, it is obvious that successive synthesis of different phage-specific proteins over a period of time implies that

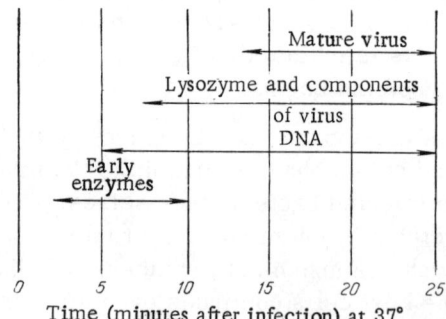

Fig. 17. Distribution of intracellular processes of bacteriophage reproduction in time (Edgar and Epstein, 1965).

the giant molecule of phage DNA synthesizes different groups of messenger RNA molecules in the early periods after infection (Khesin and Shemyakin, 1962; Khesin et al., 1963; Khesin, 1965, 1966; Hall and Spiegelman, 1961; Kano-Sueoka and Spiegelman, 1962; Hall, Nygaard, and Green, 1964; Spiegelman, 1963). It was shown by the method of hybridization of phage RNA formed at different periods after penetration of phage DNA into the bacterial cell (Hall and Spiegelman, 1961; Khesin and Shemyakin, 1962; Hall, Nygaard, and Green, 1964) that messenger RNA molecules formed at different times of phage development interact complementarily with different segments of the polynucleotides of phage DNA. This is shown by the fact that C^{14}-labeled messenger RNA molecules obtained from cells at different times after infection with phage do not compete with each other during the formation of complementary complexes with phage DNA denatured and "unwound" by thermal "melting" (thermal conjugation). This led these investigators to consider that messenger RNA is formed on different segments of phage DNA at different stages of intracellular phage development.

The same conclusion was reached as a result of experiments using a different and extremely elegant method (Kano-Sueoka and Spiegelman, 1962; Spiegelman, 1963). These workers labeled the RNA formed at different periods after infection with various isotopes. The RNA formed 3-5 min after infection was labeled with C^{14}-uridine, and the RNA synthesized between 13 and 15 min after infection was labeled with H^3-uridine. These RNA fractions were then compared with respect to certain biochemical properties (distribution of molecular size, gradient centrifugation, chromatographic fractionation). In cases when both types of label were used at the same period, the properties of the fractions labeled with C^{14} and H^3 were identical. If, however, the C^{14} and H^3 labels were used at different periods, the characteristic yield curves of phage RNA bearing these two labels clearly differed. One of these experiments is shown schematically in Fig. 18. Differences in the RNA fractions formed at different periods of phage infection, with respect to composition and other characteristics, can be seen and are associated with differences in the composition of the proteins coded by them.

Later experiments by Khesin and co-workers (1963) showed that the ability of phage DNA to synthesize early RNA molecules first is predetermined by certain features of the actual DNA structure. Phage DNA, purified by phenol deproteinization, and renatured after melting by heat *in vitro*, can act as template for synthesis of early messenger RNA fractions. To change over from the synthesis of early to late RNA in the middle of infection requires protein synthesis. If protein synthesis is inhibited in the infected cells by chloramphenicol (which inhibits protein synthesis only and not RNA synthesis), only early forms of RNA are synthesized in the cells. The change to synthesis of late forms of RNA is not observed under the conditions of chloramphenicol inhibition. If, however, chloramphenicol was added when synthesis of late forms of RNA had just begun, synthesis of early forms of RNA continued longer than the usual period, and was accompanied by synthesis of late forms. By inhibiting protein synthesis, chloramphenicol apparently fixed the process of change from one stage to another of the synthesis of phage RNA occurring at a particular time. The "switching on" of those cistrons (genes) of DNA which determine synthesis of late forms of mRNA was thus found to require protein synthesis, while synthesis of early forms of RNA is independent of protein synthesis and predetermined by the structural properties of DNA.

Phage DNA according to these workers' findings, in any situation begins by synthesizing early RNA and early proteins. They showed that the proteins necessary for the changeover to synthesis of late RNA are not special DNA-dependent RNA-polymerases. In this case RNA synthesis on DNA templates is carried out by means of RNA-nonspecific polymerase of the bacteria. Some phages, however, form a specific enzyme.

These experiments thus reveal a dual type of control, a primary determination (synthesis of early forms of RNA and passiveness of the late genes) predetermined by the actual DNA structure. The changeover to synthesis of late proteins cannot take place without preliminary synthesis of early proteins. No evidence has yet been found to suggest that the synthesis of a particular protein, such as a gene repressor or activator, is essential for this purpose.

Later work in Khesin's laboratory (Khesin et al., 1966) showed that whereas early forms of RNA can be synthesized *in vitro*, late forms require an intact cell structure for their synthesis, integrity of the membrane component of the infected bacterial cells being particularly important.

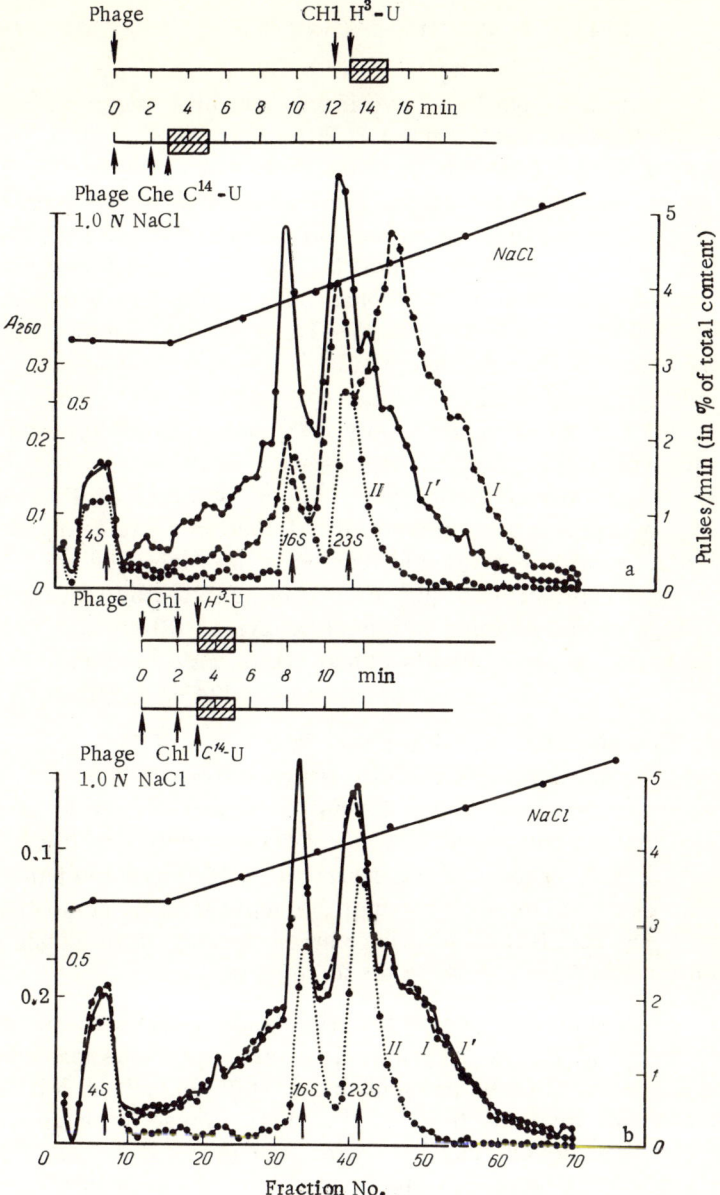

Fig. 18. Chromatography of T2-specific RNA on a methylated albumin column (Spiegelman, 1963). Legend: Chl represents chloramphenicol; H^3-U, tritium-labeled uridine. C^{14}-U, uridine labeled with C^{14}. Shaded rectangles denote periods of administration of label. Curves of optical density indicate the presence of preformed stable cell components (4S, 16S, 23S), located in the order as shown from left to right. a) Mixture of two RNA samples; H^3-labeled RNA obtained from culture labeled between 13 and 15 min; b) control mixture labeled with both isotopes at the same period; I and I') radioactivity (H^3 and C^{14}, respectively); II) absorption.

Other investigations have shown that this change of program from synthesis of early forms to late forms of RNA also requires DNA synthesis. If synthesis of phage DNA is prevented by ultraviolet irradiation of the phage, the formation of early phage-specific enzymes is not brought to an end but persists longer than the usual period, whereas the synthesis of late proteins, which are structural components of the mature phage, does not begin (Dirksen et al., 1960; Watanabe, 1957; Delihas, 1961). This was also demonstrated by the experiments of Cohen and co-workers (1963) to study infection of thymidine-deficient mutants of *E. coli* by $T6r^+$ phages. Phage reproduces in these mutants only if thymine, histidine, and uracil are present in the medium. If, however, synthesis of phage DNA is prevented by thymine deficiency, only early forms of RNA and early proteins are synthesized after infection. The changeover to synthesis of late proteins and lysozyme likewise does not take place when replication of phage DNA is inhibited by incorporation of analogs of amino acids into early enzymes (Ebisuzaki, 1963). These enzymes consequently cannot participate in synthesis of new DNA and the old DNA continues to synthesize early proteins only.

Khesin (1965) postulates that late forms of RNA are produced on newly synthesized phage DNA, which differs in some of its properties or in its connections with cell structures from the genetic DNA of mature phage and from the DNA accumulating in cells before lysis with respect to the distribution of active and passive genes. Khesin also suggests that the repression of some DNA regions may consist essentially of strengthening of the bonds between the two DNA chains in these regions.

When synthesis of late proteins begins, synthesis of early proteins ends. However, synthesis of early proteins may still continue and early RNA be present in a certain amount in cells synthesizing late proteins (Hall et al., 1964). These workers suggest that the changeover from early to late proteins is regulated not only at the DNA level, but also at the level of influences exerted on cytoplasmic processes, such as the ability of RNA molecules to compete for ribosomes. The possibility of regulation of the changeover from early to late syntheses by mechanisms located in the cytoplasm has also been postulated by Kozaka and co-workers (1963). They made observations on synthesis of an early phage-specific protein, desoxycytidine triphosphatase, during infection of bacteria with mixtures of normal phages and mutants injured by ultraviolet radiation. In certain combinations normal phage proved to be recessive, while the injured phage was dominant, resulting in absence of changeover to synthesis of late proteins even when DNA of normal phage was present.

Fresh possibilities for elucidation of the principles governing the changeover between synthesis of different forms of messenger RNA in time were suggested in Khesin's (1966) paper to the Ninth International Congress of Microbiology. He attempted to connect this changeover with the great affinity (allosteric) of RNA-polymerase for the early genes in the composition of DNA-phage. Next, at a certain moment during consecutive reading of the information contained in the early genes, the repressor region of the genome comes into operation and represses the early genes. Under these conditions, in Khesin's opinion, RNA-polymerase no longer finds its appropriate point of application on the DNA molecules and starts to attach itself to the free DNA segment and to read the late genes, although possessing less affinity for them. It should also be mentioned that a limited number of starting points for the action of RNA-polymerase is likewise found in the native DNA of T4 phages (Bremer et al., 1966).

Sueoka and Kano-Sueoka (1964) studied the biochemical and physico-chemical properties of 17 forms of transfer RNA (sRNA) of *E. coli* after infection with T2 phage. For 16 forms of aminoacyl-sRNA no difference was found between their properties in uninfected and infected cells, but with leucyl-sRNA fractions obtained at various times after infection, appreciable differences were present. In the period between 3 and 5 min after infection a new peak of leucyl-sRNA appeared, to disappear again at 8 min. Allowing for degeneracy of the code, it can be assumed that the new peak contains a leucine anticodon differing from the leucine anticodon in the sRNA of *E. coli*. According to code deciphering experiments, several codons in the template can code leucine (CUG, CUC, CUU, UUA, UUC). In this case, if leucine codons in the template (or in the DNA cistron) are represented, for example, by UUG and CUU only, only the anticodons to them in the sRNA molecule will be active, and anticodons to UUA and CUC will be inactive. If, on the other hand, cistrons in phage DNA are heterogeneous with regard to their leucine codons, and in one particular segment leucine is coded by a codon to which no anticodons are present in the sRNA of *E. coli*, this segment will be "translated" only when the necessary sRNA fraction appears in the cells. Consequently, the genetic program of phage DNA may be switched on at a certain moment (in this particular case, after 3 min) of the synthesis of an sRNA specially adapted for "translation" of a particular group of cistrons, whereas the remaining cistrons are translated by *E. coli* sRNA.

Further investigation of this phenomenon (Kano-Sueoka and Sueoka, 1966) reveals, however, that transformation of leucyl-sRNA takes place, not between the 3rd and 8th minutes, but between the 1st and 3rd minutes after infection. These workers therefore abandoned their earlier hypothesis concerning regulation of the changeover from early to late proteins, and suggested that modification of leucyl-sRNA may lead to "arrest" of protein synthesis by the host cell.

Bass and co-workers (1967) showed, however, that during the destruction of bacterial cells infected with phage T2, derepression of synthesis of bacterial DNA takes place. These workers consider that this process is evidence against the existence of specific phage repressors and they postulate the possibility of inhibition of synthesis of host DNA in the infected cells purely because of topographic separation of host DNA from DNA-polymerase.

Dove (1966), working with phage λ, showed that its DNA contains special genes which induce late syntheses. In this case proteins required for DNA replication are produced first, followed by replication of the DNA, at which stage transcription of the genome reaches the zone of genes inducing late functions, thereby putting into operation the process of maturation and synthesis of lysozyme.

Yet another interesting hypothesis of the molecular mechanism for the changeover from synthesis of early proteins (and early forms of mRNA) to late proteins in the T-phages was based on the result of experiments (Edlin, 1965) to study phenotypic reversion of certain mutants of phage T4 by the action of 5-fluorouracil (a mutagen usually incorporated in the uracil position, but counted in many cases as cytosine). Working with mutants, in some of which the mutations affected early genes, and in others late genes, i.e., genes active at the end of reproduction, Edlin found that fluorouracil, when given in the early stages, and then displaced by an excess of uracil, nevertheless caused phenotypic reversion of mutants of the late genes. This indicated that synthesis of late messenger RNA, into which the fluorouracil was incorporated, evidently took place immediately after the beginning of infection, before DNA replication. Although this conclusion is in conflict with the data discussed above, no other explanation of these facts has yet been put forward. In my opinion, there are other possible explanations, such as the absence of competition between fluorouracil and uracil and continuation of the action of fluorouracil in the late period. Reversion of the phenotype is perhaps connected with modification of anticodons in sRNA rather than with modification of codons in messenger RNA. In any case, the contribution made by the codons and anticodons to phenotypic reversion must be equal, and in one hypothesis which has been suggested to explain the mechanism of suppressor mutations (mutations restoring the phenotype, not by true restoration, but as the result of a new mutation at another locus) in bacteria this mechanism is explained in fact by changes in sRNA.

On the basis of his findings, Edlin suggests that the changeover from synthesis of early to late proteins, when both are synthesized together, takes place for the following reasons: 1) Early messenger RNA (mRNA) is formed faster than late mRNA; 2) early mRNA is rapidly renewed, while late mRNA is stable; 3) late mRNA is more intimately associated with ribosomes than early mRNA, i.e., it is more competitive for places of synthesis [item (2) may be the result of item (3) on the assumption that only free mRNA undergoes degradation].

According to this model, therefore, immediately after infection mainly early proteins, together with a little of the late group, are formed. Gradually, however, with accumulation of late forms of mRNA, the ribosomes are reprogrammed for synthesis of late proteins only.

b) Genetic Function of the Phage DNA Macromolecule during Phage Replication and Reproduction. The DNA molecule, located in the head of the phage and, as already mentioned, being of very large size through the continuity of its polynucleotide chains, is thus the material program providing information not only for the composition of proteins to be synthesized, but also for the regulatory mechanism (the "clock") predetermining the proper time for synthesis of a particular protein, and thus ensuring planned reproduction of phage particles. It is therefore essential that we examine the problem of functional and structural properties of phage DNA from this point of view.

The DNA molecule of large phages, like the genetic molecules of small phages, contains complete information for morphogenesis of the phage. Although it has not yet been possible to show that chemically purified DNA of the T-even phages, with a molecular weight of the order of 110×10^6-160×10^6, retains its infective properties, several smaller λ phages, containing DNA with a molecular weight of 33×10^6 (Caro, 1965) and a molecular length of about 17-$18\,\mu$, possess DNA whose infectivity after purification has been clearly demonstrated (Meyer et al., 1961). The infectivity of DNA from T1 phages, after purification with phenol, has also been shown to be somewhat weaker in the case of infection of bacterial protoplasts freed from their cell wall (Brody et al., 1964). The length of the DNA molecule of T2 and T4 phages is about 55-$56\,\mu$ (Thomas and MacHattie, 1964).

Because of the exceptional simplicity and the genetic "purity" of the study of mutations in phages and the development of methods for obtaining genetic recombination (hybridization) with phages, work done during the last few years on the production of a genetic map of phage T4 has led to the discovery of several important phenomena. The most interesting of these is discovery of the closed (circular) character of the genetic map of this phage (Streisinger et al., 1964; Baylor et al., 1965).

The circular character of the genetic map of the small phage ϕX174 is obvious because, as already mentioned, the DNA molecules of this phage also are circular in character. Electron-microscopic investigations have revealed the circular structure of the DNA in the larger phage λ (Ris and Chandler, 1963; MacHattie and Thomas, 1964) (Fig. 19).

The fact that the genetic map of phage λ corresponds to the parameters of its DNA was shown by Rading and Kaiser (1963). In their experiments the DNA molecule was broken into two parts, one of which, as mixed infection experiments showed, carried all the marker genes characteristic of the left side of the genetic map, while the other part contained marker genes of its right side. The DNA of this phage can change reversibly from a circular to a linear state because of the presence of complementary segments at the ends of its linear chains capable of linking with each other (Hershey and Burgi, 1965).

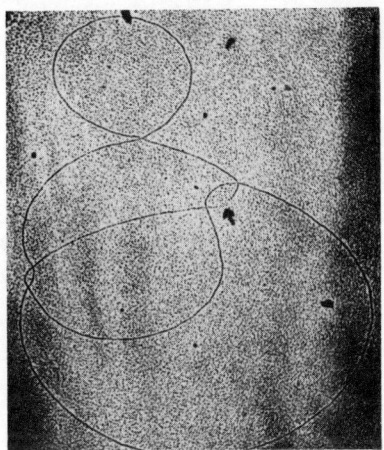

Fig. 19. Circular structure of DNA-phage (Ris and Chandler, 1963).

We must consider that linear agreement exists between the genetic picture and DNA molecule of phage T4, whose genetic map has been studied in greatest detail and whose DNA molecule also contains all genes (cistrons) of the phage. The genetic map of this phage, obtained as a result of work by many investigators studying recombination and linkage among its various mutants, is shown graphically in Fig. 20 (Edgar and Wood, 1966).

By comparing mutations undergone by particular syntheses (synthesis of early proteins, lysozyme, DNA, and so on) with the position of these mutations on the genetic map, a very important discovery was made. The position of individual genetic loci was found to correlate in many cases with the temporal order in which these loci function during intracellular phage reproduction. In addition, genes linked with one another functionally, such as the group of genes determining synthesis of early proteins or genes controlling synthesis of the structures of the phage particle, were found to be located in clusters on the genetic map.

The circular genetic map of the T-even phages would be very easy to understand if it could be assumed that the DNA molecules of these phages are circular in structure. However, native DNA molecules of phage T2 are linear in structure on electron micrographs. After denaturation and renaturation, the DNA molecules

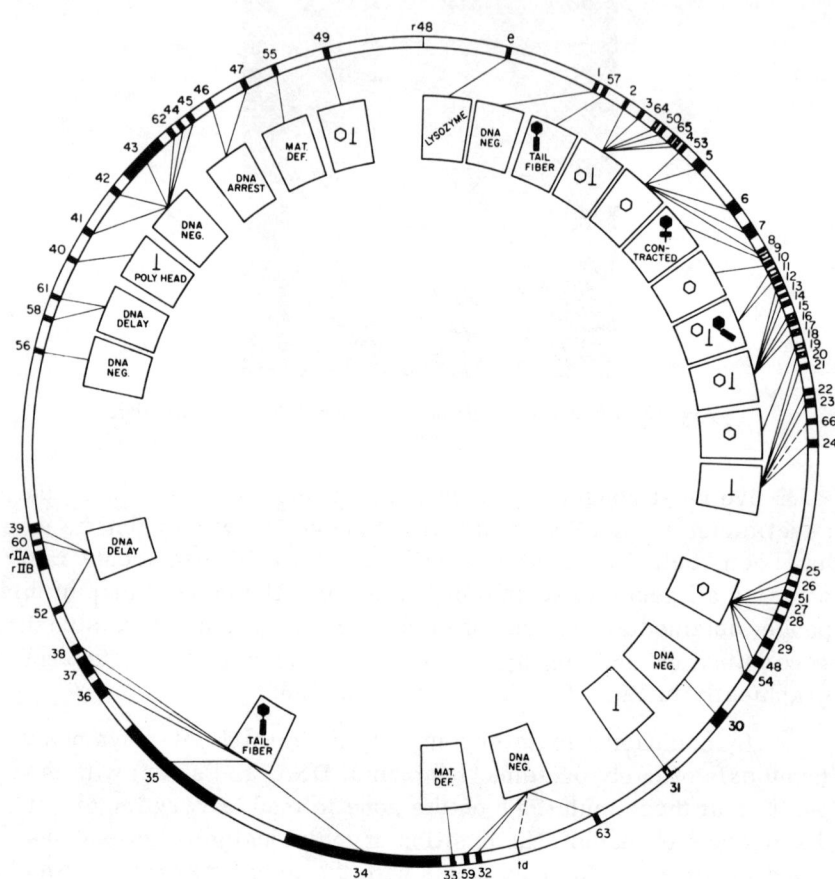

Fig. 20. Defective phenotypes of conditional mutants of T4D under restrictive conditions. Characterized genes are represented by shaded areas illustrating relative locations, and, if known, approximate map lengths. The enclosed symbols indicate defective phenotypes as follows: DNA NEG.: no DNA synthesis; DNA ARREST: DNA synthesis arrested after a short time; DNA DELAY: DNA synthesis commences after some delay; MAT DEF.: maturation defective, DNA synthesis is normal but late functions are not expressed; a hexagon indicates that free heads are produced, an inverted T, that free tails are produced; TAIL FIBER: fiberless particles produced; gene 9 mutants produce inactive particles with contracted sheaths; gene 11 and 12 mutants produce fragile particles which dissociate to free heads and free tails. (Edgar and Wood, 1966).

of this phage can form rings (Thomas and MacHattie, 1964). However, these workers claim that the circular character of the genetic map may be due, not to the circular structure of the phage DNA molecule in bacteria, but to rearrangement of linear fragments of maternal DNA during processes of replication of phage DNA in the bacterial cell. They consider that native linear DNA molecules of phage T2 are a population of molecules in which the same nucleotide sequence is cyclically heterogeneous in the sense that the sequence typical of the terminal zone of one molecule is found in another in the center of its linear structure, and vice versa, as illustrated in Fig. 21. Accordingly, during alkaline

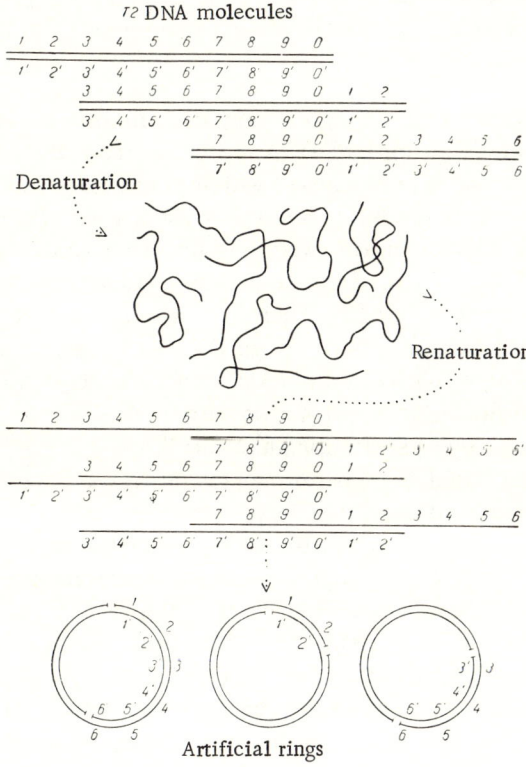

Fig. 21. Diagram showing how a set of linear DNA double helices (duplexes) with sequences consisting of circular rearrangements of the same information can form circular molecules by separation of the chains following random association and restoration of the double-helical structures (Thomas and MacHattie, 1964).

denaturation and renaturation of these molecules, the circular structures clearly observable in electron micrographs are formed.

The formation of these DNA molecules, heterogeneous with respect to the linear localization of the gene groups, can in fact take place as the result of what is known as fragmentation of the parent DNA molecule and incorporation of different fragments in different parts of the DNA molecules in the phage progeny. The discovery of fragmentation of the parent DNA molecule was the result of a series of investigations to study the fate of parent molecules during phage reproduction, i.e., to study whether they enter the phage progeny and, if so, in what form: as a DNA double helix (conservative replication), as separate polynucleotides (semi-conservative replication), fragmentarily, or dispersively.

The discovery of a fragmentary distribution of polynucleotide segments of the original DNA molecule from the phage particle (both small and large phages) which penetrates into the bacterial cell was first made by Kozinskii (Kozinskii, 1961; Kozinskii and Kozinskii, 1963). In his experiments, Kozinskii used a very ingenuous method. DNA of the host cells was labeled with 5-bromouracil and thus became "heavy." The phage DNA in T4 was labeled with P^{32}. Synthesis of phage DNA is known to take place at the expense of breakdown of host DNA and, consequently, in the case of conservative or semi-conservative replication after the second and third cycles of reproduction, the "light" P^{32}-labeled polynucleotides must have been preserved in the progeny without further dispersion of the label in the progeny DNA. However, a different pattern of distribution was observed, in which DNA molecules of the whole progeny contained fragments of parent P^{32}-labeled DNA corresponding to the dispersive method of replication. However, Kozinskii remarks that the dispersive method of replication can exist in two types. In one type distribution of the parent material in the progeny may be rather like the dissolving of alcohol in water. In the other type, dispersion may have the character of fragmentation in which progeny DNA resembles a patchwork quilt: comparatively large fragments of parent DNA are inserted in it here and there. In this case the structure of the giant RNA molecule must also be fragmentary, composed of subunits consisting of elementary discrete carriers of genetic information.

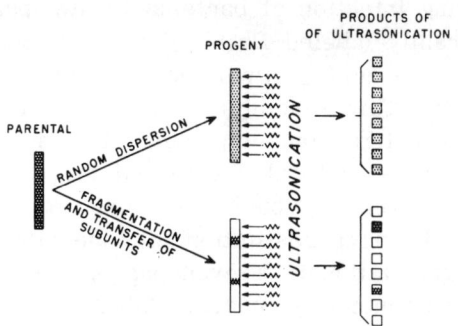

Fig. 22. Two possible methods of dispersion of parent P^{32}-DNA in progeny DNA (Kozinskii, 1961).

The two possible types of dispersive replication of DNA are shown schematically in Fig. 22.

By producing partial degradation of the progeny DNA by sonication, followed by division of the fragments into light (P^{32}-labeled) and heavy fractions, Kozinskii showed that dispersion of large fragments of parent DNA in fact takes place. The molecular weight of these fragments varied from 5×10^5 to 5×10^6, i.e., not less than the molecular weight of the RNA receiving from the DNA the information contained in it. Kozinskii considers that each discrete carrier of genetic information is reproduced in subsequent replications in accordance with the principles of complementary autoreplication, corresponding to the semiconservative model of DNA autoreplication suggested by Watson and Crick. This type of DNA fragmentation is not the rule of all phages. A semiconservative method of distribution of genetic material in the phage progeny with partial fragmentation has been found for DNA of recombinant forms of lysogenic phage (Meselson and Weigle, 1961).

Later experiments (Kozinskii, Kozinskii, and Shannon, 1963) showed that the process of fragmentary recombination of segments of the parent phage DNA molecule of phage T4 can be inhibited by chloramphenicol if added 5-7 min after infection, i.e., during the period of synthesis of early protein. On this basis, these workers postulate the existence of a special phage-specific enzyme, named "recombinase." It was also found (Kozinskii and Kozinskii, 1964) that recombination between two parent molecules by DNA segments takes place during infection of bacteria by two phage particles containing differently labeled DNA (with P^{32} in one stage, with H^3 in the other) and during simultaneous inhibition of DNA synthesis in bacteria. This recombination was not connected with depolymerization and random dispersion, for artificial fragmentation of the DNA molecules inhibited the recombination process. Many segments (subunits) labeled with P^{32} and H^3 were incorporated during restitution into the single polynucleotide chain. Fragmentation in T-even phages has also been demonstrated experimentally in other laboratories (Bresler et al., 1964; Konrad and Stent, 1964).

The possibility is not ruled out that fragmentation and recombination of this type are connected with the presence of noncoincident breaks or weak links in each of the two mutually connected polynucleotide chains of a type indicated in Fig. 23. The problem of the presence or absence of such breaks has often been discussed, but it has never been solved. It has recently been reinvestigated (Freifelder and Davison, 1963; Davison et al., 1964) with the use of DNA preparations from a large number of different phages, by studying native and denatured DNA preparations in the ultracentrifuge. When DNA is denatured, its double helix is unwound, and if breaks are present in each of its chains, heterogeneity in molecular weight must be revealed, and this in fact was established.

However, the degree of manifestation of this heterogeneity varied in different phages: in some (T7 bacteriophage) about 50% of the polynucleotides were unbroken during denaturation of DNA and about 50% were fragmented, while in phages T2 and T4 the percentage of polynucleotides with possible breaks (or weak points) was more than 90, in phage λ it was about 30, and in other phages (B3,

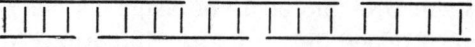

Fig. 23. Double-helical DNA chain with breaks.

D3, E79, F116, SP8, α, C, T5) it varied from 15 to 100. Irradiation with α-rays increased the number of possible breaks. The authors cited conclude that in any population of bacteriophages there is a certain number of particles containing noncoincident breaks in the structure of their DNA, or breaks in one of the two polynucleotides in the DNA. The character of the disruptions in denatured DNA shows that distribution of weak links or breaks along the polynucleotides is not random, but most of them are located mainly near the middle of the chain. At the same time, the presence of these breaks does not mean loss of infectivity, for the effective infectivity of large phages was about 100%, although only a few of their single polynucleotides was intact. The opposite possibility, i.e., that breaks in polynucleotide chains are essential for infectivity, has not been studied.

However, the possibility remains that the circular character of the genetic map of the T-even phages is also explained by the circular structure of DNA established after penetration of phage into the bacterial cell. Although the DNA in the heads of mature phage possesses a linear structure after isolation from the heads, according to Kozinskii and Kozinskii (1965) the state of partially replicated DNA at the beginning of the replication process is modified: it becomes circular in shape, but the ring is closed not by a covalent bond, but by a weaker link.

It is very important to note that, although in experiments *in vitro* both polynucleotides of double-helical phage DNA, after uncoiling during denaturation, can serve as templates for synthesis of complementary RNA molecules, during normal synthesis *in vivo* only one of the two polynucleotides is transcribed in the form of messenger RNA. For this reason, one of the paired polynucleotides was described as "coding" or genetic, and the other as "replicative" (Roth, 1964; Tocchini-Valentini et al., 1963; Womack and Barricelli, 1965).

This phenomenon has been studied particularly thoroughly by Marmur and co-workers, using phage SP8 for this purpose (Marmur and Greenspan, 1963; Marmur, Greenspan, et al., 1963). Each of the two chains of this phage differ so much from the other in its nucleotide composition (pyrimidines are predominant in one chain, but purines in the complementary chain) that, after denaturation of this DNA, the different chains can be fractionated by several methods (into light, or purine and heavy, or pyrimidine) and, in particular, by centrifugation in a cesium chloride density gradient and by MAK column chromatography.

It was found that only the "heavy" H-chain of the DNA double helix of this phage can form DNA/RNA hybrids with the RNA which is synthesized in the host cells (*Bacillus subtilis*) during infection by this phage and is detected by the pulse label method usually used for determining messenger RNA. The light (purine) L-chain can form RNA/DNA hybrids only with the RNA which was synthesized *in vitro* in a system in which this DNA-polynucleotide chain acted as primer for RNA-polymerase. Marmur and co-workers were unable to determine what structural properties of phage DNA are in fact responsible for this polarized transfer of information and synthetic differentiation between the two components of phage DNA.

However, important facts helping to understand this mechanism were obtained by work in another laboratory using phages ϕX174 and T2 for this purpose (Wood and Berg, 1962, 1963, 1964). They studied the ability of different forms of DNA to stimulate synthesis of RNA and proteins *in vitro* in an artificial system consisting of components of *E. coli* (soluble proteins + ribosomal system). They found that native DNA determines the synthesis of active RNA, i.e., that form of RNA which can act as template for protein synthesis. Meanwhile, although denatured DNA can form RNA, the product is inactive and not only does not stimulate protein synthesis, but may actually inhibit this synthesis. RNA molecules synthesized under the control of native double-helical DNA and a mixture of its two chains differed from each other in many properties. RNA from native DNA was heterogeneous in composition with respect to sedimentation coefficient (from 8 to 12S), whereas RNA from denatured DNA was more homogeneous and had a shorter chain (4S).

The DNA of phage ϕX174 is single-stranded. In this form, in the experiments carried out by Wood and Berg, it showed weak activity in relation to RNA synthesis, and this RNA did not stimulate protein synthesis. Synthesis of active RNA began only after formation of the replicative double-helical form., Meanwhile it is known (Hayashi et al., 1963) that messenger RNA induced during infection with phage ϕX174 does not yield molecular hybrids with

DNA from mature phage, but hybridizes on a basis of complementarity with double-helical replicative DNA of this phage. It must be supposed that it is complementary only to the "new" chain of this replicative DNA. The reducing effect of denaturation of DNA is not connected with destruction of or injury to its polynucleotides, because renaturation almost completely restored its activity. It has also been shown (Wood and Berg, 1963, 1964) that RNA formed on double-helical DNA interacts better with ribosomes.

Phage DNA (T4) is present inside *E. coli* cells in a double-helical form all the time after infection (Pouwels et al., 1963; Bresler et al., 1964). Disturbance of the selective transfer of information from only one chain of double-helical T2 DNA after fragmentation or denaturation was also demonstrated by Green (1964).

Clearly, therefore, the double-helical structure of DNA provides for the transfer of information to RNA from one of its chains, but in a manner which is essential for protein synthesis, i.e., based on boundaries between individual genes. Although DNA when uncoiled into separate chains is capable of transferring its nucleotide information to RNA, this process becomes "nonsense" as a result of disturbance of the boundaries between cistrons, lack of recognition of the starting and finishing boundaries of synthesis, and so on.

c) Study of the Nature of "Assembling" of the Phage Particle from its Various Components. Morphogenesis of the large DNA-phages consists in its final form of a unique process of assembling of the relatively complex phage particle (Fig. 16) from its components. Simple plant and animal viruses are assembled spontaneously, the process being predetermined by the structure of the homogeneous protein subunits. Purely spontaneous automatic assembling of the complex phage particle is impossible, because too many heterogeneous parts are included in its composition. A detailed study of the genetic map and character of various mutations of phage T4 (by the method of complementation and recombination) has shown that a large part of the genetic map and many of the genes (about 30) are in fact responsible for the correct assembling of the phage particle from its protein components, synthesized under the control of other genes (about 10) (Epstein et al., 1963).

Edgar and Epstein (1965) called the genes regulating assembly of the particle morphogenetic. Mutations of these genes lead to disturbances of assembly, the tail fibrils remain disconnected, the tail part of the phage is separated from its head, the head is improperly formed, and so on (Fig. 20). What is the mechanism of action of these genes? Other genes code synthesis of particular proteins, early or late. What are the products of the "assembling" genes? Edgar and Epstein (1965) postulate that genes of this type also produce specific proteins, but in small quantities. In their opinion these proteins play the role of special connecting links ensuring that the units are correctly assembled. Because of their small quantities, however, they have not yet been identified.

Conclusion

The material described in this chapter clearly demonstrates evolution of molecular morphogenetic mechanisms in viruses. In the simplest case this is tantamount to the synthesis of one protein. In the case of synthesis of two or three virus-specific proteins, the morphogenetic problem of determination of the linear and temporal sequence of their synthesis and of the precise regulation of their relative quantities assumes importance.

In this case regulating members appear in the genetic informational macromolecules, determining in accordance with an assigned program the polarity and sequence of processing of the information and the specific speed with which the information contained in each gene is read. With a further increase in structural complexity, genetic and protein-synthesizing functions (DNA and RNA) are separated, and functionally different groups of genes appear. Finally, functions connected with organization of the assembling of complex phage particles appear.

All these different forms of genetic instructions are recorded by nucleotide sequences. The genetic code, which first appears in this structurally simplest form of life, has persisted up to its highest forms without any significant changes.

Forms of genetic regulation of morphogenesis existing at the virus level have not disappeared during the transition to more complex forms of life, but on the contrary, has been built upon and made more complex, and the examination and study of these more complex forms of morphogenesis can only be fruitful if we can understand the simplest and most elementary methods of regulation.

Literature Cited

Allison, A. C., Pereira, H. G., and Farthing, C. P., 1960, Virology, 10:316.
Ames, B. N., and Hartman, P. E., 1963, Cold Spring Harbor Symp. Quant. Biol., 28:349.
Amman, J., Delius, H., and Hofschneider, P. H., 1964, J. Mol. Biol., 10:557.
Atabekov, I. G., and Novikov, V. K., 1966, Biokhimiya, 31:157.
August, J. T., Shapiro, L., and Eoyang, L., 1965, J. Mol. Biol., 11:257.
Baltimore, D., 1964, Proc. Nat. Acad. Sci., USA, 51:450.
Baltimore, D., 1966, J. Mol. Biol., 18:421.
Baltimore, D., Becker, Y., and Darnell, J. E., 1964, Science, 143:1034.
Baltimore, D., Eggers, H. J., Franklin, R. M., and Tamm, J., 1964, Proc. Nat. Acad. Sci., USA, 49:843.
Baltimore, D., and Franklin, R. M., 1962, Biochem. Biophys. Res. Comm., 9:388.
Baltimore, D., and Franklin, R. M., 1963, J. Biol. Chem., 28:3395.
Baltimore, D., Girard, M., and Darnell, J. E., 1966, Virology, 29:179.
Bass, I. A., Zograf, Yu. N., and Khesin, R. B., 1967, Molecul. Biol., 1:436.
Bassel, A., Hayashi, M., and Spiegelman, S., 1964, Proc. Nat. Acad. Sci., USA, 52:796.
Baylor, M. B., Hessler, A. Y., and Baird, J. P., 1965, Genetics, 51:351.
Becker, Y., and Joklik, W. K., 1964, Proc. Nat. Acad. Sci., USA, 51:577.
Ben-Porat, T., and Kaplan, A. S., 1962, Virology, 16:261.
Ben-Porat, T., and Kaplan, A. S., 1965, Virology, 25:22.
Bockstahler, L. E., and Kaesberg, P., 1962, J. Biophys., 2:1.
Bremer, H., Konrad, H., and Brunner, R., 1966, J. Mol. Biol., 16:104.
Breslei, S. E., Drabkina, L. E., Mosevitskii, M. I., and Timkovskii, A. L., 1964, Dokl. Akad. Nauk SSSR, 156:947.
Brody, E., Coleman, L., Mackal, R. P., Werninghaus, B., and Evans, E., 1964, J. Biol. Chem., 239:285.
Burdon, R. H., Billeter, M. A., Weissman, Ch., Warner, R. C., Ochoa, S., and Knight, C. A., 1964, Proc. Nat. Acad. Sci., USA, 52:768.
Burkinskaya, A. G., Gitel'man, A. K., and Vokrunova, G. K., 1964, Vopr. Virusol., No. 5:569.
Burton, A., and Sinsheimer, R. L., 1963, Science, 142:962.
Bykovskii, A. F., 1964, In: Virology and Immunology, Series: Fundamentals of Molecular Biology (ed. V. A. Engel'gardt), Izd. Nauka, p. 176.
Caro, L. G., 1965, Virology, 25:226.
Carusi, E. A., and Sinsheimer, R. L., 1961, Federated Proc., 20:438.
Chamberlin, M., and Berg, P., 1964, J. Mol. Biol., 8:297.
Clark, J. M., Chang, A. Y., Spiegelman, S., and Reichmann, M. E., 1965, Proc. Nat. Acad. Sci., USA, 54:1193.
Cohran, G. W., Dhaliwal, A. S., Welkie, G. W., Chidester, J. L., Lee, M. H., and Chandrasekar, B. K., 1962, Science, 138:46.

Cohen, S. S., Sekiguchi, M., Stern, J. L., and Barner, H. D., 1963, Proc. Nat. Acad. Sci., USA, 49:699.
Cohen, S. S., 1963, Ann. Rev. Biochem., 32:83.
Cooper, P. D., 1961, Nature, 190:302.
Cooper, S., and Zinder, N. D., 1962, Virology, 18:405.
Cornuet, M. P., and Manifacier, S. A., 1962, C. R. Acad. Sci., 255:1660.
Davis, J. E., and Sinsheimer, R. L., 1963, J. Mol. Biol., 6:203.
Davis, J. E., Strauss, J. H., and Sinsheimer, R. L., 1961, Science, 134:1427.
Davison, F. F., Freifelder, D., and Holloway, B. W., 1964, J. Mol. Biol., 8:1.
Delihas, N., 1961, Virology, 13:242.
Delius, H., and Hofschneider, P. H., 1964, J. Mol. Biol., 10:554.
Dirksen, M. L., Wiberg, J. S., Koerner, J. F., and Buchanan, J. M., 1960, Proc. Nat. Acad. Sci., USA, 46:1425.
Doi, R. H., and Spiegelman, S., 1962, Bacteriol. Proc., 97:153.
Doi, R. H., and Spiegelman, S., 1963, Proc. Nat. Acad. Sci., USA, 49:353.
Dove, W., 1966, J. Mol. Biol., 19:187.
Dulbecco, R., and Vogt, M., 1963, Proc. Nat. Acad. Sci., USA, 50:236.
Ebisuzuka, K., 1963, J. Mol. Biol., 7:379.
Edgar, R. S., and Epstein, R. H., 1965, Scient. Amer., 212:71.
Edgar, R. S., and Wood, W. B., 1966, Proc. Nat. Acad. Sci., USA, 55:498.
Edlin, G., 1965, J. Mol. Biol., 12:363.
Epstein, R. H., Bolle, A., Steinberg, C. M., Kellenberger, E., Tour, E. B., Cheralley, R., Edgar, R. S., Susman, M., Denhardt, G. H., and Lielausis, A., 1963, Cold Spring Harbor Symp. Quant. Biol., 28:375.
Erikson, R. L., and Franklin, R. M., 1966, Bacteriol. Rev., 30:267.
Fiers, W., and Sinsheimer, R. L., 1962, J. Mol. Biol., 5:408, 420, 424.
Flanagan, J. F., and Ginsberg, H. S., 1962, Federation Proc., 21:463.
Franklin, R. M., Wecker, E., and Henry, C., 1959, Virology, 7:220.
Freifelder, D., and Davison, P. F., 1963, Biophys. J., 3:49.
Gendon, Yu. Z., 1964, In: Virology and Immunology, Series: Fundamentals of Molecular Biology, ed. V. A. Engel'gardt, Izd. Nauka, Moscow, p. 86.
Gomatos, P. J., and Tamm, J., 1963a, Proc. Nat. Acad. Sci., USA, 49:707.
Gomatos, P. J., and Tamm, J., 1963b, Proc. Nat. Acad. Sci., USA, 50:878.
Goodman, H. M., and Rich, A., 1963, Nature, 199:318.
Green, M., 1962, Cold Spring Harbor Symp. Quant. Biol., 27:219.
Green, M. H., 1964, Proc. Nat. Acad. Sci., USA, 52:1388.
Green, M., and Pina, M., 1962, Bacteriol. Proc., 3:129.
Guthrie, G. D., and Sinsheimer, R. L., 1960, J. Mol. Biol., 2:297.
Hall, B. D., Nygaard, A. P., and Green, M. H., 1964, J. Mol. Biol., 9:143.
Hall, B. D., and Spiegelman, S., 1961, Proc. Nat. Acad. Sci., USA, 47:137.
Haruno, I., Nozu, K., Ohtaka, Y., and Spiegelman, S., 1963, Proc. Nat. Acad. Sci., USA, 50:905.
Haselkorn, R., and Fried, V. A., 1964, Proc. Nat. Acad. Sci., USA, 51:1001.
Haselkorn, R., Fried, V. A., and Dahlberg, J. E., 1963, Proc. Nat. Acad. Sci., USA, 49:511.
Hayashi, M., 1965, Proc. Nat. Acad. Sci., USA, 54:1736.

Hayashi, M., Hayashi, M. N., and Spiegelman, S., 1963, Proc. Nat. Acad. Sci., USA, 50:664.
Hayashi, M., Hayashi, M. N., and Spiegelman, S., 1964, Proc. Nat. Acad. Sci., USA, 51:351.
Hayes, W., 1964, The Genetics of Bacteria and Their Viruses, Blackwell Sci. Publ. Oxford.
Hershey, A. D., and Burgi, F., 1965, Proc. Nat. Acad. Sci., USA, 53:325.
Hofschneider, P. H., 1960, Z. Naturforsch., 15b:441.
Il'yachenko, V. N., 1963, Vopr. Virusol., 8:667.
Joklik, W. K., 1962a, Cold Spring. Harbor Symp. Quant. Biol., 27:199.
Joklik, W. K., 1962b, J. Mol. Biol., 5:265.
Kaerner, H. G., and Hoffmann-Berling, H., 1964, Nature, 202:1012.
Kano-Sueoka, T., and Sueoka, N., 1966, J. Mol. Biol., 20:183.
Kano-Sueoka, T., and Spiegelman, S., 1962, Proc. Nat. Acad. Sci., USA, 48:1942.
Karesek, M., and Schramm, G., 1962, Biochem. Biophys. Res. Comm., 9:63.
Kellenberger, E., 1966, In: Principles of Biomolecular Organizations, Ciba Found. Symp., Churchill Ltd., London, p. 192.
Kelly, R. B., Gould, J. L., and Sinsheimer, R. L., 1965, J. Mol. Biol., 11:562.
Kelly, R. B., and Sinsheimer, R. L., 1964, J. Mol. Biol., 8:602.
Khesin, R. B., 1965, Uspekhi Sovr. Biol., 59:12.
Khesin, R. B., 1966, In: Ninth International Congress of Microbiology, Proceedings of Symposia, Moscow, p. 14.
Khesin, R. B., Gorlenko, Zh. M., Shemyakin, M. F., Bass, I. A., and Prozorov, A. A., 1963, Biokhimiya, 28:1070.
Khesin, R. B., and Shemyakin, M. F., 1962, Biokhimiya, 27:761.
Khesin, R. B., Shemyakin, M. F., Bass, I. A., Astaurova, O. B., Kamzolova, S. G., Kiselev, N. A., and Manyakov, V. F., 1966, In: Structure, Properties, and Genetic Functions of DNA, Kurchatov Institute of Atomic Energy, Moscow, p. 167.
Khesin, R. B., Shemyakin, I. F., Gorlenko, Zh. M., Bogdanova, S. L., and Afanas'eva, T. P., 1962, Biokhimiya, 27:1092.
Kim, Y. T., and Wildman, S. G., 1962, Biochem. Biophys. Res. Comm., 8:394.
Kingsbury, D. W., 1966, J. Mol. Biol., 18:204.
Kleinschmidt, A. K., Burton, A., and Sinsheimer, R. L., 1963, Science, 14:961.
Konrad, M. W., and Stent, G. S., 1965, Z. Vererbungsl., 96:66.
Kozaka, M., Matsubara, K., and Takagi, Y., 1963, Biochem. Biophys. Res. Comm., 11:244.
Kozinski, A. W., 1961, Virology, 13:124, 377.
Kozinski, A. W., and Kozinski, P. B., 1963, Virology, 20:213.
Kozinski, A. W., and Kozinski, P. B., 1964, Proc. Nat. Acad. Sci., USA, 52:211.
Kozinski, A. W., and Kozinski, P. B., 1965, Proc. Nat. Acad. Sci., USA, 54:634.
Kozinski, A. W., Kozinski, P. B. and Shannon, P., 1963, Proc. Nat. Acad. Sci., USA, 50:746.
Landgridge, R., Billeter, M. A., Borst, R. H., Burdon, R. H., and Weissman, C., 1964, Proc. Nat. Acad. Sci., USA, 52:114.
Landgridge, R., and Gomatos, R. J., 1963, Science, 141:694.

Loeb, T., 1960, Science, 131:932.
Loeb, T., and Zinder, N. D., 1961, Proc. Nat. Acad. Sci., USA, 47:282.
Loh, P. C., and Riggs, J. L., 1961, J. Exptl. Med., 114:149.
Luria, S. E., 1958, Protoplasmologia, IV:1.
MacHattie, L. A., and Thomas, C. A., 1964, Science, 144:1142.
Maclean, E. C., and Hall, C. E., 1962, J. Mol. Biol., 4:173.
Marmur, J., and Greenspan, C. M., 1963, Science, 142:387.
Marmur, J., Greenspan, C. M., Palecek, E., Kahan, F. M., Lavine, J., and Mandel, M., 1963, Cold Spring Harbor Symp. Quant. Biol., 28:191.
Meselson, M., and Weigle, J. J., 1961, Proc. Nat. Acad. Sci., USA, 47:857.
Meyer, F., Mackal, R. P., Tao, M., and Evans, E. A., 1961, J. Biol. Chem., 236:1141.
Montagnier, L., and Sanders, F. K., 1963, Nature, 199:664.
Mora, P. T., McFarland, V. W., and Lubrorsky, S. W., 1966, Proc. Nat. Acad. Sci., USA, 55:438.
Nomura, M., Matsubara, K., Okamoto, K., and Fujimura, R., 1962, J. Mol. Biol., 5:535.
Nomura, M., Okamoto, K., and Asano, K., 1962, J. Mol. Biol., 4:376.
Nonoyata, M., and Ikeda, Y., 1964, J. Mol. Biol., 9:763.
Ohtaka, J., and Spiegelman, S., 1963, Science, 142:493.
Okamoto, K., Sugino, Y., and Nomura, M., 1962, J. Mol. Biol., 5:527.
Paranchych, W., 1963, Biochem. Biophys. Res. Comm., 11:28.
Paranchych, W., and Graham, A. F., 1962, J. Cell. Comp. Physiol., 60:199.
Penman, S., Scherrer, K., Becker, Y., and Darnell, J. E., 1963, Proc. Nat. Acad. Sci., USA, 49:654.
Palasa, H., and Green, M., 1965, Virology, 25:68.
Plagelmann, P. G. W., 1966, Bacteriol. Rev., 30:288.
Pouwels, P. H., Veldhuisen, G., Jansz, H. S., and Cohen, J. A., 1963, Biochem. Biophys. Res. Comm., 13:83.
Rading, Ch. M., and Kaisar, A. D., 1963, J. Mol. Biol., 7:225.
Ralph, R. K., Matthews, R. E. F., Matus, A. J., and Mandel, H. G., 1965, J. Mol. Biol., 11:202.
Reddi, K. K., 1966, Proc. Nat. Acad. Sci., USA, 55:593.
Reich, E., Franklin, R. M., Shatkin, A. J., and Tatum, E. L., 1961, Science, 134:556.
Reichmann, M. E., 1964, Proc. Nat. Acad. Sci., USA, 52:10009.
Rich, A., Penman, Sh., Becker, Y., Darnell, J., and Hall, C., 1963, Science, 142:1658.
Ris, H., and Chandler, B. L., 1963, Cold Spring Harbor Symp. Quant. Biol., 28:1.
Roth, J. S., 1964, Nature, 202:182.
Rubenstein, J., Thomas, C. A., and Hershey, A. D., 1961, Proc. Nat. Acad. Sci., USA, 8:111.
Rueckert, R. R., and Zillig, W., 1962, J. Mol. Biol., 5:1.
Rueckert, R. R., Zillig, W., and Doerfler, W., 1962, J. Mol. Biol., 5:10.
Salzman, N. P., Shatkin, A. J., and Sebring, E. D., 1964, J. Mol. Biol., 8:405.
Scharff, M. D., Schatkin, A. J., and Lewinton, L., 1963, Proc. Nat. Acad. Sci., USA, 50:686.
Schafer, W., 1963, Bacteriol. Rev., 27:1.
Schwarts, J. H., Eisenstadt, J. M., Brawerman, A., and Zinder, N. D., 1965, Proc. Nat. Acad. Sci., USA, 53:195.

Sekiguchi, M., Taketo, A., and Takagi, Y., 1960, Biochim. Biophys. Acta, 45:199.
Shapiro, L., and August, J. T., 1966, Bacteriol. Rev., 30:279.
Shipp, W., and Haselkorn, R., 1964, Proc. Nat. Acad. Sci., USA, 52:401.
Sinsheimer, R. L., 1959, J. Mol. Biol., 1:37, 43.
Sinsheimer, R. L., 1961, J. Chim. Phys., 58:986.
Sinsheimer, R. L., 1962, Scient. Amer., 207:109.
Sinsheimer, R. L., Starman, B., Nagler, C., and Guthrie, S., 1962, J. Mol. Biol., 4:142.
Skalka, A., 1966, Proc. Nat. Acad. Sci., USA, 55:190.
Skold, O., and Buchanan, J. M., 1964, Proc. Nat. Acad. Sci., USA, 51:553.
Spiegelman, S., 1963, In: Informational Macromolecules, ed. H. Vogel et al., Academic Press, New York.
Stent, G. S., Molecular Biology of Bacterial Viruses, W. H. Freeman and Co., San Francisco and London.
Streisinger, G., Edgar, R. H., and Denhardt, G. H., 1964, Proc. Nat. Acad. Sci., USA, 51:775.
Sueoka, N., and Kano-Sueoka, T., 1964, Proc. Nat. Acad. Sci., USA, 52:1535.
Thomas, C. A., 1963, In: Molecular Genetics, Part I, ed. J. H. Taylor, Academic Press, New York, p. 113.
Thomas, C. A., and MacHattie, L. A., 1964, Proc. Nat. Acad. Sci., USA, 52:1297.
Tocchini-Valentini, G. P., Stodolsyk, M., Aurisicchio, A., Sarnat, M., Graziosi, F., Weiss, S. B., and Geiduschek, E. P., 1963, Proc. Nat. Acad. Sci., USA, 50:935.
Tromans, W. J., and Horne, R. W., 1961, Virology, 15:1.
Tsugita, A., and Fraenkel-Conrat, H., 1960, Proc. Nat. Acad. Sci., USA, 46:636.
Tsugita, A., and Fraenkel-Conrat, H., 1962, J. Mol. Biol., 4:73.
Tsugita, A., and Fraenkel-Conrat, H., 1963, In: Molecular Genetics, Part I, ed. J. H. Taylor, Academic Press, New York.
Tsugita, A., Fraenkel-Conrat, H., Nirenberg, M. W., and Matthaei, J., 1962, Proc. Nat. Acad. Sci., USA, 48:846.
Vainshtein, B. K., and Kiselev, N. A., 1964, In: Virology and Immunology, Series: Fundamentals of Molecular Biology, ed. V. A. Engel'gardt, Izd. Nauka, Moscow, p. 7.
Van Kammen, A., 1963, Meded. Landbouwhogeschool, Wageningen (Nederland), 63(12):1.
Vinograd, J., Lebowitz, J., Radloff, R., Watson, R., and Laipis, P., 1965, Proc. Nat. Acad. Sci., USA, 53:1104.
Vittman, G. G., 1962, Fifth International Biochemical Congress, Symposium No. 1, Izd. AN SSSR, Moscow, p. 274.
Voorma, H. O., Gout, P. W., van Duin, J., Hoogendam, B. W., and Bosch, L., 1964, Biochim. Biophys. Acta, 87:693.
Warner, J. R., Knopf, P. M., and Rich, A., 1963, Proc. Nat. Acad. Sci., USA, 49:122.
Watanabe, J., 1957, J. Gen. Physiol., 40:521.
Watanabe, J., Watanabe, K., and Hinuma, J., 1962, Biochim. Biophys. Acta, 61:976.
Weissman, C., and Borst, P., 1963, Science, 142:1188.
Weissman, C., Borst, P., Burdon, R. H., Billeter, M. A., and Ochoa, S., 1964, Proc. Nat. Acad. Sci., USA, 51:682.
Weissman, C., Simon, L., and Ochoa, S., 1963, Proc. Nat. Acad. Sci., USA, 49:407.

Wilcox, W. C., and Ginsberg, H. S., 1961, Proc. Nat. Acad. Sci., USA, 47:512.
Wilcox, W. C., and Ginsberg, H. S., 1963, Virology, 20:269.
Wildy, P., and Horne, R. W., 1963, In: Progr. Med. Virol., 5:1 (Karger, Basel), New York.
Wittman, H. G., 1964, In: Informational Macromolecules, ed. H. J. Vogel et al., Academic Press, New York.
Wittman, H. G., and Wittman-Liebold, B., 1963, Cold Spring Harbor Symp. Quant. Biol., 28:589.
Womack, F., and Barricelli, N. A., 1965, Virology, 27:600.
Wood, W. B., and Berg, P., 1962, Proc. Nat. Acad. Sci., USA, 48:94.
Wood, W. B, and Berg, P., 1963, Cold Spring Harbor Symp. Quant. Biol., 28:237.
Wood, W. B., and Berg, P., 1964, J. Mol. Biol., 9:452.
Yamazaki, H., Bancroft, J., and Kaesberg, P., 1961, Proc. Nat. Acad. Sci., USA, 47:979.
Zech, H., and Vogt-Kohne, L., 1955, Naturwissenschaften, 42:337.
Zhdanov, V. M., 1966, Ninth International Congress of Microbiology, Proceedings of Symposia, Moscow, p. 397.
Zhdanov, V. M., Strakhanova, V. M., and Ershov, F. I., 1964, Vopr. Virusol., No. 5: 564.

Chapter 3

Molecular-Genetic Mechanisms of Morphogenesis and Intracellular Biochemical Differentiation of Monochromosomal Bacterial Cells

Introduction

Unicellular living organisms are represented on our planet by an extensive group of widely different microorganisms, varying widely in the complexity of their living systems and belonging to different taxonomic categories: bacteria, algae, fungi, protozoa. The most elementary of these groups from the biochemical point of view is the bacteria, among which the smallest independently living unicellular organisms have been identified: the mycoplasmas, or pleuropneumonia-like organisms (*Mycoplasma laidlawii* and *Mycoplasma gallisepticum*). The cell diameter of this microorganism is about 0.1μ, and its mass about 1-1000th of that of ordinary bacteria. The smallest mycoplasma cells are smaller than vaccinia virus particles. The total molecular weight of DNA in the cell of this tiny microbe is about 50×10^6, i.e., even lower than is typically found in the large phages (Morowitz and Cleverdon, 1959; Morowitz and Tourtellotte, 1962; Morowitz, 1966). In this case the number of DNA-cistrons determining the complete development cycle of the microorganism and maintaining its specialization apparently does not exceed 100 (at a rough calculation: 1000 nucleotides in one DNA chain per protein) so that for 100 types of proteins this figure is evidently the least possible for autonomous semiparasitic reproduction of a cell system.

The microorganism whose biochemistry has been most fully studied, *Escherichia coli* is much more complex and consists of no fewer than 1000 different types of protein. It will thus be obvious that this group of living organisms comes next to the viruses in its complexity, which explains why it must be considered in our discussion of the stages of evolution of the molecular-genetic systems regulating morphogenesis and differentiation.

§1. Types of Morphogenic Reactions in Bacteria

As will be clear from the previous chapter, even the large phages lose their apparent individuality during reproduction. They transfer to the host cell merely a molecular system of genes, a program of development in material form, determining the structural elements of the mature forms, later assembled into the morphologically specific structure of the phage particle. Bacterial cells do not lose their individuality during reproduction. The formation of new cells in most cases takes place by division of existing cells, when typical morphogenesis, in the form of execution of a program over a period of time, is practically absent. However, when compared with viruses and phages, we find that the bacteria have evolved new morphogenetic biochemical and genetic processes in response to environmental conditions. In viruses and phages synthesis of a particular substance, such as a methylated nucleotide, takes place in accordance with a timetable of successive formation of a series of enzymes coded by a successive series of cistrons in the phage DNA. In bacteria, hundreds of enzyme reactions and syntheses are taking place continuously in the cytoplasm, and each enzyme cycle is no longer an act of morphogenesis but a fragment of the ordinary living activity of the cell at all stages of its life cycle, activity associated with the maintenance of functions not present in phages, such as assimilation, energy metabolism, synthesis of systems of activation and protein formation, formation of ribosomes, and so on.

Typical morphogenetic processes in bacteria are sexual reproduction and spore-formation, although these phenomena constitute a small part of the total dynamics of reproduction of bacterial cells. Clearly, therefore, we cannot regard the morphogenesis or ontogenesis of bacteria as an independent problem, but we can speak only of morphogenetic phenomena or reactions in the vital activity and reproduction of bacteria.

The principal morphogenetic phenomena at the bacterial level of evolution of living structures are as follows:

1) Enzyme synthesis induced by substrates in the environment;
2) Sexual reproduction;
3) The process of division of cells and cell chromosomes;
4) Spore-formation;
5) Formation of ribosomes and lysosomes;
6) Cycles of reactions of synthesis and breakdown of substances taking place in successive stages and in a definite order in space and time.

From the molecular-genetic point of view these morphogenetic phenomena can be represented as follows. If the genome of a bacterium, by analogy with phage, can be represented conventionally by groups of functionally related genes (cistrons), and in many cases this is in fact so, at any given moment only some of these groups are active. In some segments of the bacterial chromosome containing a program of synthesis and breakdown, consisting of successive cycles of constitutive reactions (such as the synthesis of certain amino acids), the activation of successive genes takes place at high speed by successive linking or even by the formation of a polycistronic template. In other cases, all genes of the cycle are active simultaneously and the system works continuously rather than in waves. The third type of chromosomal segments contain genes which are activated by substrate from the environment and are inactive in the absence of substrate (substrate induction). Areas of the fourth type are responsible for the synthesis of functional proteins of a nonenzymic nature, such as ribosomal and membrane proteins, etc., and they consist of independent genes. The fifth type contains a program for cell division which operates periodically; the sixth — a program of sexual reproduction which operates infrequently; the seventh — a program for spore-formation connected with environmental conditions, and so on. The actual arrangement of these regions may be very complex. The problem of the concrete structure of the bacterial genetic map will be examined separately.

Some chromosomal regions may be redundant in the genome, others unique; some regions consist of many genes and others of only one or two, and so on.

Before trying to draw up actual schemes to illustrate the purposive coordination and interaction of all these morphogenetic phenomena, we must look at the basic facts concerning molecular mechanisms of different types of morphogenetic processes in bacteria. At this stage it is convenient to do so not in the order reflecting the general biological significance or relative frequency of each particular process, but in an order based on the completeness of their molecular-genetic study.

§ 2. Substrate-Induced Synthesis of Bacterial Enzymes as a Morphogenetic Process

We shall begin our analysis of morphogenetic phenomena in bacteria with the examination of induced protein synthesis for many reasons, but primarily because it was in fact the study of syntheses of this type which gave birth to the molecular genetics of bacteria, the theory of the operon and genetic induction and repression, the well-known model of regulation of genetic activity developed by Jacob and Monod (1961), and a number of other principles at present used in the analysis of morphogenetic phenomena at both lower and higher levels.

It is not our purpose to undertake a comprehensive examination of all aspects of induced enzyme synthesis and its genetic regulation, for this is a far too extensive field in modern biochemistry and genetics. We shall merely consider those few aspects which are important to our analysis of other ontogenetic problems.

a) The Principle of Enzymic Induction. Induced synthesis of enzymes is a purposive adaptive reaction of bacteria to changing environmental conditions. Obviously if bacterial cells are able to use several substances, each of which is present only periodically in the environment, it is more economical for the cell to employ an enzyme system capable of processing each of these substances periodically, as the need arises, and in the necessary amount, rather than to form several enzymes continuously under conditions when they cannot be used. Biochemical adaptive reactions of this type are found in all cells, under all levels of organization, and by now more than 100 enzyme systems induced by environmental substrates have been described in different types of cells of multicellular organisms and microorganisms. The extent to which the various forms of these systems have been studied differs, most attention having been paid to the induced synthesis of β-galactosidase, an enzyme formed by *E. coli* cells if the carbohydrate lactose and other β-galactosides hydrolyzed by this enzyme are present in the growth medium. The velocity of synthesis of β-galactosidase is increased 10,000 times if lactose is present, and in bacteria grown on lactose, β-galactosidase accounts for about 6-7% of the total protein.

The system of induced synthesis of β-galactosidase has served as basic model for years of research in several laboratories studying the mechanisms of this synthesis. The most widely known studies are those of Jacob and Monod, resulting in the creation of an experimentally based concept of regulation of genetic activity, which we shall describe briefly on the basis of the publications summarizing their conclusions (Jacob and Monod, 1963, 1964).

b) <u>Concept of the Operon, a Group of Linearly and Functionally Connected Genes, Exemplified by the Regulation of Synthesis of β-Galactosidase. Regulator Genes and Repressor Genes. Scheme of the Lactose Operon.</u> Biosynthesis of β-galactosidase, a protein consisting of one or several specific polypeptides, in *E. coli* cells is effected through a gene, i.e., a segment of DNA in the form of nucleotide triplets coding the essential sequence of amino acid residues for these polypeptides. This gene is called the structural gene or structural cistron, and it must possess colinearity with the protein to be synthesized, i.e., precise linear matching of the nucleotide regions of the DNA with corresponding peptide groups. Colinearity of the structural gene with protein synthesized on its bases has been demonstrated most clearly by comparison of the parameters of the A-protein of tryptophan synthetase and the A-gene of *E. coli* responsible for its synthesis (Yanofsky et al., 1964). The same rule certainly must apply to other structural genes also.

In some cases structural genes are constantly active (constitutive synthesis), but in the case of the β-galactosidase gene it has been showed that when lactose is absent from the medium the activity of this gene is inhibited (repressed) in the sense that it synthesizes virtually no template for protein synthesis (messenger RNA). This repression disappears immediately with the addition of lactose, and the experiments of Jacob and Monod were based on the study of the mechanism of this repression.

The essence of their experiments was to study mutants of *E. coli* in which the mutation affected this system (for example, the isolation of cells with constitutive synthesis of β-galactosidase) and to determine the location of these mutations on the genetic map of *E. coli*. We shall examine techniques for the location of genes on the genetic map of *E. coli* a little later on.

An extensive series of investigations showed that most probably the genetic system of *E. coli* contains regulator genes which form special substances—repressors of structural genes. By their action the structural gene is "repressed," while fixation of the repressor by substrate by a process of specific steric interaction supposedly brings the structural gene from a passive state back into an active state.

Jacob and Monod's concept of the mechanisms and systems of genetic regulation contains very many hypotheses, and before discussing their general pattern we must first examine the conclusions which can now be regarded as firmly established experimentally.

The first important point to note is the discovery of the precise relationship between several functionally interconnected structural genes, in the sense that they are induced or repressed together, in association with changes in the intensity of synthesis of this group of proteins, each of which is formed independently and at its own speed; the relative velocities of their synthesis remain unchanged (Jacob and Monod, 1961, 1964). Such a group of genes has been called the o p e r o n. The following genes have been discovered in the lactose operon:

Z-gene, the gene for β-galactosidase, consisting of several structural cistrons forming polypeptide subunits of the protein, β-galactosidase;

Y-gene, the gene for synthesis of galactoside-permease; and also, it seems,

Ac (or X) gene, the gene for synthesis of galactoside-transacetylase.

Besides these structural genes, mutation analysis has revealed the presence of a special region in the operon through which the regulator gene exerts its effect on the operon. This area is called the operator. The position of the regulator gene itself, represented by the letter i has been accurately marked on the genetic map, and its product is the repressor of the lactose operon.

The conclusion to be drawn from the results of this experimental work is that if a series of reactions catalyzed by several enzymes in succession is necessary for a particular biochemical process, the genes determining them show a tendency to be arranged in the bacterial chromosome in a single cluster, the operon. The operon can be in an active or repressed state depending on the presence or absence of substrate (inducer).

It is considered that the regulator gene under normal conditions constantly forms a special macromolecular compound, the repressor, which has at least two highly specific receptor groups, one "identifying" the operator segment and bringing about repression of the operon, while the other determines fixation of the repressor with inducer and its resulting inactivation (or, conversely, its activation). The use of indirect genetic methods has shown (Bourgeois et al., 1966) that the regulator i gene of the lactose operon actually codes the formation of a protein, while the operator 0 gene of this operon evidently forms no products (is not "translated" into protein).

What is the molecular mechanism by means of which information is read simultaneously from different cistrons of the same operon? Jacob and Monod postulated that the operon can function as original template for synthesis of the polycistronic RNA template, which can form polyribosomal systems and provide for the

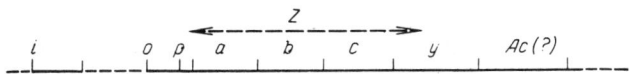

Fig. 24. Scheme of lactose operon of *Escherichia coli* (Jacob and Monod, 1964).

synthesis of several different proteins at once in accordance with the same principle of polycistronic protein synthesis as is typical of virus RNA. In their opinion the discovery of messenger RNA molecules similar in size to virus RNA molecules among a heterogeneous system supports this hypothesis of a polycistronic template.

Mapping of the operator of the lactose operon and establishing its linear autonomy from the structural genes was demonstrated with a series of mutants in which deletions of different magnitude were present in this operon. Observations on the position of these deletions showed that yet another special area may exist between the operator region and the strucural genes, known as the promotor region (P). Deletion of the operator involves a change to the constitutive synthesis of enzymes, while deletion of the promotor causes inactivation of the operon.

We shall not make a detailed analysis in this section of the extensive factual material in support of the validity of Jacob and Monod's concept of the structure and principles governing regulation of the lactose operon. This material has recently been examined in many excellent surveys of problems associated with substrate induction of enzymes and the operon (Ames and Martin, 1964; Stent, 1964; Shapot, 1965).

Genetic and biochemical studies of the lactose operon have now become extensive and work is going on in many laboratories. Meanwhile, the basic principles of Jacob and Monod's scheme have received precise experimental confirmation. The work of Kiho and Rich (1964) is particularly interesting from this point of view. They obtained experimental confirmation of Jacob and Monod's hypothesis of a polycistronic messenger RNA template, containing the integral information of the lactose operon and providing for the synthesis of several proteins determined by this operon on polyribosomes at the same time.

They found that during induction of synthesis of β-galactosidase in *E. coli* this enzyme can be found to be present only in the fraction of very large polysomes containing up to 40 individual ribosomes. With polysomes of this size, the molecular weight of the messenger RNA lies between 1,200,000 and 1,300,000, which is adequate for coding either a very large protein molecule with a molecular weight of 140,000, or several smaller protein molecules. The molecular weight of β-galactosidase is about 500,000, but it consists of four subunits each built of three similar polypeptides, each with a molecular weight of about 40,000. Consequently, synthesis of this enzyme in fact requires the coding of a linear structure with molecular weight of 40,000. An "excess" of information, if present, in the opinion of Kiho and Rich can code the structure of the other enzymes of the lactose operon.

Alpers and Tomkins (1965) studied the timetable of synthesis of two enzymes of the lactose operon—β-galactosidase and thiogalactoside-transacetylase—during induction and deinduction, and showed that during induction of the operon, transacetylase synthesis does not begin until 2-3 min after synthesis of β-galactosidase begins, while conversely, after removal of the inducer from the medium, transacetylase synthesis continues for 2-3 min after the end of β-galactosidase synthesis. In their discussion of this time sequence of syntheses from the point of view of integral polycistron and mRNA, these workers suggest that it reflects a linear sequence of translation of information. The polycistron template marks the boundaries of the cistrons by special start and finish codons. These codons are responsible for the delineation of separate templates for several polypeptides. The lactose mRNA has been shown to start synthesis of galactosidase with the first cistron, even before the process of transcription of its transacetylase cistron has ended (Leive and Kollin, 1967). The whole transcription process takes about 3.5 min.

Studies of the lactose operon in *E. coli* have stimulated similar approaches to investigation of the synthesis of other inducible enzymes in various microorganisms. Here, too, in several cases it was found that enzymes which are metabolically related and induced together are usually located side by side on the chromosome map, forming a gene cluster. Typical in this respect is the galactose operon, consisting in *E. coli* of three enzymes: galactokinase, uridyl-transferase, and epimerase (Morse, 1962; Buttin, 1961, 1962; Wu and Kalckar, 1966). This operon has the same structure also in *Saccharomyces* (Douglas and Pelroy, 1963). However, structural integrity of the operon is by no means a universal rule.

§3. The Principle of Combination of Successively Functioning Genes, Responsible for Synthesis of Successively Active Enzymes Concerned Jointly with the Formation of Various Metabolites, into Linear Structures, or Operons

Discovery of the fact that the synthesis of each polypeptide subunit is specified by a definite structural gene (cistron) together with the genetic mapping of a circumscribed region of the DNA demonstrated that enzymes which are related to each other for the synthesis, breakdown or modification of a particular compound are usually clustered together in operons of self-regulating type. This means that the end product of the reaction, if it accumulates above a specified level, turns into a repressor of the cycle as a whole. As a result of work of this type, the study of metabolic cycles from the standpoint of elucidation of coordination of all reactions has been placed on a precise genetic basis.

Many multicomponent biochemical conversions are now being studied from this standpoint, but we shall mention only a few cases.

a) **The Histidine Operon.** Histidine synthesis in *Salmonella typhimurium* consists of 10 successive reactions converting the 5-carbon chain of phosphoribosyl pyrophosphate into the 5-carbon chain of histidine (Ames and Hartmann, 1962).

The successive reactions of this synthesis and the enzymes catalyzing them are shown in Fig. 25 (Ames, 1965; Ames and Hartmann, 1963). The study of more than 1000 mutants unable to synthesize histidine because of blocking of one or another reaction enabled Ames and Hartmann (1963) to draw a map of the histidine loci of the *Salmonella* chromosome showing that all are connected in a linear manner into a structural unit of operon type (Fig. 26), in which, however, the order of the reactions does not coincide with the order of the corresponding genes in the linear sequence. The operator of the histidine operon lies on the right of the map (the G-gene). Analysis of the character of polarity of the mutations (mutation of the middle cistrons leads to reduction of synthesis of cistrons located on the left, while synthesis of enzymes determined by cistrons on the right is unchanged) shows that

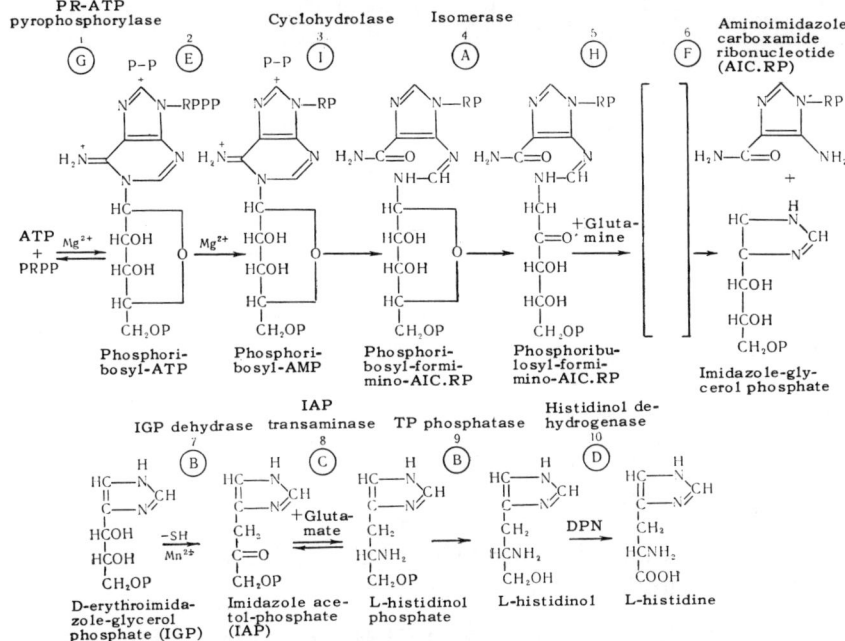

Fig. 25. Order of reactions in histidine biosynthesis (Ames, 1965). The structure of compounds given in parentheses is unknown. Stages F and H can take place also in the opposite order. P) phosphate, R) ribose, AIC) aminoimidazolecarboxamide.

transcription of the operon begins at the right of the operon and takes place linearly from right to left.

The study of orientation of the histidine operon on the genetic map of S. typhimurium (Hartmann et al., 1965) showed that if this operon is located on the bacterial chromosome, it lies in a clockwise direction between the following genes (operons): gal−try− −H1−His0−G−D−C−B−H−A−F−I−E−metG−purG−str−metA. It is usually assumed that the histidine operon is transcribed into one molecule of messenger RNA, which is then transcribed by the polyribosome system from one end to the other. The histidine operon and corresponding mRNA consist of about 13,000 nucleotides. The possibility that such a polycistronic histidine messenger RNA exists was demonstrated by the work of Martin (1963). The RNA fraction, labeling of which took place differently in histidine-constitutive and histidine-deficient strains, had a sedimentation

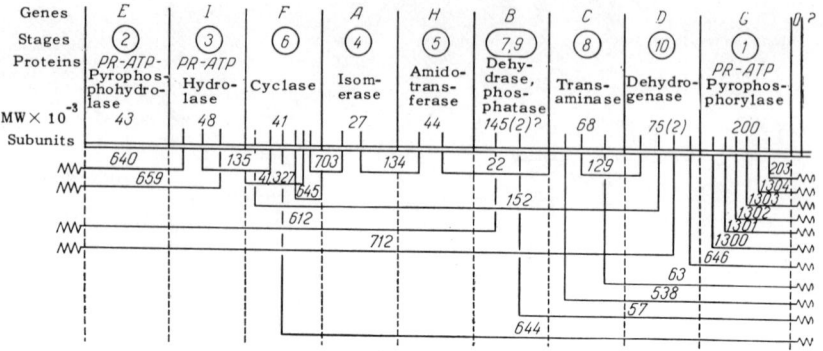

Fig. 26. General map of the histidine operon from data of Ames and Hartmann and collaborators (Ames, 1965). Capital letters between diagrams refer to gene loci. Stages of reactions successively controlled by each gene are circled. Usual names of enzymes are given. Below each name of an enzyme its approximate maximal molecular weight ($\times 10^{-3}$) is given. Extent of certain mutations (deletions) is shown at the bottom of the diagram.

constant of 34S, corresponding to a molecular weight of 4×10^6, very close to that calculated for the whole operon.

Whitfield and co-workers (1964) determined the molecular weight of nearly all the enzymes of the histidine operon. Adding these weights together and allowing for the dimerism of some enzymes, they calculated that the histidine operon must code the position of 4500 amino acid residues. If the size of the operon (13,000 nucleotides) is taken into account, the code ratio (nucleotides/amino acid) in this case is 2.9/1, in good agreement with the theoretical ratio of 3/1 for a triplet code.

Goldberger and Berberich (1965) carried out some interesting work to determine the time sequence of activity of the cistrons of the histidine operon based on the kinetics of repression and derepression of five (of the ten) enzymes of this operon from the beginning, middle, and end regions of the histidine operon of S. typhimurium. The specific activity of these enzymes was measured during a period soon before and after removal of histidine from the growth medium of the "leaky" mutant of this microorganism.

They found that the enzymes are derepressed (histidine synthesis begins) in a time sequence corresponding to the linear sequence of the genes in the histidine operon. About 20 minutes elapses between derepression of the first and the last enzyme (cistron). If, however, an excess of histidine is added to the derepressed culture, the sequence of repression of the five enzymes is similar in character and so also are the time intervals. This may be explained either by successive synthesis of individual molecules of the mRNA-operon or by successive "translation" of the polycistron template, starting at one end (by ribosomes). However, in this case there is no possibility of simultaneous synthesis of proteins from different cistrons of the template, previously accepted for virus RNA.

b) The Tryptophan Operon. The study of the tryptophan operon of *E. coli* is mainly associated with a series of investigations by Yanofsky and co-workers, summarized by Yanofsky (1963). The relative position of five cistrons is now known.

1	2	3	3b	4	Cis 4

Cistrons 1 and 2 correspond to A- and B-polypeptide subunits of tryptophan-synthetase, while cistrons 3A, 3B, and 4 determine the synthesis of enzymes converting anthranilic acid into indoleglycerophosphate. The order of these cistrons corresponds to the sequence of the synthetic reaction.

Synthesis of tryptophan-synthetase is repressed by exogenous tryptophan, and derepression is observed if the tryptophan concentration in the medium is restricted. The A-protein subunit of tryptophan-synthetase may constitute, in the course of these variations, from 0.01 to 1% of extractable cell protein (Yanofsky, 1960).

The operator locus of the tryptophan operon in *S. typhimurium* is linked with the gene for anthranilate-synthetase (Margolin and Mukai [1964]). In *E. coli*, an important role in the regulation of all activities of the tryptophan operon is also played by the gene for anthranilate-synthetase (Somerville and Yanofsky, 1965), the mutation of which is reflected in the other activities of this group of genes. Analysis of the character of these mutations and of their effect on the other genes of the operon, recently undertaken by Matsushiro and co-workers (1965), led them to suggest that in

E. coli also the operator locus is directly linked with the anthranilate-synthetase cistron.

However, Somerville and Yanofsky (1964) showed that synthesis of A-protein can take place under conditions when no B-protein is synthesized.

This can be explained either by the formation of individual forms of mRNA for each cistron, or by the ability of the integral operon (polycistron) RNA to translate the information with some degree of selectivity, i.e., by the phenomenon of regulation at the level of interaction between messenger RNA molecules and ribosomes.

The work of Imamoto and co-workers (1965) showed that a common (polycistronic) RNA is usually formed in *E. coli* for the whole tryptophan operon. It has a sedimentation constant of 33S (mol. wt. higher than 2×10^6). However, in some mutants, this messenger RNA macromolecule is fragmented into monocistronic units. The operator of the tryptophan operon of *E. coli* also is connected with the anthranilate locus. As a result of the method of isolation of the tryptophan operon together with a segment of the bacterial chromosome carried by bacteriophage $\phi 80$, this polycistronic RNA could also be synthesized *in vitro* (Okamoto et al., 1965). In *S. typhimurium* (Bauerle and Margolin [1966]) the tryptophan operon consists of two suboperons: one consisting of three, the other of two genes. Each suboperon is transcribed independently, so that during its activation the operon produces not one, but two different forms of polycistronic messenger RNA.

c) Other Integrated Single-Operon Structural Combinations of Genes. The study of localization of individual genes in systems of functionally connected biochemical reactions and of their corresponding localization on the genetic map of individual microorganisms have now become the main trend in biochemical genetics. Research in this field has led to the discovery and detailed study of several other biochemical systems with genetic organization of the operon type. The best known of these are: the leucine operon of *Salmonella*, consisting of four successive cistrons, the arabinose operon of *E. coli*, also consisting of four cistrons, and the isoleucine—valine operon of *Salmonella* and *E. coli*, consisting of five connected cistrons, etc. (Ames and Martin, 1964). These results suggest that the organization of a series of functionally connected metabolic reactions in a genetically determined system of the operon type is apparently the simplest method of programmed regulation. Since an intermediate link in this regulation is the synthesis of messenger RNA, the problem of translation of information contained in the operon into concrete processes is naturally bound up primarily with the investigation of which forms of messenger RNA are synthesized in connection with a particular operon.

§ 4. Aspects of the One Operon–One Molecule of Messenger RNA Theory, Modulators of the Velocity of Synthesis

I mentioned above that there are two views regarding the character of transcription of information from polycistronic loci of DNA. According to one view, every gene (cistron) synthesizes its own molecule of RNA (the one gene–one mRNA molecule theory); according to the other, an individual messenger RNA is synthesized by a complete operon and carries information for several proteins (the one operon–one mRNA molecule theory Fig. 27).

Goldberger and Berberich (1965) represented these two possible alternatives of transcription of genetic information as applied to the histidine operon in the form of the following scheme (Fig. 28). To the two variants, these workers added a third, the formation of polycistronic RNA, in which the protein synthesis of each

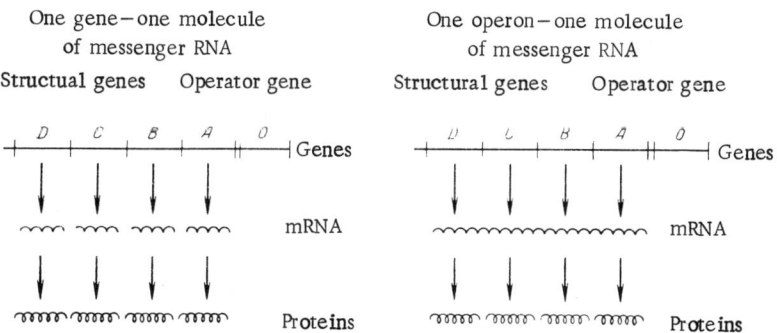

Fig. 27. Two biochemical models of transcription of the information of an operon.

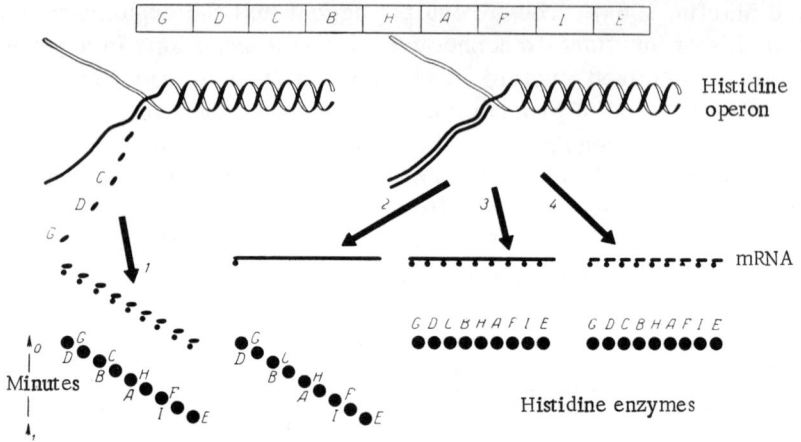

Fig. 28. Four possible models of protein synthesis under the control of polycistron operons. The histidine operon is shown at the top of the scheme. Capital letters denote genes (cistrons) controlling histidine biosynthesis. DNA is represented as a double helix; messenger RNA as straight lines of different lengths; ribosomes as small black circles; and proteins as larger black circles. The time scale is indicated in the bottom left corner of the scheme (Goldberger and Berberich, 1965).

cistron may be independent, and a fourth, in which polycistronic RNA is first formed and then breaks up into individual cistrons before interacting with ribosomes.

Polycistronic forms of messenger RNA, carrying information for several proteins, as we have seen are typical of viruses and are found in bacteria during the study of synthesis of proteins coded in one operon. However, in this case also we are forced to ask whether this RNA functions in the stage of protein synthesis on polysomes as a single linear system or as a sum of independently working cistrons, producing different numbers of protein molecules.

As we have seen in the case of virus polycistronic RNA templates, despite their integral structure, they produce different proteins in sharply different amounts. Each cistron of such a template functions at its own intensity and different cistrons produce different numbers of molecules per unit time depending on the amount of them required for virus reproduction. In this situation it is easier to assume that each cistron is not functionally connected with each other, but works independently at its own individual speed.

§ 4] THE ONE OPERON—ONE MOLECULE OF MESSENGER RNA THEORY

In the polycistronic operons controlling the multistage synthesis of a metabolite such as histidine, at every stage different numbers of molecules of enzymes must be formed because the specific activity of the enzymes differs and depends both on protein structure and on the character of the reaction it catalyzes. From this aspect, the linear polarity of transcription of information from polycistronic messenger RNA molecules of a polysome system raises the question of the mechanism responsible in this case for differences in the velocity of protein synthesis by different cistrons running their course in the polysomes. Ames and Hartmann (1963) proposed a very interesting explanation of this phenomenon and undertook its preliminary experimental verification.

According to their theory, as the polycistronic messenger RNA moves in relation to the polysome system, the velocity of protein synthesis in its various parts is slowed. They postulated that the sequence of the genes in the histidine operon (which does not correspond to the biochemical sequence of reactions) is connected with the number of molecules of each enzyme synthesized. By analyzing the frequency of mutations of polarity, they concluded that many triplets (of the 64 possible) can retard the transcription and translation of information. The essence of the matter is that if any nucleotide triplet (codon) XYZ requires an anticodon in the molecules of acceptor sRNA for its translation into a protein "text," a lowered content of this fraction of sRNA with the corresponding anticodon may act as modulator of the velocity of translation, which is reduced at this locus in connection with a decrease in the number of codon—anticodon interactions.

Ames and Hartmann assume that in normal cells this class of modulator coding triplets provides for the transcription of polycistronic mRNA at different speeds for its different cistrons, thereby ensuring the level of production from each cistron necessary for balanced synthesis of a particular metabolite. In their opinion the modulator triplets may be found near the beginning of every gene.

To verify this hypothesis they attempted to purify the enzymes of histidine biosynthesis and to determine their relative amounts in cells. They assumed that enzyme D may be present in an amount equal to or less than that of G; enzyme C may be present in an amount equal to or less than that of D; and so on.

Preliminary results confirmed their theoretical expectations. Two of the series of enzymes produced by genes on the right side of the operon (histidinol-dehydrogenase and pyrophosphorylase) (genes D and G) are present in large amounts. Two of the series of enzymes formed by the left side of the operon, on the other hand, were present in much smaller amounts.

It has also been shown for the lactose operon (Zabin, 1963a, 1963b) that the β-galactosidase cistron, which is nearer to the operator, synthesizes its protein ten times faster than the transacetylase cistron of the same operon. For every 25 moles of β-galactosidase subunits, only one mole of transacetylase is synthesized.

Ames and Hartmann (1963) point out that in formulating their hypothesis of modulator triplets they used the ideas of Itano (1963), who first suggested that degeneracy of the code and differences in the relative amounts of each of the 64 possible sRNA fractions could explain the unequal amounts of normal and abnormal hemoglobins synthesized in heterozygotes containing the abnormal hemoglobin gene in one of the genomes. Hemoglobin mutation, changing the sequence of amino acids, in every case reduces the quantity of hemoglobin synthesized. Tens of abnormal hemoglobins are now known, and in every case the allele gene (of heterozygotes) of abnormal hemoglobin transcribes its information to a lesser degree than the normal hemoglobin allele.

It is interesting to note that certain other connected groups of enzymes are formed in equimolar amounts (Mier and Cotton, 1966). In this case, these workers suggest, each cistron of the polycistronic RNA works independently with equal speed in the polysome system.

§ 5. Genetic Determination of Functionally Interconnected and Successive Reactions by Structurally Separate Genes and by Groups of Genes Located in Different Parts of the Bacterial Chromosome. Arginine Synthesis by *E. coli* and Cysteine Synthesis by Salmonella

It is necessary to point out, however, that the time sequence of biochemical reactions does not always by any means correspond to the linear sequence of cistrons on the genetic map of the bacterial chromosome, as we have seen in the case of lactose, histidine, tryptophan, and other operons. The regulation of a group of functionally interconnected reactions by a single operon is the simplest case of genetic structural regulation. Fragmentation of loci and the appearance of polyoperon regulation is a more complex situation.

The location of genes determining synthesis of enzymes for the successive stages of arginine synthesis in *E. coli* is typical in this respect. Synthesis of this amino acid by *E. coli* consists of the following reactions (the numbers denote the enzymes catalyzing each reaction): Glutamate $\xrightarrow{1}$ Acetylglutamate $\xrightarrow{2}$ Acetylglutamate-semialdehyde $\xrightarrow{3}$ Acetylornithine $\xrightarrow{4}$ Ornithine $\xrightarrow{5}$ Citrulline $\xrightarrow{6}$ Arginine succinate $\xrightarrow{7}$ Arginine.

The study of the arrangement of arginine loci in the *E. coli* genome by the method of interrupted conjugation (Gorini et al., 1961) showed that loci of the structural genes of enzymes participating in arginine biosynthesis do not lie side by side but are dispersed along the length of the chromosome, forming four groups (Fig. 29).

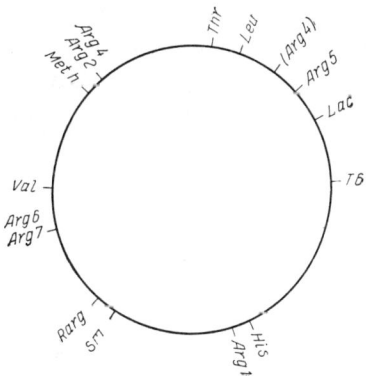

Fig. 29. Arrangement of loci of enzymes of arginine biosynthesis in the chromosome of *Escherichia coli* K-12 (Gorini et al., 1961). The crossover time for movement of the chromosome from one cell to another is 22-23 min for ARG5, 25-28 or 34-35 min for ARG4, 36-37 min for ARG2, 55-59 min for ARG6 or 7, and 68 min for ARG1. LAC markers are transferred after 18 min and HIS after 70 min.

Arginine, the end product of the reaction, represses 6 of the 7 enzymes. However, mutants exist in which none of the enzymes of the arginine cycle is repressed by arginine. It has also been shown that repression of the enzymes of this cycle is not successive (when No. 1 is repressed, absence of No. 1 influences No. 2, and so on), and under the influence of arginine simultaneous repression of all enzymes participating in its biosynthesis takes place (Gorini and Maas, 1957, 1958). Genetic mapping showed that the responsibility for the whole sequence of the arginine cycle is controlled by one particular locus (R arg) located in the chromosome of strain K-12 between the Val and Sm loci. Transfer of this locus by genetic recombination to strain B, which is not repressed by arginine, transmits to its progeny a repressibility of the same type as in K-12. This locus is thus the regulator gene of the cycle.

Gorini and co-workers (1961) postulate that simultaneous repression of genes in different locations may occur for two reasons: either the enzymes which they synthesize are dimers and one of the subunits of each of them is common to all (which is improbable), or each enzyme contains a common sequence which serves as "identification mark" on the chromosome in the nucleotide text, thus enabling the repressor to recognize enzymes which differ in other respects.

This dispersal of loci of arginine synthesis on the genetic map of *E. coli* has also been demonstrated by other investigations (Maas, 1961; Maas and Clark, 1964). The gene determining the ability of arginine loci to be repressed by arginine (the R gene), as this investigation showed, is located separately and at some distance from the arginine loci.

If the o p e r o n model is applied to this system, in Maas's opinion it has to be assumed that each structural gene of arginine synthesis possesses its own operator, and that the repressor acts on all operators at once along the lines indicated in Fig. 30. Experimental evidence in support of this type of regulation of the cistrons of arginine synthesis was obtained in work by Vogel and co-workers (Baumberg et al., 1965). Since some genes of the arginine cycle are side by side, however, it may be considered that each group acts separately as an operon, and the whole system becomes a p o l y o p e r o n system.

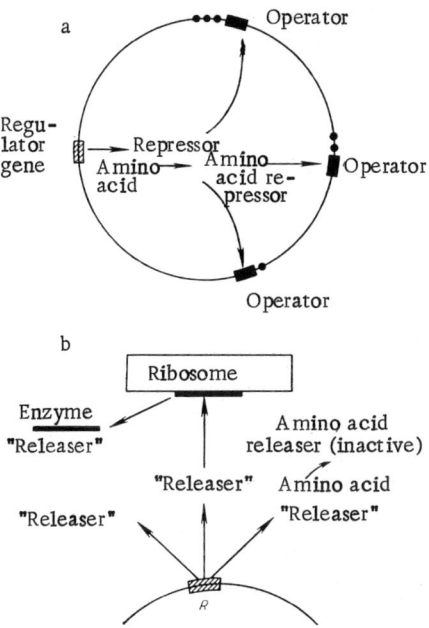

Fig. 30. Hypothetical models of the repression mechanism with a "broken" operon. a) Model with participation of an operator; b) model with participation of a "releaser" (Maas et al., 1961).

The genetic regulation of cysteine synthesis with utilization of sulfates by *Salmonella* have the same type of polyoperon structure (Demerec et al., 1963). This system consists of 14 cistrons combined into five linked groups (operons) at different locations (with 3, 3, 5, 2, and 1 cistron per operon). Such polyoperon systems of regulation of metabolic cycles are more common in microorganisms at a higher level of evolution than in the simplest bacteria (Demerec, 1965; Horowitz, 1965).

§ 6. Possible Nature of Genetic Repressors and Mechanisms of Repression of Operons and Cistrons of the Bacterial Chromosome

The repressor and the repression of definite genes constitute a special category of molecular-genetic phenomena appearing in a primitive form in the genetic system of phages, but most marked as a result of evolution of a more complex genetic system in bacteria. This type of regulation of gene activity appeared when it became necessary for the genes or groups of genes to work only periodically, or to work only under certain influences, such as the composition of the external or internal environment, or under the influence of other factors.

The genetic apparatus of the cell controls not only the morphogenesis and specific characteristics of the cell through the synthesis of enzymes and other proteins, but also the routine functions of the cell and even the dynamics of these functions, and the character of its interaction with the environment. Accordingly, of, let us say, 1000 genes combined into 140 operons, at any given moment not all the operons are active, and those which are active are not so to the same degree. Some operons, whose function is not required at that moment, are inhibited (repressed), but they can be brought back into an active state by the action of certain factors (activators). It can naturally be assumed that repression of a particular operon is brought about by some form of substance, and since this repression depends on a special regulator gene, whose location has been precisely mapped, it is also natural to assume that the repressor substance is formed by the regulator gene and that the external substrate somehow either blocks the repressor or stops its formation. As we have seen, this is essentially Jacob and Monod's hypothesis of the mechanism of induced enzyme synthesis.

Under ordinary conditions, according to this scheme, the repressor, by its affinity for the operator, blocks synthesis of messenger RNA by the operon, and this makes the synthesis of a particular group of proteins impossible. The substance combining with the repressor (R) is known as the effector (F) and the reaction is represented by $R + F = RF$. If a system is capable of induction, only the pure repressor R can block the operon.

On the other hand, in metabolic systems such as the synthesis of tryptophan, histidine, and many other substances, when the internal or external metabolite (usually the end product of the reaction) inhibits instead of inducing the system (feedback inhibition), it is postulated that this end product combines with a prorepressor, converting it into active repressor. Such inhibition is necessary when the temporary appearance of a utilizable substance (an amino acid, for example) makes its complete synthesis from precursors uneconomic, so that this synthesis is either inhibited or stopped.

The concept of "repressors" and "prorepressors" appeared initially as a model (hypothesis) based on the analysis of genetic data for localization of regulator genes and operators, and discovery of the invalidity of other schemes of induction and repression assuming the direct action of metabolites on structural genes.

Of course, as soon as this concept was formulated, attempts were made to study the possible nature of repressors and the mechanism of their action on genetic structures experimentally.

Arguments based on the molecular conformation properties which the repressor must possess (ability to recognize the substrate molecule and affinity for the operator group), its action on the operator, its hypothetical transfer through the cytoplasm from the regulator gene located in another part of the chromosome, and various other considerations led to the suggestion that the repressor is protein in nature (Jacob et al., 1962).

In this and other investigations concerned with the theoretical analysis of the influence of conformational changes in protein structure on their function (allosteric effects), these workers advance the view that the hypothetical repressor protein contains two specific receptor sites in its molecule (Jacob and Monod, 1964; Monod et al., 1963). One site interacts with the inducer (the effector), the other with the operator. During this interaction with the effector, influencing the conformation of the molecule, the degree of affinity of the second site for the operator is modified, being either reduced (inducible system) or increased (repressible system).

It must be pointed out, however, that some metabolic systems have now been studied with feedback inhibition by the end product, in which the end product reacts directly with the enzyme of the first reaction and inhibits it (Datta et al., 1964; Kennell and Magasanik, 1964; Nierlich and Magasanik, 1965). It has been shown in the case of histidine synthesis that the inhibitor of synthesis is not free histidine, but the compound histidinyl-sRNA, i.e., a more direct product for incorporation into protein synthesis (Schlesinger and Magasanik, 1964). During valine biosynthesis by

E. coli, the end product repressing the enzyme of this synthesis is not valine, but valyl-sRNA (Eidlic and Neidhardt, 1965).

Although by now many investigations have been carried out by workers who, on the basis of certain indirect properties of repressors, postulate that the repressors in bacteria are specific proteins (Hiatt et al., 1963; Sypherd and Strauss, 1963; Maas and Clark, 1964; Müller-Hill et al., 1964; Gallant and Stapleton, 1964; Gallant and Spottwood, 1964), a specific protein produced by the regulator gene has been isolated only in two investigations. In the first case this was a protein formed by the regulator gene for the alkaline phosphatase of *E. coli* and somehow linked with repression of this enzyme (Garen and Otsuji, 1964).

This protein was identified by electrophoretic analysis of the protein composition of different *E. coli* mutants, some of which were very active with respect to the inducible synthesis of alkaline phosphatase (up to 6% of their dry weight) while others were active with respect to constitutive phosphatase synthesis (deletion of the regulator gene). Mutation of the regulator gene locus led to changes in the electrophoretic and immunochemical properties of this protein.

The amino acid composition of the regulator protein was studied. It possessed acid properties due to a very high content of glutamic and aspartic acids. However, this investigation still did not demonstrate that this protein itself is the operon repressor, but merely indicated that it was somehow connected with repression.

Sadler and Novick (1965) carried out a convincing and important investigation to obtain evidence of the existence of a repressor of the lactose (LAC) operon and to study its properties (without physical separation of the repressor). According to genetic analysis of recombinations of *E. coli*, as mentioned previously, the regulator gene of the lactose operon (I) forms a repressor, hypothetically a special allosteric protein, capable of reacting both with the inducer (substrate) and the operator (or other locus of the operon). According to the classical scheme, union of the repressor with inducer modifies its properties and prevents its union with the operon.

It had been shown earlier that the regulator (I) gene may undergo mutations modifying its specificity relative to the inducer only, and not to the operon (Wilson et al., 1964). Sadler and Novick also used this principle of studying mutations of the I-gene to study the nature of the repressor, but they used a very broad spectrum of mutations of this gene, largely temperature sensitive mutations. In certain mutants of *E. coli* K-12 the product formed by the I-gene (repressor) is thermolabile, and warming the culture from 30 to 41.5° for a short time causes its inactivation; in other mutants the repressor itself is stabile, but synthesis of the repressor, i.e., activity of the I-gene, is sensitive to temperature. If a bacterial line with a thermolabile repressor (i^{TL}-allele) is heated to 41.5°, derepression of the operon soon takes place (just as during induction), and synthesis of β-galactosidase is activated. If, on the other hand, inducer is added before heating, the thermolability of the repressor is modified.

From experiments of this type, in different combinations and with different mutants of *E. coli* Sadler and Novick concluded that the repressor is a protein composed of subunits (the polymer itself is stable but the monomers are thermolabile), which actually takes part in a reaction with inducer and operator. In another investigation in the same laboratory (Novick et al., 1965) it was shown that the formation of the repressor itself is not an inducible reaction but a constant process, so that it is unnecessary to assume yet another regulating system controlling the work of the regulator genes.

The first report of actual isolation and identification of the genetic repressor was published at the end of 1966 (Gilbert and Müller-Hill, 1966). These workers used a very simple and ingenious method. They isolated the repressor of the lactose operon of *E. coli*, the product of the i-gene. According to the theory, this repressor unites with the substrate, i.e., with galactose. For its isolation by dialysis, a labeled substrate (isopropylthiogalactoside) also was used. The repressor turned out to be a protein with molecular weight 150,000-200,000. Strains with deletion of the i-gene did not form repressor.

§ 7. Functioning of the Bacterial Chromosome as Carrier of the Entire Program of Morphogenesis and Metabolism of Bacterial Cells

a) Genetic Map of the Bacterial Chromosome. We shall not consider the methods by which the genetic map of the bacterial chromosome is compiled. These methods have long since acquired classical status and they are described in standard textbooks of genetics (Lobashev, 1963; Muntzing, 1963). They are also fully described in a special monograph on bacterial genetics (Jacob and Wollman, 1961). It is only necessary to point out that now that it is possible to map the loci by three different and independent methods — by studying genetic recombination, by the method of transfer of loci by prophage (transduction), and by interruption of conjugation at different phases of transfer of the donor chromosome into the recipient cell — we now know the very precise details of the nature of the genetic map, especially of *E. coli*.

Both the chromosome and the genetic map of *E. coli* have the character of a closed circle, which can, however, be broken at a certain point during conjugation, whereupon the chromosome is slowly passed as a linear structure from one cell into the other at the point where they unite. The point of rupture differs in different donors.

In 1961, Jacob and Wollman collected together all known information concerning the mapping data for *E. coli* in the circular chromosome of this microorganism, and in their monograph (Jacob and Wollman, 1961) they presented the first complete circular map of the linkage group of characteristics for *E. coli* K-12. The loci for 60 genes were shown on this map. By 1964 about 100 genes had been located on the map (Fig. 31). At the present time the position of no fewer than 150 genes could be included on the general map. There is still far to go before all the genes of *E. coli* (about 1500) are mapped, but the task will certainly be completed within the lifetime of the present generation of geneticists. After the phages, clearly *E. coli* will be the first cell system for which this complex natural crossword puzzle — the genetic map — will be completely solved.

b) **Biochemical Structure of the Chromosome of *E. coli* and Its Replication.** For various technical reasons, the study of the biochemical organization of the bacterial chromosome has so far made very little progress. The colinearity, i.e., coincidence of the linear parameters of the A gene of tryptophan-synthetase and of the corresponding polypeptide (Yanofsky, Carlton, et al., 1964), and the precise linear demarcation of individual cistrons revealed by mapping by the interrupted conjugation method show that the basis for the bacterial chromosome of *E. coli* consists of a linear informational polymer, i.e., a DNA double helix, formed in various ways. Some authors accept that the bacterial chromosome may be monomolecular in character, i.e., composed of one giant molecule of double-helical circular DNA of the same type as is found in bacteriophage. The molecular weight of such a molecule for the *E. coli* chromosome is about 2×10^9.

It has been shown more recently, however, that the chromosome of *E. coli* and of *Bacillus subtilis* apparently consists, not of one continuous DNA molecule, but of a series of molecules with a molecular weight of about 250×10^6, held together by special protein cross-links (Massie and Zimm, 1965).

The most typical objects for the study of DNA of the bacterial chromosome and its replication are two bacteria: *E. coli* and *B. subtilis*.

Morphologically, as Cairns (1963a, b) clearly showed, the chromosome of *E. coli* is a closed ring, in full agreement with the character of its genetic map. Replication of this integral structure, connected by a definite system with cell division, is a complex morphogenetic process, for which the speed and the time and place of onset is regulated in accordance with a specific genetic program. The experimental study of replication of the *E. coli* chromosome has been carried out very intensively in recent years and a theoretical analysis of the results obtained have been made in a number of surveys (Kuempel and Pardee, 1963; Jacob, Brenner, and Cuzin, 1963; Lark, 1963; Sibatani and Hiai, 1964; Maaløe, 1961; Jacob and Monod, 1964), and this relieves us of the necessity for a detailed examination of the process, so that we can rest content with the basic facts and explanations.

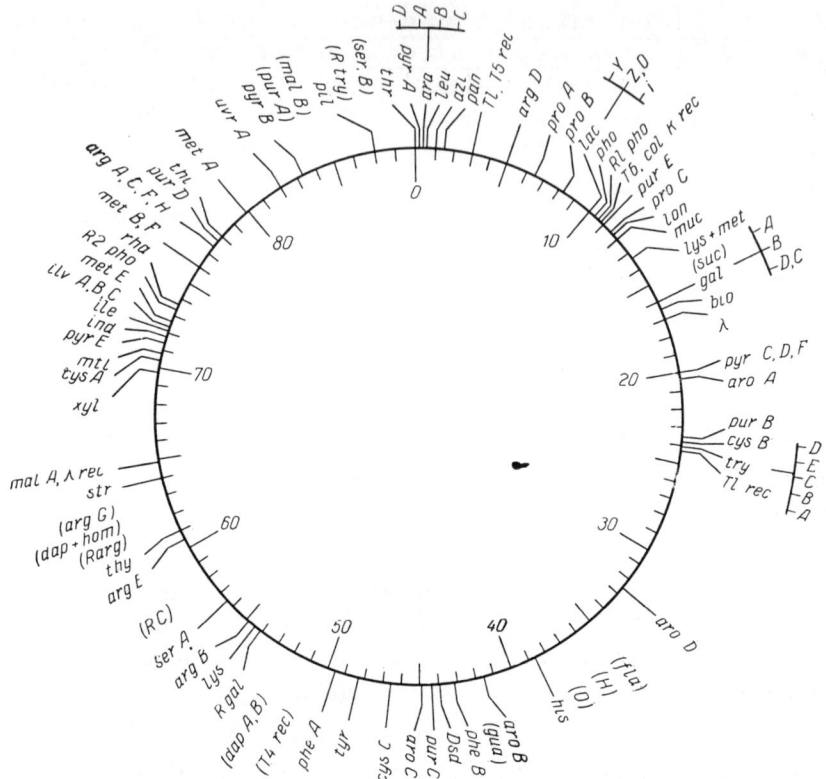

Fig. 31. Genetic map of *Escherichia coli* (Taylor and Thoman, 1964). The map is graduated in intervals of 1 minute and has a total length equal to the time taken for transfer of the chromosome during conjugation (89 min). The symbols denote: ability to synthesize threonine (thre), leucine (leu), proline (pro), lysine (lys), arginine (arg), biotin (biot), pyrimidines (pyr), etc. The capital letters for some polygenes (try, ara, etc.) denote the order of the cistrons within the operon.

1) **Onset of replication of the *E. coli* chromosome and its connection with the cell cycle.** In synchronized cultures of *E. coli* with rapid cell cycles (45 min or less) DNA synthesis, as determined from the rate of incorporation of labeled compounds, takes place throughout the cycle, whereas in slowly growing cultures of *E. coli* DNA synthesis in the chromosome (i.e., its replication) takes place only during a certain part of the cell cycle (Maaløe, 1961; Kuempel and Pardee, 1963).

In other bacteria, notably *Alcaligenes faecalis*, DNA synthesis also occurs during only a part of the cell cycle (Lark, 1963). This indicates that a period of rest occurs between two periods of DNA replication, and that the onset of replication at a particular moment of the cell cycle is connected with a definite stimulus. The nature of this stimulus is not yet known, but it evidently is not connected with absence of essential conditions for DNA synthesis. This follows from experiments (Brunfaut-Steux and Thomas, 1961), which showed that *E. coli* cells infected with phage λ could synthesize phage DNA immediately after their transfer to fresh medium before the onset of synthesis of bacterial DNA, which began later in accordance with the program of the *E. coli* cycle. The nucleotide composition of DNA of phage λ is almost identical with that of *E. coli* DNA. The fact that DNA synthesis began independently in this case does not indicate that the specific stimulus for the onset of replication cannot be transmitted through the cytoplasm. A characteristic example is shown by another organism, the mold *Physarium polycephalum*, which has multinuclear cells. Many of the nuclei in its cytoplasm start their cycle of division at the same time, indicating some form of interconnection through the cytoplasm (Nygaard et al., 1960).

2) **Rate of replication of DNA of the bacterial chromosome and its regulation.** If two types of genetic structures, such as the chromosome and phage λ, are present in the cytoplasm of *E. coli*, and if these are in a state of separation, DNA synthesis in these structures takes place at different rates. If, on the other hand, the phage DNA (in the form of prophage as a result of lysogeny) is incorporated into the chromosome of *E. coli* and has become part of it, the rate of synthesis of phage DNA will become the same as that of synthesis of other parts of the chromosome. A similar case is observed when synthesis of the chromosome and an episome is compared in the separate and combined form. In male *E. coli* cells (F^+) the episome (the circular DNA structure determining the sex type of the bacterial cell) is autonomous and replicates independently of the chromosome, sometimes at a very fast rate. In male cells of another type (Hfr) the episome is attached to the bacterial chromosome like a prophage, and in this case its replication is subordinated to the regulator mechanisms of the chromosome (Jacob and Monod, 1964). On the other hand, if lactose or galactose loci (operons) of

the *E. coli* chromosome are attached by genetic recombination to DNA of the sex factor (episome), these loci will begin to replicate at the rate characteristic of the episome, regardless of the rate of replication of the bacterial chromosome. However, these loci are incapable of independent replication unless incorporated into a genetically organized structure (the chromosome or episome), and they remain in the cell as an isolated fragment.

It was accordingly postulated that structures such as chromosomes or episomes are the elementary units of replication *in vivo*, and the term "replicon" was given to them (Jacob, Brenner, and Cuzin, 1963). It was also postulated that their ability to behave as a single entity and to replicate at a certain speed are determined by the activity of specific loci present in this replicon. The presence of such loci (determinants) was demonstrated genetically by the production of mutants in which tese loci were damaged. Mutations inhibiting the replication system in most cases are lethals; however, some mutations which are manifested only under certain conditions (for example, at a raised temperature), made it possible to localize these regulator genes in the bacterial chromosome.

Cairns (1963b), on the basis of his autoradiographic experiments with *E. coli* cells rapidly labeled with tritiated thymidine, calculated that the rate of replication of DNA was about 20-30 μ/min. In a period of 30 min this amounts to 600-900 μ, which is approximately the length of the chromosomal DNA, assuming that it is all present in the form of a giant double-helical molecule.

3) **Molecular mechanism and linear sequence of the process of replication of the *E. coli* chromosome from the starting point.** The molecular mechanism of DNA replication, based on the Watson and Crick DNA model (semiconservative type of replication) assumes that during uncoiling of the double-helical polynucleotide, each chain forms its complementary "replica" (Fig. 1). This model of DNA replication has been confirmed experimentally for many cases, and with particular clarity in the classical experiments of Meselson and Stahl (1958) when studying the process of DNA replication in *E. coli*. Having obtained an original "heavy" N^{15}-labeled form of DNA, these workers then grew bacteria on a medium containing the light nitrogen isotope (N^{14}). Heavy and light forms of DNA could be

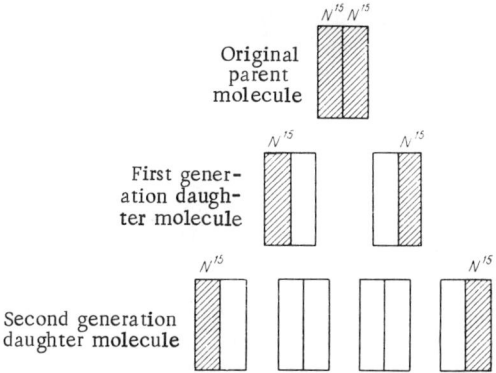

Fig. 32. Scheme showing results of experiment by Meselson and Stahl (1958) demonstrating semiconservative replication of DNA (explanation in text).

separated by prolonged centrifugation in a cesium chloride gradient at 140,000 g. Precise agreement between the replication process and the Watson and Crick model was observed in the experiment (Fig. 32).

By incorporating two prophage markers (prophage λ and 424) into the bacterial chromosome and then studying the kinetics of replication of the prophages during the DNA replication cycle in various sexual forms of *E. coli*, Nagata (1962, 1963a, b) showed by some demonstrative experiments that in the sexual form of *E. coli*, in which the episome is attached to the chromosome, the replication wave not only moves successively, but also moves cyclically in the sense that replication of the circular chromosome starts at a certain fixed point and moves continuously in one direction around the circle until the end. Each locus of the chromosome replicates once only in this period. Autoradiographic data (Cairns, 1963a) show that in Hfr-chromosomes, after completion of one round of replication, a new cycle begins at exactly the same starting point and proceeds in the same direction. A schematic model of replication of the circular chromosome, taken from Cairns (1963a), in the form of an interpretation of his autoradiographic data, is given in Fig. 33. The place at which replication begins is located near the attachment of the episome (F-factor), i.e., it corresponds to the "end" of the chromosome, using the term "end" to describe its locus which is transferred last into the F'-form during conjugation.

Fig. 33. Two stages of replication of a circular chromosome (Cairns, 1963a).

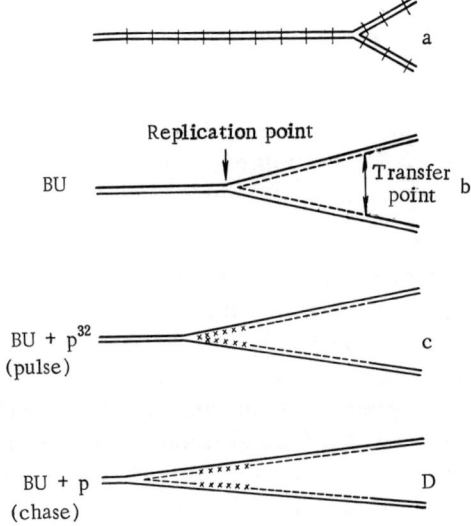

Fig. 34. Scheme showing a segment of a bacterial chromosome containing the growth point. a) Fragmentation of DNA in the course of extraction. DNA strands shown by continuous lines. Short intersecting lines indicate lateral breaks and may be assumed on the three lower diagrams. b) The same chromosome segment after labeling with 5-bromouracil (5BU). The DNA strand containing 5BU is shown by a broken line. c) The same segment after brief pulse labeling with P^{32} to mark the fragment of the growth point during cultivation with 5BU. P^{32}-containing zone into hybrid fragments during subsequent growth with 5BU and with unlabeled phosphate (Hanawalt and Ray, 1964).

Meanwhile, in cells of *E. coli*, not containing episomes, replication of the chromosome was random, i.e., in each new cycle it could begin at a new starting point. In this case there was no precisely fixed starting point of replication on the genetic map although, regardless of where it started, the replication wave always traveled around the complete circle, so that the linear sequence of replication was preserved.

The presence of only one growth point in the chromosome of *E. coli* was also demonstrated by the experiments of Bonhoeffer and Gierer (1963).

A very interesting method of isolating the "growth point" of the bacterial chromosome was developed by Hanawalt and Ray (1964). By pulse labeling of the replicating chromosome by different methods in turn, with bromouracil (a thymine analog and P^{32}) and isolating the labeled fragments after fragmentation of the chromosome into loci (bromouracil-DNA separates as the heavier fraction during gradient centrifugation), these workers were able to follow the movement of the growth point along the chain (Fig. 34).

4) **Models of regulation of the replication process of the *E. coli* chromosome.** Jacob and co-workers (Jacob, Brenner, and Cuzin, 1963, 1964), when discussing the existing factual material on replication of the *E. coli* chromosome, postulate that like other regulator mechanisms, in bacteria there is a molecular-genetic system controlling replication and consisting of at least two chromosomal loci, one of which determines and produces a diffusible factor, evidently protein in nature, while the other is a receptor for the first factor, is mounted in the DNA chain and like an operator gene determines the point where replication begins. They proposed the term "initiator" for the first, diffusible factor and the "replicator" for the second. The whole scheme is shown in Fig. 35. The mechanism of action of the initiator is unclear, and the authors of the model submit that it is either specific DNA-polymerase or some enzyme which can break the circular structure of DNA at a certain point near the replicator, thus enabling DNA-polymerase to begin to act from this site, and thereafter to code each polynucleotide of the chromosome chain. Butler (1963, 1965) puts forward the very interesting hypothesis that DNA-polymerase is a double ring, which can move along the two DNA chains, separating them and causing replication.

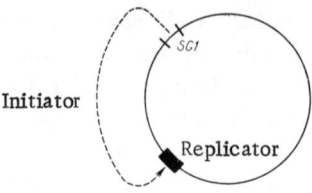

Fig. 35. Simplest model of positive regulation of the replicon. According to the model the replicon is a circular structure carrying two specific determinants. The structural gene determines synthesis of a diffusible active substance, the initiator. The initiator acts on the replicator, starting replication which then continues along the entire circular structure (Jacob, Brenner, and Cuzin, 1963).

However, we have seen that according to Nagata's findings, the onset of replication is not programmed in the F'-type of *E. coli*, and the cycle of DNA replication can begin at various points around its circular structure. To explain this situation, Nagata (1963b) suggests that there are many potential replicators in the *E. coli* chromosome. In F'-cells, at the end of cell division, one of these replicators, because of its random distribution, receives a specific signal from the newly formed bacterial membrane and is activated into a state in which it can react to the initiator, and a new cycle of DNA replication begins from this starting point. The replicator in the Hfr line, with a fixed starting point near the end of the chromosome, is activated by the episome (F-factor), or the F-factor introduces its own replicator into the system and this plays the dominant role in reception of the signal from the cell surface.

So far as the time sequence of these events in the cell cycle of *E. coli* is concerned, this is also hypothetically explained at present, by assuming the existence of successive trigger mechanisms operating in the corresponding order. In the case of another organism, *B. subtilis*, electron microscopy has revealed protrusions of the cell membrane, known as mesosomes, to which the chromosomes are attached. If we accept that the chromosomes in *E. coli* are somehow connected to the membrane, we can devise a model providing for coordination between growth of the bacterial cell and replication (Fig. 36; Jacob, Brenner, and Cuzin, 1963). The signal determining the onset of the DNA replication cycle, according to

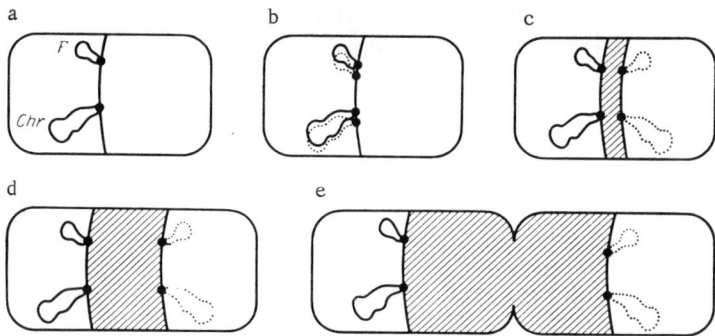

Fig. 36. Model of regulation of DNA synthesis and of the even distribution of DNA between daughter bacteria (Jacob, Brenner, and Cuzin, 1963). The bacterium (F^+) carries two independent self-reproducible structures: a chromosome (Chr) and fertility factor (F). Both structures are replicons and are attached independently to specific sites on the bacterial surface along the equatorial perimeter. At a certain stage of the bacterial cycle the cell surface transmits a signal through the replicons for replication to start. Elements of the bacterial membrane to which daughter replicons are attached gradually separate from the initial membrane; a, b, c, d, e—stages of division of the bacterial cell.

this scheme, originates at the cell membrane level. In each cycle of replication, synthesis of the membrane takes place between the points of attachment of the newly formed nuclear structures, thus insuring their gradual separation. The new replication cycle remains inhibited until growth of the membrane (coordinated, in turn, with other syntheses), reaches a certain limit. The membrane then gives the signal triggering the new replication cycle. The possibility cannot be ruled out that the beginning of DNA replication is coordinated with cell size (Koch and Schoechter, 1962) or with its surface/volume ratio.

With regard to molecular changes in DNA associated with the start of replication, variants other than local opening of the ring suggested by Jacob and Monod may also be postulated. This may be local denaturation, a wave of uncoiling, and so on. Intensive experimental studies of all these interactions are at present going on, but as in the case of the repressor, the nature of the proteins responsible for the trigger mechanisms of replication is still unknown. Investigations have clearly shown, however, that for replication to start normally the synthesis of a certain protein is necessary, and that there is another protein which accumulates only in the presence of replication (Lark et al., 1963; Pritchard and Lark, 1964; Lark and Lark, 1964).

A hypothetical model of replication of the *E. coli* chromosome, worked out in very great detail, has recently been published by Sibatani and Hiai (1964). Since this is purely hypothetical at the present time, I shall not analyze all its aspects in detail, although I should mention that these workers have presented other possible explanations of the precise coordination between the individual acts of replication.

c) Replication of the Chromosome of *B. subtilis*.
Like *E. coli*, *B. subtilis* has become a classical subject for the study of the biochemical genetics of microorganisms. Although *B. subtilis* is far removed from *E. coli* systematically, replication of its chromosome takes place more or less in accordance with the same principles as in the case described above. Studies of this replication by different methods (Wake, 1963; Sueoka and Yoshikawa, 1963; Yoshikawa et al., 1964; Oishi et al., 1964) showed that division of the chromosome of *B. subtilis* is polar in character and takes place in a linear manner from one end of the chromosome to the other. In rapidly growing cultures and germinating spores this division is dichotomous in character (Fig. 37).

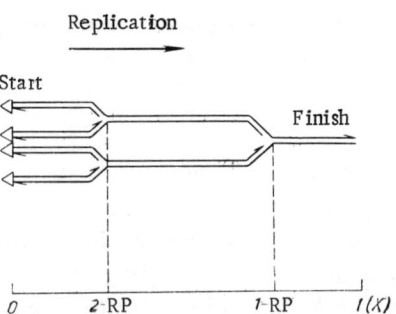

Fig. 37. Model of dichotomous replication of the chromosome of *Bacillus subtilis* (Oishi et al., 1964). Before the first replication, starting at point 1 RP reaches the end, the second cycle of replication (2 RP) begins.

d) Transcription of Genetic Information from DNA of the Bacterial Chromosome. Transcription of genetic information contained in DNA takes place through an intermediary, i.e., through synthesis of mRNA, molecules of which interact with the cytoplasmic systems of protein synthesis and act as direct templates for protein synthesis. The nucleotide genetic code is transcribed as a specific sequence of amino acids, and the resulting proteins, by their coordinated activity, provide for the complex metabolism of the cell, its reaction to the environment, its morphogenetic changes and growth, and so on.

Consequently, RNA synthesis by the bacterial chromosome is a basic element in the transcription of genetic information, and the distribution of sites of this synthesis along the chromosome, as well as in time during the cell cycle, is a mechanism controlling the vital activity of the cell.

The first point which must be mentioned is that only one polynucleotide chain in the DNA molecule, the basis of the bacterial chromosome, carries actual genetic information, and the other complementary chain plays no part in protein synthesis but simply serves for the replication of "genetic" DNA (it is sometimes called the "nonsense" or "reference" strand). It has been shown experimentally that not only in phages, as was pointed out in the preceding chapter, but also in *E. coli* the total messenger RNA of the cell is complementary to only half of the possible nucleotide sequences of the total cell DNA (McCarthy and Bolton, 1964). For this purpose the very ingeneous method of DNA-agar columns (columns containing denatured DNA), through which RNA is passed, was used (Britten, 1963; Bolton et al., 1963).

Interaction between DNA and RNA in agar showed that a population of cell RNA contains molecules homologus to only half the possible nucleotide sequences of cell DNA. If the cell DNA is separated mechanically into two parts, and one part, in the form of single-stranded polynucleotides, is applied to the agar while the other part of the DNA is passed through it in denatured form, the "agar" DNA is completely transformed into a double-stranded form through interaction with the external DNA. If, however, cell RNA is passed through such a column, DNA/RNA helices are formed by only half of the "agar" DNA. The mRNA molecules formed in the chromosomes are thus complementary to only one strand of each double-helical DNA molecule and identical to the other. That is why, as McCarthy and Bolton consider, this RNA does not form double helices spontaneously in the cells.

Meanwhile, if synthesized *in vitro*, complementary RNA forms secondary double helices (Geiduschek et al., 1962). Later work in the same laboratory (Geiduschek et al., 1964) showed that RNA synthesis *in vitro* (on DNA templates) is symmetrical (i.e., it copies both DNA strands) if the DNA is partially degraded. If the DNA molecules are intact, RNA synthesis is asymmetrical, i.e., in this case only one strand acts as template. This evidently accounts for the results obtained by Bresler and co-workers (1964), who obtained artificial double-helical DNA hybrids, introduced them into *B. subtilis* cells as transforming agent, and found that both strands possess genetic transforming significance. Definite evidence of transcription of information from only one strand of chromosomal DNA was obtained in the experiments of Guild and Robinson (1963) studying pneumococcal DNA from this standpoint.

However, from the point of view of analysis of the controlling function of the bacterial chromosome, it is essential that we now consider, not the very general problem of DNA-dependent RNA synthesis, but how the vast circular polyoperon structure of this chromosome carries out synthesis (transcription) of RNA at each operon at its own speed and with the required coordination. Some operons (especially the induced), as McCarthy and Bolton (1964) demonstrated and calculated, can synthesize messenger RNA from 100 to 200 times faster than the mean rate of synthesis of this RNA by the whole chromosome. An interesting model of the possible individualization of DNA operons was devised by Jones and Truman (1964). They postulate that one DNA strand, i.e., that carrying genetic information, at the border of each operon or independent cistron, possesses either a break or a segment capable of being ruptured by RNA-polymerase. In this case the operon locus of DNA moves away (uncoils) a short distance from the second DNA strand and becomes a template for RNA synthesis. The exposed segment of the second DNA strand, by reacting during that time with desoxyribonucleoside-triphosphates, synthesizes a new complementary chain for itself (this model is shown in Fig. 2).

However, there is evidence which supports quite different views on which DNA strand contains information. Facts have been found and theories put forward which indicate that during formation of complementary RNA, the first stage of synthesis is the exposure of one segment of one of the polynucleotide strands of the DNA, and this exposed area (not the zone which had branched off) becomes the template for RNA synthesis (Chamberlin and Berg, 1964; Sinsheimer and Lawrence, 1964; Warner et al., 1963). The first stage of this synthesis is formation of a DNA-RNA hybrid. Free RNA does not appear in this system until the ratio between DNA and RNA in the hybrid reaches a certain value. Judging from the known polarity of four of the operons (arabinose, leucine, lactose, and galactose), the orientation of these operons is the same as regards transcription, i.e., the direction of RNA synthesis is the same along the whole chromosome (Ames and Martin, 1964). In inducible systems, the inducer brings about a sharp increase in the rate of synthesis of the messenger RNA which is complementary to a particular operon (Gray et al., 1964; Hayashi et al., 1963; Imamoto et al., 1964).

§ 8. Differences in Development of Genetic Information of the Bacterial Chromosome Connected with the Cell Cycle and Morphogenetic Phenomena

The literature devoted to changes in biochemical properties of bacteria (enzymes, RNA, DNA, etc.) associated with the age of the cultures is extensive. However, the information applies to the special problem of aging of bacterial cultures, and we shall not consider it now. In the present case we are concerned with the much less thoroughly investigated problems of changes in biochemical activity of bacterial cells in the shorter interval between cell divisions, i.e., in the course of a life cycle of bacterial cell, usually measured in minutes. Investigations of this type are possible only on synchronized cultures. In bacteria, synthesis of RNA and DNA usually takes place throughout the cell cycle, i.e., cells newly formed after division gradually increase (double) their content of both RNA and DNA by the time of the next division, and this corresponds to the process of replication of the bacterial chromosome. Nevertheless, because of the replication of the chromosome in a linear sequence, it may be postulated that the synthesis of messenger RNA may vary in the course of the cell cycle, both quantitatively and qualitatively. This phenomenon has been studied in detail in Chargaff's laboratory (Rudner et al., 1964) on *E. coli*. By analyzing fast-labeled RNA (mRNA) of *E. coli* at 10-minute intervals of the life cycle of this organism, these workers found that both in the first cycle (10-100 min) and after division (110-150 min) the rate of synthesis and composition of the mRNA varied distinctly. The nucleotide ratio $(A + G)/(C + U)$, for example, varied (as radioactivity) during the cycle from 0.98 to 1.25.

Halvorson and co-workers (1966) found that during the cell cycle in yeasts and *E. coli* changes take place not only in the nucleotide composition of RNA, but also in the types of synthesis. DNA synthesis took place only in the earliest stages of cell division. Synthesis of ribosomal, mRNA and sRNA reached a maximum at different phases of the cycle, and the peaks of these syntheses replaced one another — ribosomes were first formed, then they were "charged" with mRNA, after which synthesis of sRNA, the amino acid carrier, increased in intensity. As these workers found, transcription of the genome in yeasts and bacteria takes place consecutively, and the order of enzyme synthesis is determined by the order of the structural genes in the chromosome. These results indicate possible differences in the information given out by bacterial DNA during the cell cycle.

A result similar in principle, but obtained by comparison of other phases of development, was obtained by Doi and Igarashi (1964) who studied forms of mRNA from *B. subtilis* during sporulation. Experiments on hybridization of this RNA with DNA showed that the RNA (associated with sporulation) originates from particular genetic loci that are inactive during ordinary cell reproduction.

This line of research into biochemical differentiation of the life cycle of bacteria was continued by Aronson (1965) who studied mRNA formed, also during sporulation, by *Bacillus cereus*. Aronson also studied in this case the qualitative characteristics of the mRNA of *B. cereus* by examining its hybridization with DNA fixed on agar. The character of saturation of the DNA with RNA taken from stationary, growing, or sporulating cultures differed very considerably.

In later work (Aronson, 1965), an mRNA specific for the sporulation period only was identified. It belongs to the group of stable forms of RNA and is connected with the cell membrane. This RNA is responsible for synthesis of the structural protein of the spores.

It is interesting to note that during spore-formation in bacteria, not only the composition of mRNA, but also the fractional composition of transfer sRNA varies. Comparative study of the composition of fractions of aminoacyl-sRNA from vegetative and sporulating cells of *B. subtilis* by chromatography on methylated albumin (Kaneko and Doi, 1966) showed that, at the beginning of sporulation, the ratio between the two different types of valyl-sRNA changed sharply (on account of the sRNA and not of the aminoacyl-sRNA-synthetase). By the end of sporulation, the changes observed had returned to normal. In another investigation (Lazzarini, 1966), on the same microorganism, a marked increase in lysyl-sRNA was found in the spores, attributable to new types of lysyl-sRNA not present in vegetative cells.

Some particularly interesting results have recently been obtained by the study of the dynamics of synthesis of certain enzymes in the course of the cell cycle. If the gene determining synthesis of a particular enzyme is active (during constitutive synthesis or during induced synthesis in the presence of substrate), the process of linear replication of DNA during the cell cycle may double this activity as a result of synthesis of another daughter gene at various periods of the cycle depending on the distance of this gene from the starting point of replication of the bacterial chromosome. In this case activity of the enzyme rises abruptly at a certain moment of the cycle. In this way it has been possible to determine the number of genes for a particular enzyme, because if two genes are present in one chromosome there must be two of these abrupt increases in activity.

The first results showing this type of discontinuous enzyme synthesis in *E. coli* were obtained by Masters and co-workers (1964), who found that different enzymes of *E. coli* double their activity at different moments of the cell cycle. In experiments by Halvorson and co-workers (Halvorson et al., 1964; Gorman et al., 1964), a series of enzymes (glucosidase, invertase, alkaline phosphatase) was studied from this point of view in synchronized cultures of different species of yeasts. The "jumps" in activity of the different enzymes occurred at different points of the cycle, and the experiments showed that in *Saccharomyces cerevisiae* there are two structural genes for alkaline phosphatase and invertase (two periods of synthesis during the cycle).

Similar investigations with another series of enzymes were carried out on *B. subtilis* (Donachie, 1965). He studied yet another group of enzymes (aspartate-transcarbamylase, ornithine-transcarbamylase, alkaline phosphatase) from this point of view, comparing their activity with the dynamics of DNA synthesis. Well-defined "steps" in enzyme synthesis were observed in the case of aspartate-transcarbamylase (Fig. 38). The dynamics of synthesis of the other enzymes was less clear because of their rapid breakdown after synthesis. All this shows that activity of synthesis of a particular enzyme is definitely dependent upon the number of active genes.

A new and interesting line of research into the biochemical aspects of development of bacterial cells is the study of the significance of certain specific syntheses for induction of particular phases of development. For example, it has been shown that synthesis of alanine-dehydrogenase induces processes leading to germination of spores in *B. subtilis* (Freese et al., 1964). Work by Zaitseva and Vedenina (1965) has shown that during chromosome replication the

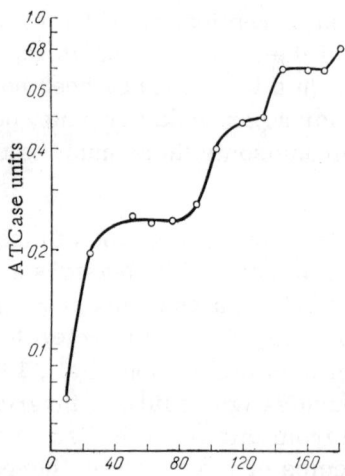

Fig. 38. Increase in enzyme activity in a culture of W23 inoculated from the stationary phase into fresh solution at 0 min (Donachie, 1965).

number of electrophoretically distinguishable proteins in the cytoplasm increased in a synchronized culture of *Azotobacter vinelandii*.

We can thus see that the genetic system of bacteria provides for a number of phenomena of regulation which are new in principle compared with the genetic system of viruses. The most important of these is ability to adapt to changing conditions of the external and internal environment, flexibility of the programmed control of internal metabolism and functions of the morphogenetic processes. This complex of new properties enables bacteria to exist independently in the external environment. As a result of these properties, life within the limits of the unicellular organism underwent complex evolution, resulting in the creation of a vast variety of forms.

However, the morphogenetic possibilities of unicellular forms of life were still restricted. True development became possible only as a result of the appearance of cell differentiation and specialization under the conditions of the multicellular organism. This step in the evolution of life implied radical changes both in the structure of the genome and in its external manifestation (the phenotype), as well as the possibility of a great leap forward in the volume of information which can be collected, reproduced, and translated by the genome. However, most systems of genetic regulation which appeared at the bacterial stage of development, persisted with certain modification, in multicellular biological systems.

Literature Cited

Alpers, D. H., and Tomkins, G. M., 1965, Proc. Nat. Acad. Sci., USA, 53:797.
Ames, B. N., 1965, Genetika, No. 5:3.
Ames, B. N., and Hartmann, P. E., 1963, In: The Molecular Basis of Neoplasma, Univ. of Texas Press, Austin, p. 322.
Ames, B. N., and Hartmann, P. E., 1963, Cold Spring Harbor Symp. Quant. Biol., 28:349.
Ames, B. N., and Martin, R. G., 1964, Ann. Rev. Biochem. 33:235.
Aronson, A. J., 1965, J. Mol. Biol., 11:576.
Aronson, A., 1965a, J. Mol. Biol., 13:92.
Bauerle, R. H., and Margolin, P., 1966, Proc. Nat. Acad. Sci., USA, 56:111.
Baumberg, S., Bacon, D. F., and Vogel, H. J., 1965, Proc. Nat. Acad. Sci., USA, 53:1029.
Bolton, E. T., Britten, R. J., Byers, T. J., Cowie, D. B., Hoyer, B., McCarthy, B. J., McOuillen, K., and Roberts, R. B., 1962, in: Carnegie Institution of Washington, Year Book 62 for the Year July 1-June 30, 1963, p. 303.
Bonhoeffer, F., and Gierer, A., 1963, J. Mol. Biol., 7:534.

Bourgeois, S., Cohn, M., and Orgel, L. E., 1966, J. Mol. Biol., 14:300.
Bresler, S. E., Kreneva, R. A., Kushev, V. V., and Mosevitskii, M. I., 1964, Biokhimiya, 29:477.
Britten, R. J., 1963, Science, 142:963.
Brunfaut-Steux, M., and Thomas, R., 1961, Arch. Int. Phys. Biochim., 69:103.
Butler, J. A. V., 1963, Nature, 199:68.
Butler, J. A. V., 1965, Nature, 207:104.
Buttin, G., 1961, Cold Spring Harbor Symp. Quant. Biol., 27:389.
Buttin, G., 1962, C. R. Acad. Sci., 255:1233.
Cairns, J., 1963a., J. Mol. Biol., 6:208.
Cairns, J., 1963b, Cold Spring Harbor Symp. Quant. Biol., 28:43.
Chamberlin, M., and Berg, P., 1964, J. Mol. Biol., 8:297.
Datta, P., Gest, H., and Segal, H. L., 1964, Proc. Nat. Acad. Sci., USA, 51:125.
Demerec, M., 1965, In: Evolving Genes and Proteins, ed. V. Bryson and H. J. Bogel, Academic Press, New York.
Demerec, M., Gillespie, D. H., and Mizobuchi, K., 1963, Genetics, 48:997.
Doi, R. H., and Igarashi, R. T., 1964, Proc. Nat. Acad. Sci., USA, 52:755.
Donachie, W. D., 1965, Nature, 205:1084.
Douglas, H. C., and Perloy, G., 1963, Biochim. Biophys. Acta, 68:155.
Eidlic, L., and Neidhardt, F. C., 1965, Proc. Nat. Acad. Sci., USA, 53:539.
Freese, E., Park, S. W., and Cashel, M., 1964, Proc. Nat. Acad. Sci., USA, 51:1164.
Gallant, J., and Stapleton, R., 1964, J. Mol. Biol., 8:431, 442.
Gallant, J., and Spottwood, T., 1964, Proc. Nat. Acad. Sci., USA, 52:1591.
Garen, A., and Otsuji, N., 1964, J. Mol. Biol., 8:841.
Geiduschek, E. P., Moor, J. W., and Weiss, S. B., 1962, Proc. Nat. Acad. Sci, USA, 48:1078.
Geiduschek, E. P., Tocchini-Valentini, G. P., and Sarnat, M. T., 1964, Proc. Nat. Acad. Sci., USA, 52:486.
Gilbert, W., and Müller-Hill, B., 1966, Proc. Nat. Acad. Sci., USA, 56:1891.
Goldberger, R. F., and Berberich, M. A., 1965, Proc. Nat. Acad. Sci., USA, 54:279.
Gorini, L., Gundersen, W., and Berger, M., 1961, Cold Spring Harbor Symp. Quant. Biol., Vol. 26.
Gorini, L., and Maas, W. K., 1957, Biochim. Biophys. Acta, 25:208.
Gorini, L., and Maas, W. K., 1958, In: Symp. on the Chemical Basis of Development, ed. W. D. McElroy and B. Glass, Johns Hopkins Press, Baltimore, p. 469.
Gorman, J., Taruo, P., La Berge, M., and Halvorson, H., 1964, Biochem. Biophys. Res. Comm., 15:43.
Gray, E. D., Haywood, A. M., and Chargaff, E., 1964, Biochim. Biophys. Acta, 87:397.
Guild, W. R., and Robinson, M., 1963, Proc. Nat. Acad. Sci., USA, 50:106.
Halvorson, H., Gorman, J., Tauro, P., Epstein, R., La Berge, M., 1964, Federation Proc., 23:1002.
Halvorson, H., Tauro, P., Epstein, R., and Smith, D., 1966, In: Ninth International Congress on Microbiology, Symposia, Moscow, p. 11.
Hanawalt, Ph. C., and Ray, D. S., 1964, Proc. Nat. Acad. Sci., USA, 52:125.
Hartmann, P. E., Rusgis, C., and Stahl, R. C., 1965, Proc. Nat. Acad. Sci., USA, 53:1332.

LITERATURE CITED

Hayashi, M., Spiegelman, S., Franklin, N. C., and Luria, S. E., 1963, Proc. Nat. Acad. Sci., USA, 49:729.

Hiatt, H. H., Gros, F., and Jacob, F., 1963, Biochim. Biophys. Acta, 72:15.

Horowitz, N. H., 1965, In: Evolving Genes and Proteins, ed. V. Bryson and H. J. Vogel, Academic Press, New York.

Imamoto, F., Morkawa, N., and Sato, K., 1965, J. Mol. Biol., 13:169.

Imamoto, F., Yamagishi, H., and Nozu, K., 1964, J. Biochem., 55:303.

Itano, H. A., 1963, Proc. Symp. Abnormal Hemoglobins (Ibadan, Nigeria), Blackwell, Oxford.

Jacob, F., and Wollman, E., 1961, Sexuality and Genetics of Bacteria, Academic Press., New York and London.

Jacob, F., Brenner, S., and Cuzin, F., 1963, Cold Spring Harbor Symp. Quant. Biol., 28:329.

Jacob, F., and Monod, J., 1961, Cold Spring Harbor Symp. Quant. Biol., 26:193.

Jacob, F., and Monod, J., 1963, In: Cytodifferentiation and Macromolecular Synthesis, Academic Press, New York, p. 30.

Jacob, F., and Monod, J., 1964, In: Molecular Biology, Problems and Perspectives, Collection to Celebrate Academician V. A. Engel'gardt's Seventieth Birthday, ed. A. E. Braunshtein, Izd. Nauka, Moscow, p. 14.

Jacob, F., Sussman, R., and Monod, J., 1962, C. R. Acad. Sci., 254(24):4215.

Jones, K. W., and Truman, D. E. S., 1964, Nature (London), 202:1264.

Kaneko, I., and Doi, R. H., 1966, Proc. Nat. Acad. Sci., USA, 55:564.

Kennell, D., and Magasanik, B., 1964, Biochim. Biophys. Acta, 81:418.

Kiho, Y., and Rich, A., 1964, Proc. Nat. Acad. Sci., USA, 51:111.

Koch, A. L., and Schaechter, M., 1962, J. Gen. Microbiol., 29:435.

Kuempel, P. L., and Pardee, A. B., 1963, J. Cell and Comp. Physiol., Suppl. 1, 62(2):15.

Lark, K. G., 1963, In: Molecular Genetics, Vol. 1, ed. J. H. Taylor, Academic Press, New York, p. 153.

Lark, C., and Lark, K. G., 1964, J. Mol. Biol., 10:120.

Lark, K. G., Perko, T., and Hoffman, E. J., 1963, Biochim. Biophys. Acta, 76:9.

Lazzarini, R. A., 1966, Proc. Nat. Acad. Sci., USA, 56:185.

Leive, L., and Kollin, V., 1967, J. Mol. Biol., 24:247.

Lobashev, M. E., 1963, Genetics, Izd. LGU, Leningrad.

Maaløe, O., 1961, Cold Spring Harbor Symp. Quant. Biol., 26:45.

Maas, W. K., 1961, Cold Spring Harbor Symp. Quant. Biol., Vol. 26.

Maas, W. K., and Clark, A. J., 1964, J. Mol. Biol., 8:365.

Maas, W. K., 1964, J. Mol. Biol., 8:365.

Margolin, P., and Mukai, F. H., 1964, Bact. Proc., p. 87, Ann Arbor: American Society for Microbiology.

Martin, R. G., 1963, Cold Spring Harbor Symp. Quant. Biol., 28:357.

Massie, H. R., and Zimm, B. H., 1965, Proc. Nat. Acad. Sci., USA, 54:1636.

Masters, M., Kuempel, P. L., and Pardee, A. B., 1964, Biochim. Biophys. Res. Comm., 15:38.

Matsushiro, A., Sato, K., Ito, I., Kida, S., and Imamoto, F. J., 1965, Mol. Biol., 11:54.

McCarthy, B. J., and Bolton, E. T., 1964, J. Mol. Biol., 8:184.
Meselson, M., and Stahl, F. M., 1958, Proc. Nat. Acad. Sci., USA, 44:671.
Mier, P. D., and Cotton, D. W. K., 1966, Nature, 209:1022.
Monod, J., Changeux, J. P., and Jacob, F., 1963, J. Mol. Biol., 6:306.
Morowitz, H. J., and Cleverdon, R., 1959, Biochim. Biophys. Acta, 34:578.
Morowitz, H. J., and Tourtelotte, M., 1962, Scient. Amer., 206:117.
Morowitz, H. J., 1966, In: Principles of Biomolecular Organization, ed. G. E. W. Wolstenholme, Churchill, London, p. 446.
Morse, M. L., 1962, Proc. Nat. Acad. Sci., USA, 48:1314.
Müller-Hill, B., Rickenberg, H. V., and Wallenfels, K., 1964, J. Mol. Biol., 10:303.
Muntzing, A., 1962, Genetic Research, Stockholm.
Nagata, T., 1962, Biochem. Biophys. Res. Comm., 8:348.
Nagata, T., 1963a, Proc. Nat. Acad. Sci., USA, 49:551.
Nagata, T., 1963b, Cold Spring Harbor Symp. Quant. Biol., 28:55.
Nierlich, D. P., and Magasanik, B., 1965, J. Biol. Chem., 240:358.
Novick, A., Lennox, E. S., and Jacob, F., 1963, Cold Spring Harbor Symp. Quant. Biol., 28:397.
Nygaard, O. F., Güttes, S., and Rusch, H. P., 1960, Biochim. Biophys. Acta, 38:298.
Oishi, M., Yoshikawa, H., and Sueoka, N., 1964, Nature, 204:1069.
Okamoto, T., Imai, M., and Yura, T., 1965, Biochim. Biophys. Acta, 103:520.
Pritchard, R. H., and Lark, K. G., 1964, J. Mol. Biol., 9:288.
Rudner, R., Prokop-Schneider, B., and Chargaff, E., 1964, Nature (London), 203:479.
Sadler, J. R., and Novick, A., 1965, J. Mol. Biol., 12:305.
Sibatani, A., and Hiai, S., 1964, J. Theoret. Biol., 7:393.
Shapot, V. S., 1965, In: Biosynthesis of Protein and Nucleic Acids, ed. A. S. Spirin, Izd. Nauka, Moscow, p. 171.
Sinsheimer, R., and Lawrence, M., 1964, J. Mol. Biol., 8:289.
Schlesinger, S., and Magasanik, B., 1964, J. Mol. Biol., 9:670.
Somerville, R. L., and Yanofsky, C., 1964, J. Mol. Biol., 8:616.
Somerville, R. L., and Yanofsky, C., 1965, J. Mol. Biol., 11:747.
Stent, G. S., 1964, Science, 144:816.
Sueoka, N., and Yoshikawa, H., 1963, Cold Spring Harbor Symp. Quant. Biol., 28:47.
Sypherd, P. S., and Strauss, N., 1963, Proc. Nat. Acad. Sci., USA, 49:400.
Taylor, A., and Thoman, M., 1964, Genetics, 50:659.
Wake, R. G., 1963, Biochem. Biophys. Res. Comm., 13:67.
Warner, R. C., Samuels, H. H., Abbott, M. T., and Krakow, J. S., 1963, Proc. Nat. Acad. Sci., USA, 49:533.
Whitfield, H. J., Smith, D. W., and Martin, R. G., 1964, J. Biol. Chem., 239:3288.
Willson, C., Perrin, D., Cohn, M., Jacob, F., and Monod, J., 1964, J. Mol. Biol., 8:582.
Wu, H. C. P., and Kalckar, H. M., 1966, Proc. Nat. Acad. Sci., USA, 55:622.
Yanofsky, C., 1963, In: Symp. Informational Macromolecules, ed. G. Vogel et al., Academic Press, New York, p. 195.
Yanofsky, C., 1960, Bact. Rev., 24:221.
Yanofsky, C., Carlton, B. C., Guest, J. R., Helinski, D. R., and Henning, U., 1964, Proc. Nat. Acad. Sci., USA, 51:266.

Yoshikawa, H., O'Sullivan, A., and Sueoka, N., 1964, Proc. Nat. Acad. Sci., USA, 52:973.
Zabin, I., 1963a, Federation Proc. 22:27.
Zabin, I., 1963b, J. Biol. Chem., 238:3300.
Zaitseva, G. N., and Vedenina, I. Ya., 1965, Mikrobiologiya, 34:945.

Chapter 4

Molecular-Genetic Systems Controlling Morphogenesis and Differentiation in Multicellular Organisms. Molecular Structure of Chromosomes of the Cell Nucleus and Structural and Biochemical Features of Genetic and Functional Differentiation of Chromosomes

Introduction

The largest number of genes in the genetic system of the most complex viruses is not more than 100, and such a system can be represented by a single polycistron DNA macromolecule. In bacteria, more than 1000 genes may be present and their genetic apparatus consists of a new structure, a special type of chromosome consisting of a small or large number of DNA molecules connected into a circular (occasionally linear) system. With the transition to multicellular organisms, the volume of genetic information and the way in which it is translated into action during reproduction differ considerably and a number of fundamentally new genetic problems concerned with differentiation and morphogenesis arise. The number of genes of different types controlling development and providing for differentiation and protein synthesis increases from several thousands in the simplest members of this group of living organisms to several hundreds of thousands or, according to some highly arbitrary calculations, to values exceeding a million (Vogel, 1964) in man. Information equivalent to a million genes can be coded in polynucleotides having a total length equivalent to one billion nucleotides (with a molecular weight of about 600 billions if calculated as double-helical DNA). Such a maximal genetic system must be built from at least 60,000 DNA molecules, each having a molecular weight of 10×10^6, or 6000 molecules with a molecular weight of 100×10^6.

The quantity of DNA contained in the ordinary nucleus usually exceeds these calculated values considerably. The spermatozoon of the trout contains 6.4×10^{-9} mg DNA with a mean molecular weight of 6×10^6 (Sadron et al., 1957), equivalent to 640,000 DNA molecules. Because of the possible existence of surplus DNA, there is no direct relationship between the mass of DNA in the nucleus and the animal's position in the scale of evolution. The DNA content in the human spermatozoon, for example, is less (about 2.8×10^{-9} mg) than in the spermatozoon of the trout. Du Praw (1965) has calculated that if all DNA molecules of one diploid nucleus of a human cell were to be joined together into a single chain its length would be about 180 cm.

Since these tens and hundreds of thousands of genes function together in a strictly coordinated manner and in various combinations, depending on the types of specialization, in the cells of the developing organism it is clear that they must be joined together into relatively stable structures in strictly definite groups and in a strictly definite order within the limits of each group. This is the only way in which automatism and stability can exist in their differential activity in the course of their regulatory function. The structures responsible for coordinated and differentiated activity of the genes and their functional groups are the cell nuclei and chromosomes.

There is no need in this chapter to examine all aspects of the biochemistry and biochemical morphology of the cell nucleus and chromosome, for much of this task has already been undertaken at the present level of our knowledge in general textbooks of cytology and genetics (Lobashev, 1963; de Robertis et al., 1962, 1965; Sager and Ryan, 1964) and in numerous surveys (Ris, 1961; Mirsky and Osawa, 1961; Taylor, 1962; Zbarskii and Georgiev,

1962; Prokof'eva-Bel'govskaya and Bogdanov, 1963; Brodskii, 1965; Gall, 1963; Moses and Coleman, 1964).

We must, however, examine the biochemical structure of the nucleus and chromosomes from one particular aspect: how does this structure provide for and reflect the regulatory function of the cell nucleus during morphogenesis and differentiation?

§1. Structural Organization of Genetic Elements of the Nucleus and Chromosomes

Groups of genes (of cistrons) are joined together in DNA molecules, and as the example of phage DNA has shown, the order and mutual arrangement of the genes and the special properties of the boundary areas determine the required morphogenetically purposive character of translation of the genetic information. The same rule must also apply to the system of combination of polygenetic DNA molecules into chromosomes, and the structure of genetic maps of the chromosomes, as we know, demonstrates the strict linear sequence of genes and groups of genes relative to each other in the chromosomes. The internal structure of chromosomes thus assumes a precisely fixed position of individual DNA molecules in the chromosomes and maintenance of their specific mutual arrangement during reproduction of the chromosomes in cell division. As recent investigations have shown, chromosome structure in the various phases of the cell cycle is based on comparatively simple principles of construction which, above all, insure ease of linear replication of this structure. In some cases (polytene chromosomes) this linear aggregation of chromosome strands leads to the formation of special structures consisting of tens and hundreds of strands lying in close proximity to one another, in which repeatedly replicating identical genetic zones lie next to one another transversely.

Chromosomes consist of two (or a multiple of two) strands (chromosomes or chromatids), capable of spiralization and despiralization, and in some cases visible under the optical microscope. For this reason the physical state of the chromosomes may vary in different stages of mitosis and interphase. Toward the moment of division they become compact and thickened in shape; during formation of the interphase nucleus, despiralization is observed accompanied by increased intensity of functional processes in the chromosomes.

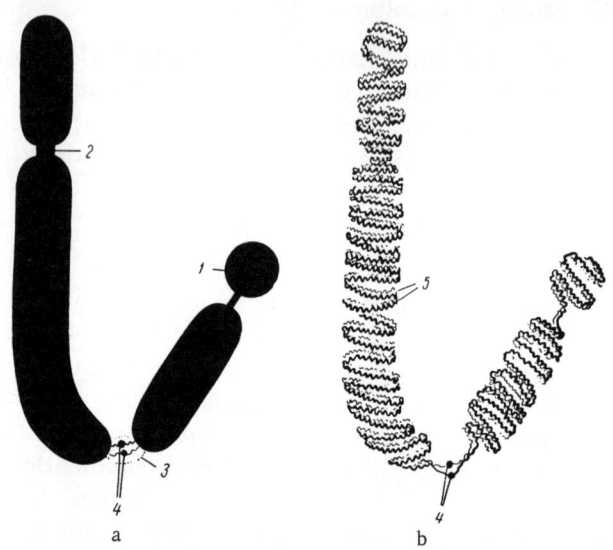

Fig. 39. Diagram of a metacentric chromosome. a) External appearance; 1) satellite; 2) secondary constriction; 3) centromere; 4) fibril of spindle; b) internal structure of the same chromosome with two chromosomes (5); a large and a small helix can be seen (de Robertis et al., 1962).

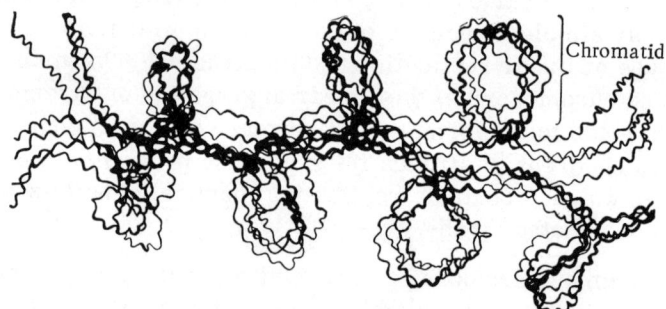

Fig. 40. Model of chromosome in early prophase of mitosis. The bracket indicates a chromatid (consisting of four subunits) coiled into a loop (Ris, 1961).

A typical model of a chromosome is shown in Fig. 39. Each chromoneme in turn according to the electron-microscopic data, has a fibrillary structure similar to that of a multistranded cable, as illustrated in Fig. 40 (Ris, 1961). Individual fibrils consist of polygenetic DNA molecules organized into a serially connected system. The character of molecular organization of the chromosome fibrils is not yet finally settled. As a result of experimental investigations so far undertaken, many structural models have been created and discussed, each of them purporting to reflect the mechanisms of helical transformation and linear replication of these fibrils, their structural stability and integrity, their ability to take part in crossing over and in various other phenomena lying at the

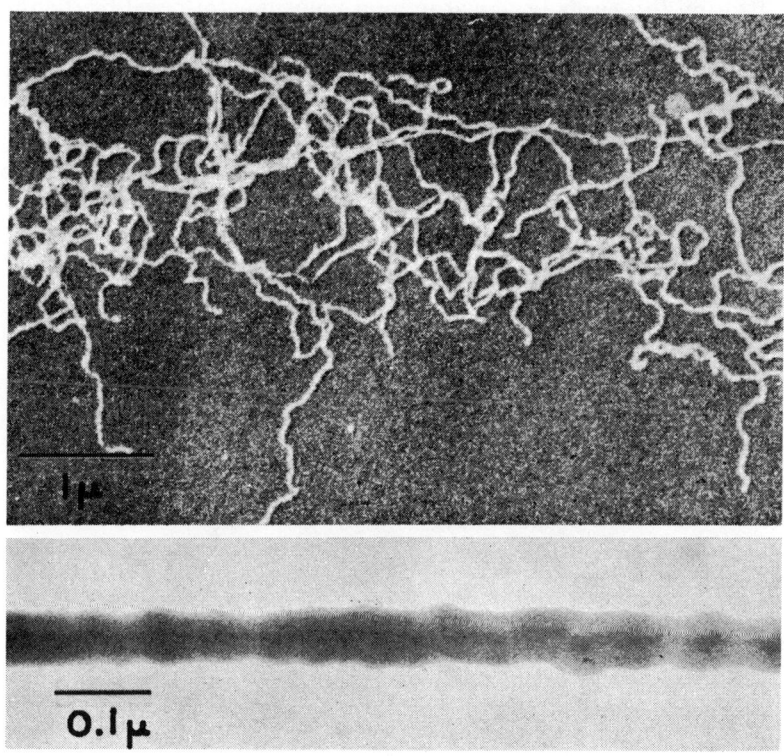

Fig. 41. Above: Fibrils from erythrocyte nuclei of the triton (*Triturus*) distributed on a water surface (dried in air, shadowed with platinum–palladium, 19,000×). Below: segment of a single fiber showing denser core (stained with uranyl, not shadowed; 140,000 ×) (Gall, 1963a).

basis of the genetic functions of chromosomes, and the transmission of their exact replicas or mutations which have arisen from one generation to the next for many thousands and millions of years.

Many old and new investigations have shown that the structure of chromosomes, in their different states (during mitosis and interphase), is based on fibrillary formations whose diameter varies from 100 to 600 Å. These investigations have been conducted on widely different objects: cell nuclei of *Tradescantia* and of bees, liver cell nuclei, oocytes and erythrocytes of *Triturus*, and so on.

Some typical micrographs of these fibrils, usually obtained by spreading homogenized nuclear material in a thin monomolecular film on the surface of water or a solution, followed by drying and conversion into a suitable preparation for electron-microscopic study, are shown in Fig. 41. The complex interweaving of the fibrils in this case is an artifact, but the existence of fibrils of this type is conclusively proved. The diameter of the fibrils varies considerably in different objects, and it may also vary appreciably

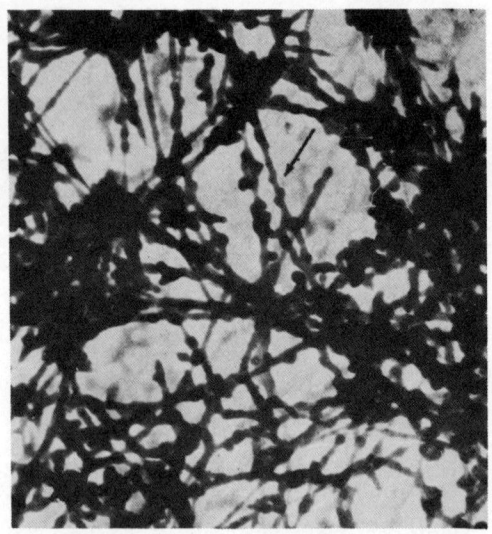

Fig. 42. Interphase chromatin fibers as seen at higher magnification (61,900×). Thickenings are visible on the fibers which in some places (arrow) have helical twists (Du Praw, 1965a).

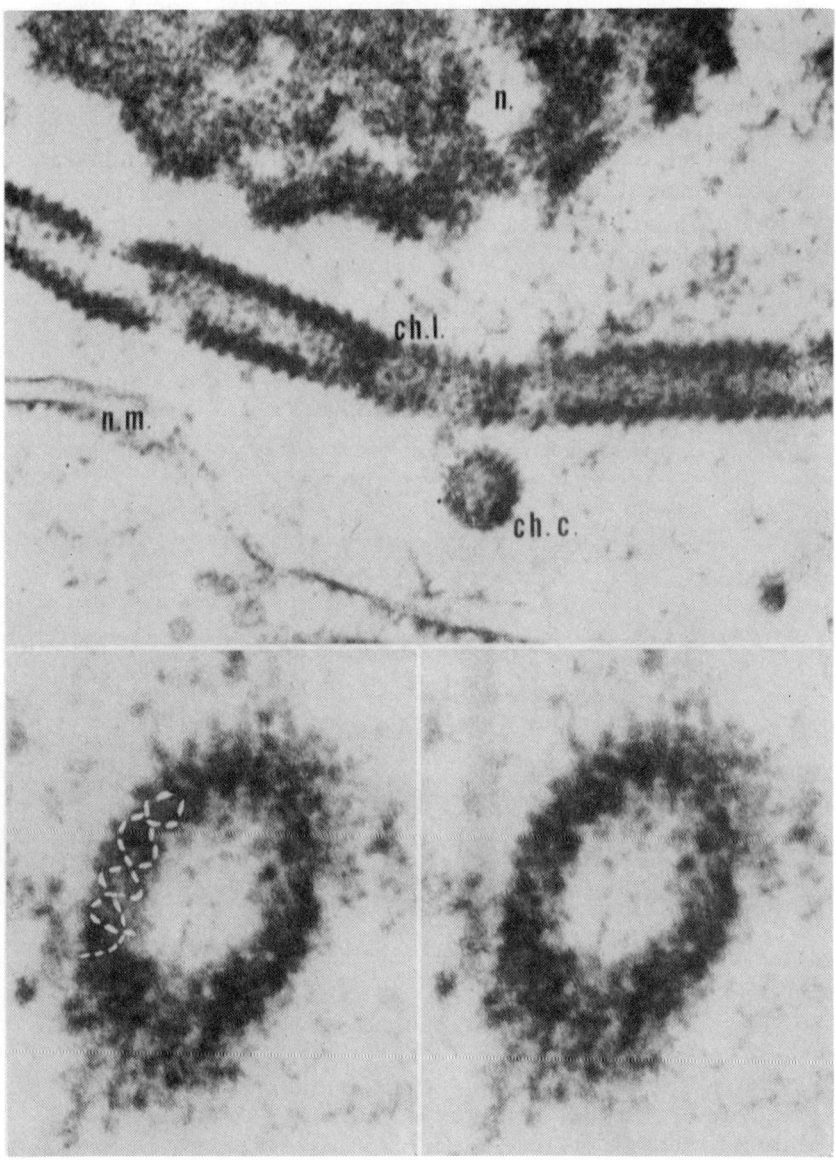

Fig. 43. Above: ch. l) longitudinal section through interphase chromosome; ch. c) cross section; n) nucleolus; n.m.) nuclear membrane. Below: oblique section through hypothetical interphase chromosome. On the left micrograph an interpretation is given of the possible arrangement of the double helix on the right of the same micrograph (Nørrevang, 1963).

in the same object. According to Du Praw (1965a, 1966), fibrils in interphase nuclei of bee embryos have parameters which lie at the limits permissible for double-helical DNA molecules allowing for the protein envelope of the DNA helix (Fig. 42).

In some cases when the state of these fibrils is studied in interphase nuclei, their secondary coiling can be seen, resulting in the formation of helical tubes about 0.2 μ in diameter (Fig. 43) (Nørrevang, 1963).

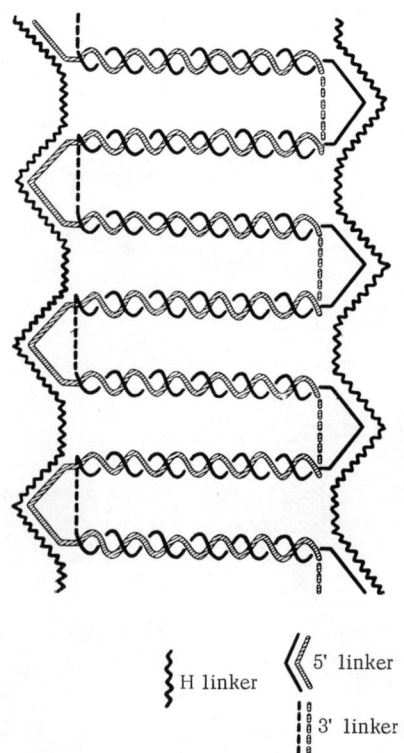

Fig. 44. Model of DNA arrangements in a chromosome showing how before or during replication the chromosome is assumed to be stabilized by the addition of H linkers to form a double axis, each of which with its attached DNA chains will become a chromatid by the time replication is complete. The stabilizing material is represented as a polymer attached to alternate 5' linkers. Although shown as a continuous chain for simplicity in illustration, it probably would consist of many small subunits attached in some way as yet unknown.

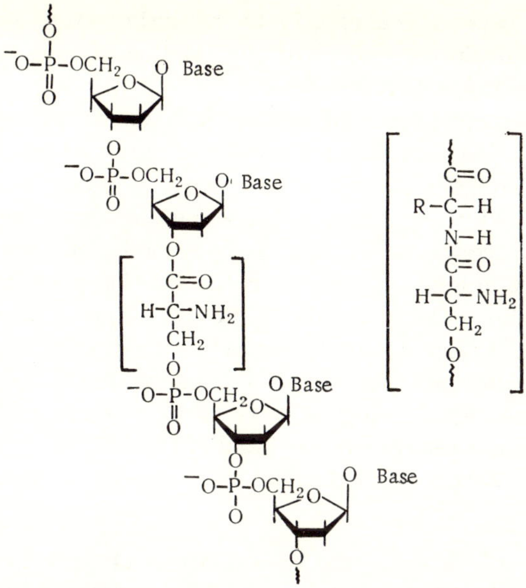

Fig. 45. Hypothetical arrangement of serine (or serine-containing peptide) in main polynucleotide chain of desoxyribonucleic acid (Bendich and Rosenkranz, 1962). On the segment of the structure an ester bond between an amino acid and the 3'-carbon atom of a nucleotide is illustrated. It is assumed that rupture of the chain by the action of hydroxylamine takes place at the site of this ester bond; as a result a hydroxamate is formed. It has been shown (see text) that hydroxylamine splits DNA into fragments retaining their double-stranded structure; hence it may be postulated that amino acid ester bonds in both DNA chains are located one opposite the other.

An interesting model of the internal structure of chromosome fibrils of this type, explaining the possibility of their lengthening and contraction (condensation), accompanied by thickening and compaction of the fibrils, has been suggested by Taylor (1962), and Freese (1958).

The compact state of a chromosome fibril is shown in Fig. 44 (after Taylor's model). The nature of the linkers joining the double strands of DNA at the bends is not yet known, although most probably they are peptides of a special type. Bendich and Rosenkranz (1962) have found that sperm DNA contains about one phosphoserine residue per 500-1000 nucleotides, and the evidence suggests that it is incorporated into the main chain. As a result of their study of the character of degradation of DNA molecules into subunits with molecular weight of about 500,000 by the action of hydroxylamine, which ruptures ester bonds of the amino acid–sRNA type but does not rupture phosphodiester internucleotide bonds, these workers postulate that the DNA molecule consists of polynucleotide units with molecular weight of about 500,000 joined together by hydroxyamino acids (or by simple peptides composed of these amino acids), one serine residue being present for every 1000 nucleotides. These amino acid ester bonds (Fig. 45) give the rigid chain considerable flexibility, allowing it to bend and coil. These serine linkages, according to Bendich and Rosenkranz, give the DNA molecules that degree of flexibility and foldability which is observed, for example, in the head of phage T2. In this case, a DNA molecule 600 times longer than the phage head itself is packed into the internal cavity.

In Taylor's opinion, the possibility of bridges (linkers) of this type between DNA molecules in chromatids also can be accepted because there is evidence (Callan and Macgregor, 1958; Gall, 1958; Macgregor and Callan, 1962) to show that proteases and ribonucleases do not rupture the loops of lampbrush chromosomes, although they are quickly broken by desoxyribonucleases.

Nevertheless, despite the fact that other recent investigations have confirmed the polyfibrillary structure of chromosomes experimentally (Dodge, 1963; Georgiev and Chentsov, 1963; Osgood et al., 1964; Jacob and Sirlin, 1963; Du Praw, 1965a, 1966) not all workers accept the polyfibrillary model. In particular, Painter (1964) has recently adduced fresh evidence in support of the earlier chromosome model based on the existence of a protein axis to which DNA molecules are attached by one end only.

Finally, two recently proposed chromosome models deserve special attention. One of them regards the chromosome as consisting of a complex fiber (Du Praw, 1965b, 1966). This model is based on an interpretation of his own electron-microscopic data concerning the chromosome structure of bee embryo cells (Fig. 46). The other model, suggested by Bogdanov and co-workers (1965) is an interpretation of their own findings and data in the literature, and it regards the chromosome as a polyfibrillary structure in which the fibrils have the special type of arrangement indicated in Fig. 47.

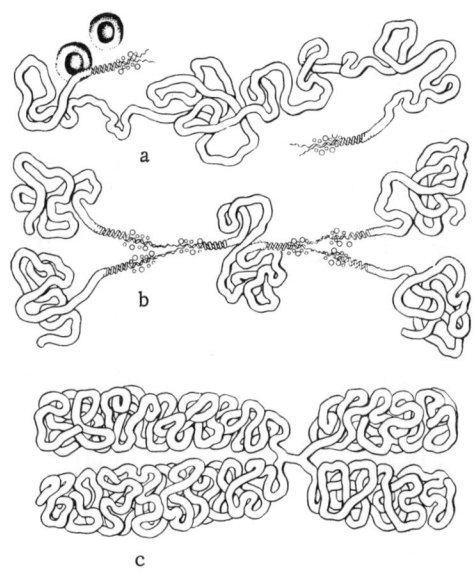
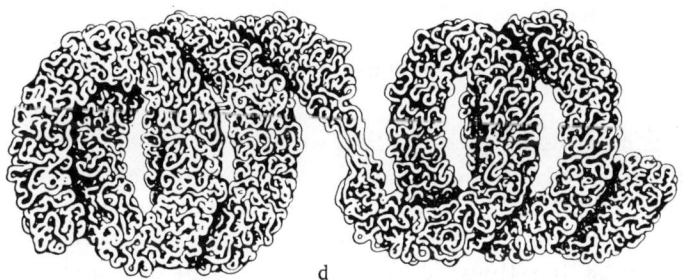

Fig. 46. Folded fiber model of the chromosome (Du Praw, 1965a, 1965b, 1966). a) Each interphase chromatin fiber consists of a DNA molecule held in the form of a secondary helix by means of its protein envelope; b) replication of the fiber takes place successively from both ends toward the middle; c) after replication the daughter fibers are folded into a condensed, compact chromosome. It is postulated that this folding is brought about by contractile protein molecules in the membrane; d) scheme of chromatid structure.

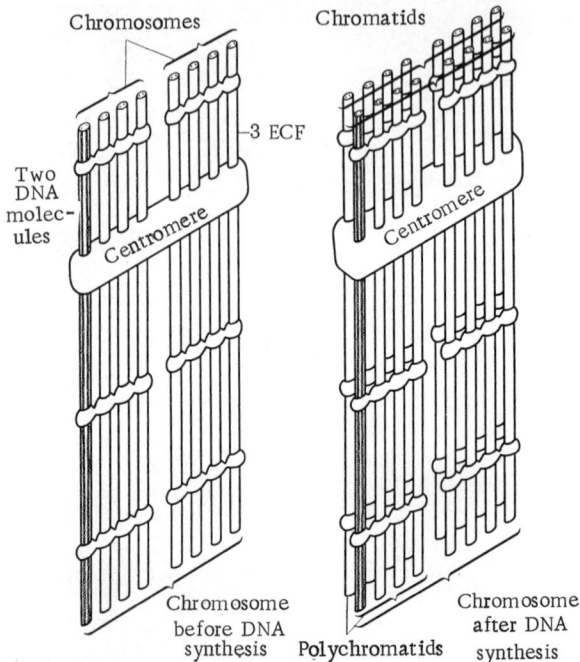

Fig. 47. Scheme of the polyfibrillary model of a chromosome. Possible arrangement of fibrils in an interphase chromosome is shown; on the left in period G_1, on the right in period G_2; ECF denotes elementary chromosomal fibril (Bogdanov et al., 1965).

It must be pointed out that we have mentioned only a very small proportion of the experimental data and theoretical models relating to molecular structure of the chromosomes of the cell nucleus. Even from the purely morphological standpoint this problem is exceptionally wide and varied, and for our purpose we have been interested only in the general principles governing the organization of genetic material into complex polymolecular structures. In the simplest forms — viruses, phages, bacteria — the boundaries between the genes and polycistron segments are not structurally distinguishable, but are nothing more than special nucleotide groups located along the general polynucleotide chain (nonsense and modulator triplets). In multicellular organisms, two new types of boundaries appear between the gene segments: protein connecting links between the DNA molecules and spatial demarcation of the chromosomes (linkage groups). This feature opens up wide prospects for intergenome regulation, activation, and repression of genes and for other processes determining functions of the chromosomes as a regulatory system of morphogenetic development.

§ 2. Reproduction (Replication) of Genetic Structures of the Cell Nucleus

In the introduction we examined the general principle governing replication of DNA, the chief information polymer of genetic structures. In the case of virus reproduction we see this synthesis in its pure form, as replication of individual DNA molecules in complete agreement with the classical scheme of Watson and Crick (Fig. 1). In the chromosomes of the cell, reproduction of genetic material solves two main problems: reproduction of DNA molecules and reproduction of the exact sequence of different DNA molecules (groups of genes) in the chromosomes. As a result, the whole process consists of precise replication of the chromosome as a definite genetic entity. Usually this replication is timed to take place at a definite initial phase of mitosis, but in some cases, in specialized cells, replication of chromatid strands is not accompanied by separation, and as a result of a series of replication, giant polytene chromosomes (such as the salivary gland chromosomes of insects) are formed.

Replication of chromosomes is a much more complex process than simple linear synthesis of DNA on other DNA templates. Replication of chromosomes is essentially a micromorphogenetic process, for besides synthesis of individual DNA molecules, specific protein linkages between them and protein (histone) molecules enveloping the DNA must also be reproduced. Proteins, however, are synthesized in special systems.

Consequently, replication of chromosomes combines synthesis of DNA molecules, the "switching on" of loci controlling synthesis of specific chromosomal proteins, the "assembling" of these proteins at the required places for linking together the DNA molecules into a single functional system, condensation (spiralization) of the chromatids, and their replication into two identical complete sets with subsequent divergence to different poles of the cell. This last process also requires a number of special syntheses, but I do not propose to discuss the analysis of the macromolecular events of mitosis but to concentrate attention instead on the mechanism of chromosome replication. All aspects of mitosis and cell division have recently been thoroughly examined in two special monographs (Mazia, 1961; Tsanev and Markov, 1964), and this enables us to concentrate on those problems of cell and chromosome division which are directly connected with mechanisms of functioning of chromosomes as systems controlling development of the organism.

The molecular mechanism of division of chromosomes, i.e., the way in which division of the uncoiled chromosome in prophase

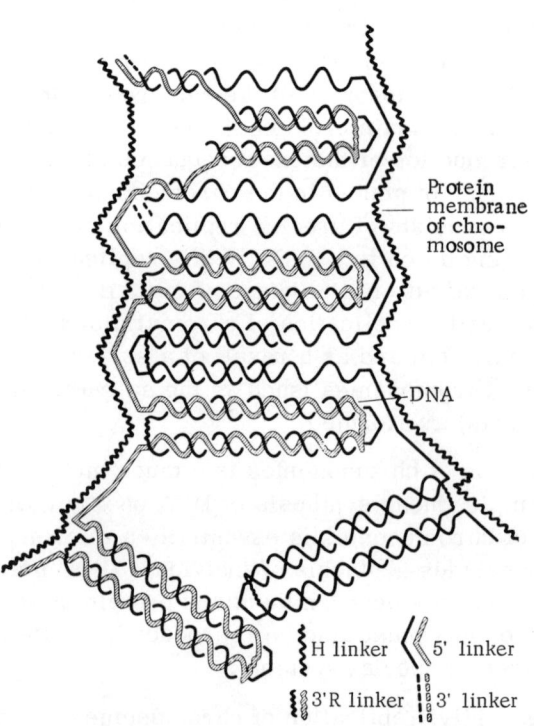

Fig. 48. Scheme showing beginning of replication (Taylor, 1963). As each pair of chains begins replication, a new 3'R linker is assumed to be inserted, which is resistant to the attack of enzyme that opens the regular 3' linkers. When the pairs of chains finish replication their growing 3' ends are assumed to be closed by the addition of another 3'R linker. The two new chromatids would consist of these pairs of DNA molecules attached to a single axial element. Another axial element (half-chromatid) would be added to each during the succeeding late interphase or prophase.

is combined with replication of each of its component DNA molecules, is still largely unknown. Nevertheless, sufficient facts on the distribution of labeled C^{14}- and H^3-nucleotide components, incorporated during replication of chromosomes, between chromatids in a series of division have now been accumulated to enable hypothetical models to be created. The best known model is that developed by Taylor (1962, 1963), which is shown in Fig. 48.

According to this scheme, the beginning of replication of chromosomes is preceded by enzymic opening of the 3' linkers in the DNA chain, starting the processes of uncoiling of the polynucleotide chains in individual double-helical DNA molecule. At the same time, closing of the 3' linkers of a particular chromatid during its replication is accompanied by insertion of new 3' linkers between the old and new polynucleotide. Taylor assumes that these temporary 3' linkers differ from those which can be broken by the enzyme, and he designates them 3'R linkers. As a result of this process the replicating structure must again consist of DNA double-helices attached to one axis of the original chromatid. The complementary chains attached to the other axis also replicate and form other DNA double-helices attached to one axis. Besides Taylor's scheme, several other models of chromosome replication are known, some more complex (Cole, 1962), others simpler (Deepesh, 1964), (Fig. 49).

Although there is no doubt that the actual mechanism of chromosome replication may differ from these models in details, on the whole the basic principle of replication is evidently faithfully reflected in them. This principle consists of the performance of a single successive replication in the chromosome of each messenger DNA molecule followed by separation of the newly formed replicas as mother and daughter chromosomes. Evidence that there is a definite and specific sequence of replication of genetic loci during chromosome replication is given by results of the microautoradiographic study of DNA synthesis in chromosomes during division.

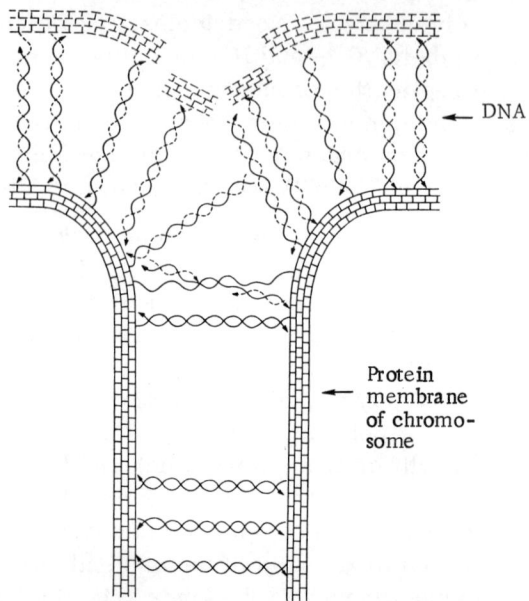

Fig. 49. Model of chromosome replication (Deepesh, 1964). Histone surrounding the DNA segments is not shown for greater simplicity of the model. Arrows on the DNA chains indicate free ends. Broken lines show newly synthesized components. Each daughter chromoneme is a hybrid of DNA and protein components.

§ 3. Asynchrony of DNA Replication in Different Parts of Chromosomes and in Different Chromosomes and an Indication of Linear Genetic and Functional Heterogeneity of Chromosomal DNA

As I stated in the introduction to this book, the main problem requiring explanation in the analysis of mechanisms of morphogenesis and differentiation is the nature of the selective activity of groups of genes in cells which have differentiated differently and the nature of the purposive change of combinations (or pattern) of active genes in time and space in precise agreement with a genetic project or plan of development. The genetic basis (in its overall significance) of most somatic cells of multicellular organisms is the same, but in different cells the work is done (i.e., the various forms of messenger RNA are synthesized) by different groups of genes and a change in these combinations precedes and predetermines morphogenetic changes in cells and cell groups. However, before analyzing the mechanism of this genetic differentiation, we must naturally examine the concrete forms in which it is expressed at the genetic level. The fact that different cells of the specialized organism differ in their morphology, protein composition, functions, and so on is well known and requires no explanation. Evidence that these cells have a common material genetic basis is given by the preservation (in most cases) of the typical chromosome number for the species, the identical (for a diploid or haploid set of chromosomes) content of DNA, the absence of tissue specificity of DNA, ability to undergo dedifferentiation (in some cases), the possibility of vegetative reproduction, and several other facts of a cytological, biochemical, and genetic order.

However, besides the evidence demonstrating the common genetic basis of cells with different specializations (within the same organism), there is considerable evidence to show the existence of definite differences between these cells at the molecular-genetic level, i.e., evidence of the actual existence of molecular-genetic and cytogenetic differentiation, sometimes reversible, sometimes irreversible, of cells with different specialization, evidence of differences between them at the nucleus-chromosome and molecule-gene levels exhibited in relation to various characteristics. Obviously this evidence must be examined first. One portion of the evidence relates to the phenomenon known as asynchronism of DNA synthesis both in individual chromosomes and in the whole set of chromosomes. This phenomenon is essentially as follows. If the chromatin fibril of the chromosome were a structure without functional and biochemical linear differentiation (in the sense of biochemical manifestations represented in its complement of genes), replication of the chromatid DNA, whether starting at one end or taking place at random, would be synchronized with respect both to different chromosomes (simultaneous beginning of replication) or to movement of the replication wave along the chromatid. However, this is not observed, but different chromosomes of diploid and haploid sets replicate asynchronously and different parts of the same chromosome differ individually in the periods, duration, and beginning of replication. Since different cells of the organism and cells of different species of animals possess specific features in this respect, this phenomenon has been regarded as a manifestation of differentiation within the genome, and it has thus been very intensively studied.

One of the results of its study has been the discovery of an inverse relationship between replication of DNA (by synthesis of the DNA → DNA → DNA → etc. type) and the active messenger function of genetic loci of the DNA (synthesis of messenger RNA): loci at which DNA synthesis is taking place at a given moment are inactive as regards synthesis of messenger RNA, and vice versa (Goldstein and Brown, 1961; Stockdale and Holtzer, 1961; Prescott, 1962).

a) Asynchrony of Linear Replication of DNA at Different Loci of the Individual Chromosome. Movement of the DNA replication wave along the genetic system has been most clearly demonstrated by the work of Prescott (Prescott, 1962; Kimball and Prescott, 1962), who studied DNA synthesis in the long macronucleus of the protozoon *Euplotes edrystomus* during amitotic reorganization of the macronucleus. DNA synthesis was studied by the use of H^3-thymidine incorporated in the DNA over short periods of time, followed by microautoradiographic determination of the zones of DNA synthesis. The experiments clearly showed that DNA synthesis in the macronucleus takes place as a narrow wave spreading toward the center from each end with a high degree of synchronism (Fig. 50a). It is interesting to note that histone synthesis in the macronucleus also takes place in the form of a wave intimately connected with the DNA wave (Fig. 50b) whereas RNA synthesis, on the contrary, stops as the wave of DNA replication draws nearer. RNA behaved as if expelled from the macronucleus into the cytoplasm. However, this particular case must be regarded more as a model of synchronism of replication, due both to the unique way in which the macronucleus functions and also to the fact that it must not be regarded as an analog of the cell chromosome of the multicellular organism. However, asynchrony of DNA synthesis is present in this organism if the macronuclei and micronuclei are compared from this standpoint. Each of these nuclei synthesizes DNA at different phases of the cycle of cell division, DNA replication taking place initially in the macronucleus and later in the micronucleus (Prescott et al., 1962).

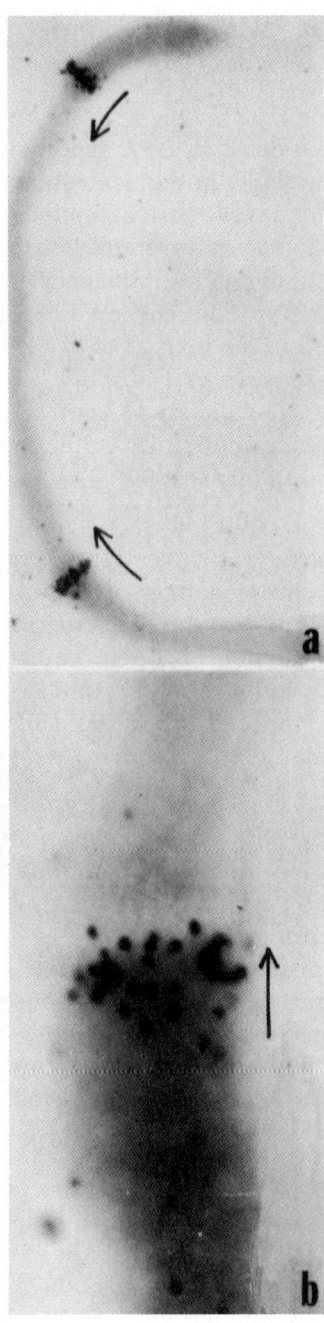

Fig. 50. a) Microautoradiograph of macronucleus isolated from *Euplotes* incubated in solution with H^3-thymidine for 20 min; denser arrangement of silver grains represent zone of DNA synthesis. Arrows indicate direction of movement of wave of DNA synthesis; b) Incorporation of H^3-histidine into *Euplotes* macronucleus during incubation for 40 min; zone of protein synthesis follows a short distance behind zone of DNA synthesis (Prescott, 1962).

In multicellular organisms, asynchrony (heterogeneity) of DNA synthesis along linear structures of chromosomes is now being studied by microautoradiographic methods (incorporation of H^3-thymidine) following the work of Taylor (1958, 1960). In the first of his investigations, Taylor observed delay in DNA synthesis (late synthesis relative to the phases of mitosis) in the centromere zone in root tip cells of *Crepis*. In the other investigation replication of chromosomal DNA was studied in tissue culture cells of the Chinese hamster. It was found that the number of simultaneously coded DNA particles varied in different parts of the chromosomes, but each DNA particle was coded only once in the course of one mitotic cycle. In some chromosomes there were loci in which DNA replication was delayed. Delay in DNA synthesis in centromere sites of certain chromosomes of human cells (in culture) has also been observed by Moorhead and Defendi (1963).

Asynchrony, that is to say different rates of DNA synthesis in different parts of the chromosomes, was investigated by incorporation of labeled thymidine into chromosomes of root cells of *Tradescantia* (Wimber, 1961) and *Vicia faba* (Peacock, 1963; Woodard et al., 1961), into chromosomes of HeLa cells (Stubblefield and Mueller, 1962; Mueller and Kajiwara, 1966), into chromosomes of cultures of leukocytes and other cells of the Chinese hamster (Prescott and Bender, 1963), and in investigations on certain other organisms.

A very clear picture of asynchrony and local heterogeneity of DNA synthesis in the chromosomes was obtained in studies of DNA synthesis in the polytene chromosomes of insect salivary glands. We shall describe the characteristics of synthesis of DNA, RNA, and proteins in these chromosomes in greater detail later in connection with our examination of chromosomal puffs. At this point, however, I shall mention the results of experiments (Plaut, 1963; Plaut and Nash, 1964) in which asynchrony of DNA synthesis in different parts of individual chromosomes was demonstrated very clearly.

These workers incubated the salivary gland of larvae of various species of *Drosophila* in a solution containing H^3-thymidine for a period of about 10-15 min, which is between one-fortieth and one-fiftieth of the total DNA replication time in polytene chromosome (10-12 h). In this case DNA synthesis was not observed

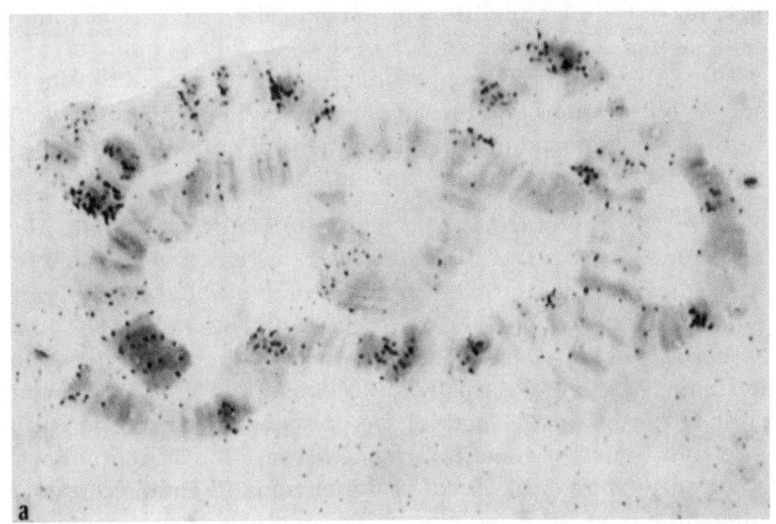

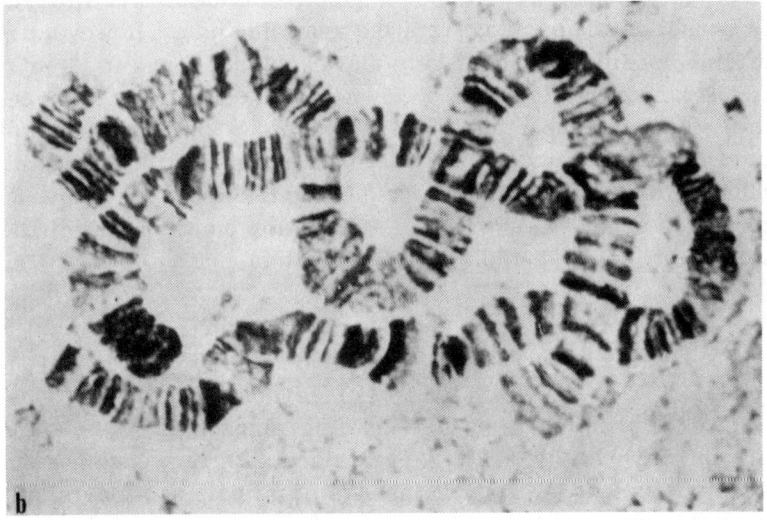

Fig. 51. a) Microautoradiograph of two chromosomes from salivary gland nuclei of *Drosophila* after incorporation of H^3-thymidine (15 µCi/ml, 10,000 µCi/µM) for 5-8 min; b) phase contrast of the same field after staining with aceto-orcein and before application of autoradiographic emulsion (× 1900). Absence of absolute correlation between density of bands and presence or absence of label will be noted (Plaut and Nash, 1964).

along the whole chromosome and it took place not successively but asynchronously in many different parts of the separate chromosomes. Moreover, the character of distribution of these sites of synthesis differed in different cells from the same part of the gland. In cases when label was administered to the *Drosophila* larva for a long period of time (several replication cycles), the chromosomes were uniformly labeled.

The typical picture of heterogeneity of DNA synthesis along chromosomes of this type is illustrated in Fig. 51 (Plaut and Nash, 1964). To explain their findings, they postulate that DNA of the chromatid strands of chromosomes is distributed along the long axis as a series of independently replicating loci. According to their findings the smallest number of these loci is about 50, and given that the maximal length of the polytene chromosome is 450 μ, this gives a figure of about 9 μ in one direction. Whether these independently replicating links of the chromatid chains of the chromosome consist of individual DNA molecules, of series of DNA molecules, or of predetermined segments of one longer DNA molecule cannot yet be decided from the available data. However, Plaut and Nash consider it most likely that each such locus in the chromatin strand consists of one DNA molecule. In polytene chromosomes, their thickness and giant size are due to the presence of hundreds and thousands of parallel chromatids, but just as DNA synthesis takes place asynchronously along the chromosome, it takes place synchronously in "cross section" in the parallel DNA molecules carrying homologous genetic information.

b) Asynchrony of DNA Replication in different Chromosomes of a Haploid and Diploid Set. Along with the phenomenon of asynchrony of DNA synthesis in different parts of individual chromosomes, in recent years the asynchrony of DNA synthesis in different chromosomes of the cell nucleus through the phases of mitosis and meiosis, and its possible functional significance, has been widely studied. These investigations have been carried out mainly with the use of tritium-labeled thymidine, selectively incorporated into DNA during its synthesis. Because of the very low energy of the β-rays emitted by tritium, the β-track produced during radioactive breakdown of tritium is so small (about 1 μ) that they enable the precise localization and velocity of DNA synthesis to be determined in segments of chromosomes by the number of black grains formed in a special photographic microemulsion. It was the use of tritium-labeled compounds and microautoradiography which stimulated the intensive study of metabolism in the chromosomes in relation to gene activity. The most interesting investigations in this field have been those of asynchrony of DNA synthesis in the chromosomes. The character and sequence of DNA synthesis in different chromosomes of the nucleus have proved to show specific differences for animals of different species, for different tissues, and for certain chromosomal abnormalities. Naturally, therefore, the possibility of explaining some of the problems connected with the manifestation of the genetic program at this level has stimulated a flood of interesting investigations.

For a more lucid discussion of the results of studies of the asynchrony of DNA synthesis in the chromosomes we should begin by briefly describing the typical layout of experiments in this direction. DNA synthesis in the nuclei takes place mainly during the S-phase of the cell cycle, which in mammalian cells lasts for between 6 and 8 h (Defendi and Manson, 1963). However, mitosis does not arise immediately after the S-phase, but after an intervening G2-phase, which lasts from 2 to 4 h. Because of the comparatively long duration of these phases, asynchrony of DNA synthesis in the chromosomes has been studied mainly by two methods: pulse labeling and continuous labeling. In pulse labeling, using synchronized cell cultures or *in vivo*, labeled thymidine is introduced into the medium for a short time (10-15 min), and then washed out of the medium and diluted at the same time with unlabeled thymidine. After a few hours the material is fixed and autoradiographs obtained. From the times between the period of incorporation and metaphase it is possible to determine what portion of the S-phase (early, middle, late) corresponds to a particular picture of radioactive segments of the chromosomes.

In the continuous labeling method, which is more frequently used, the cells are in contact with radioactive thymidine from a certain point of the S-phase (first, second, third, etc. hours) until mitosis, in some cases loci synthesizing DNA at the end of the S-phase being labeled, in other cases those synthesizing DNA at the end and in the middle of the S-phase being labeled, in a third group the chromosomes are totally labeled — for the whole period of DNA synthesis, and so on in other variations. Partial incorporation of H^3-thymidine into chromosomes on the autoradiograph in this case does not mean early DNA synthesis, but late synthesis. The fullest account of the theoretical basis and techniques used in these methods has been given recently by Hsu and co-workers (1964).

The great interest aroused by the study of asynchrony of DNA synthesis in chromosomes can evidently be attributed mainly to the work of Taylor (1960), who showed that each of the two paired X-chromosomes in female cells of the Chinese hamster, grown in culture, behaves differently in relation to DNA synthesis. One of them carries out late DNA replication throughout its length whereas the other is typified by late replication on only the longer limb of the X-chromosome (away from the centromere). In male cells one X-chromosome also carried out late DNA replication only on the long arm of the chromosome, whereas the Y-chromosome had late replication throughout its length. Similar results have been obtained in other investigations on various objects. Late DNA replication on only one of two X-chromosomes throughout its length and on the Y-chromosome was found in cell cultures of human leucocytes (Morishima et al., 1962; German, 1962, 1964a, b; Gilbert et al., 1962; Kikuchi and Sandberg, 1964), in a culture of mouse cells (Galton and Holt, 1965), in cells of the Syrian hamster (Galton and Holt, 1964), and in other organisms.

Sex in animals is determined by the distribution of the special X and Y sex chromosomes in the progeny. In most species an XX combination determines the female sex and an XY combination the male. In some inborn anomalies of sex development which have been thoroughly studied in man (Klinefelter's syndrome, Turner's syndrome, and so on), as a result of failure of the chromosomes to separate, individuals with anomalous sets of sex chromosomes are formed (XXY, XXXY, XXX, XXXX, XXXXX, and XXXXY). In all cases with an increase in the number of X chromosomes, the supernumerary X chromosomes were found to be of the late replicating type (Grumbach et al., 1963; Rowly et al., 1963; Atkins et al., 1963). The number of X-chromosomes with late replication in every case was one less than the total number of X-chromosomes, and it was on account of these chromosomes that the sex chromatin of the interphase nucleus was formed. On the basis of these facts the hypothesis was put forward (Hsu, 1964a; Taylor, 1964) that late replication of DNA in the chromosomes correlates or is directly connected with an inactive state of the genes (heteropycnosis and physical condensation of the chromosomes).

In this connection the interesting experiments of Beutler and co-workers (1962) must be mentioned. They showed that synthesis of an enzyme (glucose-6-phosphate dehydrogenase), determined by a gene located in the X-chromosome, takes place equally in cells with XX- and X-chromosome combinations. One of the X-chromosomes of the female combination is inactive in relation to synthesis of this enzyme, and from indirect data, it is this which is the late replicating chromosome. The organism is thus a mosaic: in some cells the paternal X-chromosome, and in others the maternal X-chromosome replicates late. They found this to be true in elegant experiments on heterozygotes containing the mutant gene of this enzyme in only one of the X-chromosomes. Another investigation (Hsu, 1962) showed that heteropycnotic chromosomes do not synthesize RNA in interphase. In some animals, because of peculiarities of their sex chromosomes, some differences from this replication pattern are observed (Hsu et al., 1964), but in one form or another asynchrony of DNA replication in different chromosomes is the rule.

A similar approach was used to study DNA synthesis in somatic chromosomes of cells under normal conditions and in chromosomal diseases (in trisomy, for example). Much evidence has now been accumulated in this direction also. The typical picture of asynchrony of DNA synthesis in the somatic chromosomes of the Chinese hamster is illustrated in Fig. 52 (Hsu, 1964b). In nearly every chromosome, late-replicating loci are present at different places. The pattern of late-replicating loci is species-specific and can be used as an additional criterion for identification of different pairs which are hard to differentiate by morphological criteria. Homologous chromosomes of a diploid set usually have the same DNA replication pattern in the S-phase period, but the use of a new method, that of repeated autoradiography, has enabled fine specific differences to be discovered in the linear and time dynamics of DNA synthesis in homologous chromosomes also (Stubblefield, 1964; Hsu et al., 1964; German, 1964b; Bianchi and Bianchi, 1965).

It is interesting to note that despite definite specificity in the sequence of DNA synthesis in the different chromosomes, one

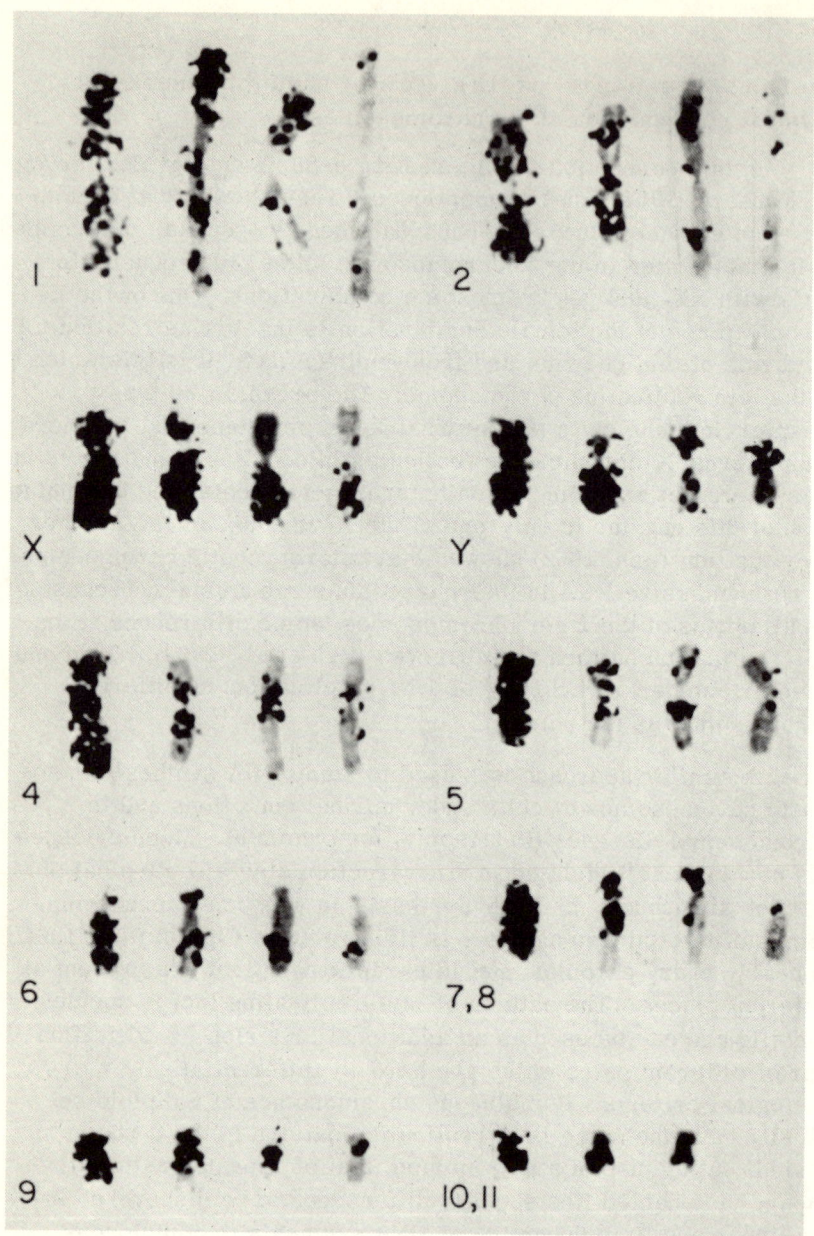

Fig. 52. Asynchronous DNA replication in chromosomes of the Chinese hamster (Hsu, 1964b). In each group the phases of replication are given in order of time, being arranged from left to right for each chromosome. One phase in one particular chromosome must correspond precisely in time to the analogous phase in another series (the numbers are the conventional numbers of the chromosomes and X and Y the sex chromosomes).

degree of variability is found in this respect (German, 1964). Sometimes in dividing cells lying next to each other this worker found differences in the distribution of late-replicating zones in the chromosome.

Since a definite reciprocal correlation is found between RNA synthesis and DNA replication in the chromosomes, it is evident that RNA synthesis in the chromosomes must also possess a definite specific organization, differentiated in time and along the length of the chromosome.

§ 4. Functional (RNA and Protein Synthesis), Morphological, and Genetic Differentiation of Chromosomes. The Study of Puffs on Polytene Chromosomes and Dynamics of Distribution in Connection with Morphogenesis and Specialization of Zones of Active RNA Synthesis in Chromosomes

The ordinary chromosomes of sex cells and of somatic cells capable of dividing contain a definite linear system of genes in a haploid or diploid set: the total complement of genes typical for the species. As we have already seen more than once, in different cells different genes or gene groups from the total set are active. Under these conditions of total genetic potency, every possible specialized function (synthesis of pepsin, of trypsin, or of hormones, formation of myosin, excretion of urea, and so on — hundreds and thousands of similar functions), every genetic characteristic, and there are thousands of them, are represented in the haploid set of chromosomes by a comparatively small number of genes, sometimes by only one gene. But the gene, as we know, is a fixed segment of DNA determining the synthesis of templates for a particular protein. Consequently, a certain number of identical genes must determine a certain possible intensity (or mass) of synthesis of certain proteins. In many cases, for most functions, the number of genes of a diploid set determining them is sufficient to produce specialization if the corresponding genes are in an active state. However, sometimes a specialized cell, in the course of its functions, must synthesize so much more of certain products that the templates of a diploid set of genes are insufficient to provide for these syntheses, and specialization of the cell is accompanied by a unique type of specialization of its nucleus and chromosomes, by the formation of giant or polytene chromosomes. In such chromosomes, as a result of endomitosis, repeated duplication of the chromatids of the interphase chromosome takes place without their separation and without condensation as a result of which the uncoiled chromosome begins to thicken by the lateral accretion of more and more new homologous chromatids, the number of which may reach 1000. Such chromosomes may be 100-250 times longer than metaphase chromosomes. Because of their giant size, these chromosomes are easy to study from different aspects, and the literature on them is exceptionally extensive. The most readily available for study are the giant chromosomes of the salivary glands of insects. It is important to note that the concept of giant chromosomes as polytene structures, each fiber of which is equivalent to an ordinary prophase chromosome, was established originally by Koltzoff (1934) and confirmed soon after by Bridges (1935) and Bauer (1935).

In giant chromosomes the number of identical genes increases in accordance with the degree of polyteny, and because of failure of separation of the chromosomes, each gene has a series of homologs of itself, so that all the structural and chemical features distinguishing the interphase monochromosome are repeated in hundreds of homologous strands and are manifested morphologically in the giant chromosome as a distinctive, specific pattern of dark and light bands and discs (when stained) corresponding to zones of chromosomes with different contents of DNA varying in their degree of helicity, shown by a distinctive alternation of thickenings and constrictions (Fig. 53, Beermann and Clever, 1964). As a result of this structure the number of identical genes increases in accordance with the number of chromosomes, and if they are in an active state the cell can synthesize relatively much more of a given product, because the chromosome can supply much more RNA to the cytoplasm. Because of the complex yet specific pattern of dark and light discs (in the four chromosomes of *Drosophila* as many as 6000 discs have been identified), one particular zone of these chromosomes has been discovered and defined as a genetically unique zone. Since the genes are arranged in linear order in the chromosomes and since the study of frequencies of crossing over in crosses of mutant forms enables the drawing of genetic maps objectively reflecting the mutual arrangement and linear sequence of the genes, a tendency has naturally arisen for

§ 4]   FUNCTIONAL, MORPHOLOGICAL, AND GENETIC DIFFERENTIATION   155

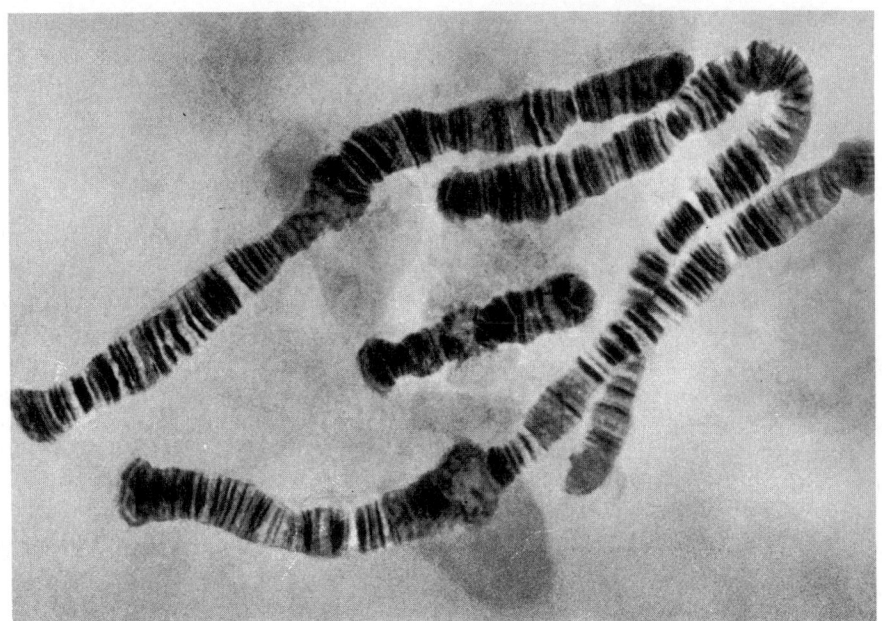

Fig. 53. Chromosome set of salivary gland cells of *Chironomus tentans* (Beermann and Clever, 1964).

symbolic genetic maps of chromosomes with the heterogeneous linear morphological structures of giant chromosomes to be compiled in order to localize a particular gene not merely on the conventional map, but also along the actual chromosome, associating it with a particular disc or group of discs.

It has been possible to do this principally because certain changes in characteristics of *Drosophila* associated with unequal crossing over (when an unequal exchange of segments of chromosomes takes place and two or three identical loci are found in one chromosome while the other loses a locus) are accompanied by corresponding changes in the disc pattern of the giant chromosomes, by the appearance of paired discs. These are all classical observations obtained many years ago and usually described in standard textbooks of genetics, but it is worthwhile examining this material briefly as an introduction to our subsequent consideration of modern trends. A particularly clear example of this type of morphological identification of the gene on polytene chromosomes is the thoroughly investigated case of unequal crossing over of the dominant Bar (B) gene which, in *Drosophila*, determines the development of narrow eyes with a decreased number of facets in *Drosophila*. In the wild-type flies (B^+) there is a single set of several distinct discs, while in flies of Bar type these discs are double and in "double-Bar" flies they are triple (Fig. 54).

In all these investigations on various classical genetic objects (*Drosophila*, *Chironomus*, corn, etc.) many cases were studied and, as a result, genetic maps with morphological details (with definite discs) of giant chromosomes were compiled. Examples of these maps can be found in scores of papers. These investigations clearly revealed the lateral congruence of the genes in giant chromosomes in each individual chromoneme.

However, a particularly important discovery from the point of view of the study of the genetic control of morphogenesis was the observation that an active (as regards RNA synthesis) state of

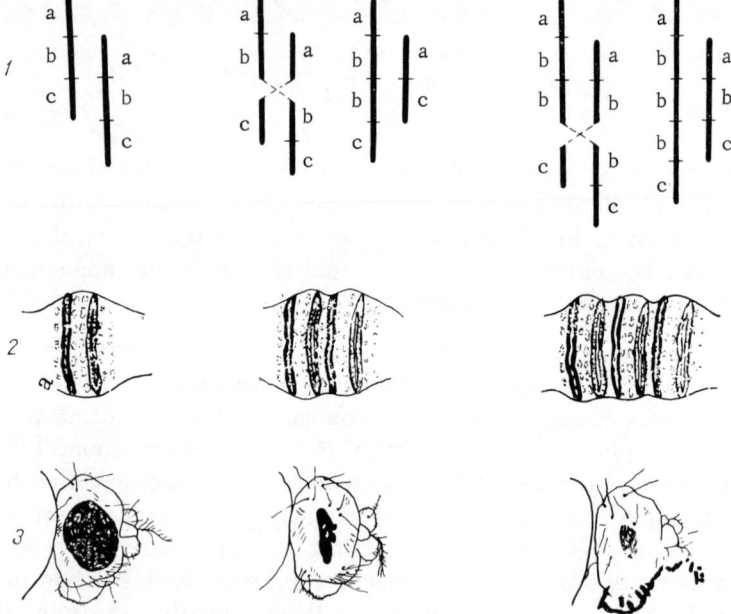

Fig. 54. Scheme of unequal crossing over in the *Drosophila* chromosome (Lobashev, 1963). 1) Mechanism of unequal crossing over; 2) changes in structure of discs in the Bar locus; 3) phenotype of insects.

homologous genes is specifically reflected in the morphology of giant chromosomes, as the formation of special outgrowths known as puffs. The formation of puffs as active genetic zones of the chromosomes was first found during the study of salivary gland chromosomes of the larvae of two insects: *Chironomus tentans* (Beermann, 1952, 1956) and *Rhynochosciara angelae* (Pavan and Breuer, 1955). These investigations also revealed that the pattern of puffs on chromosomes is not a stable criterion of their morphology, as is the disc pattern, but the distribution, size, and number of puffs vary with the type of tissue and the stage of its differentiation, and also with the period of development of the insect larvae.

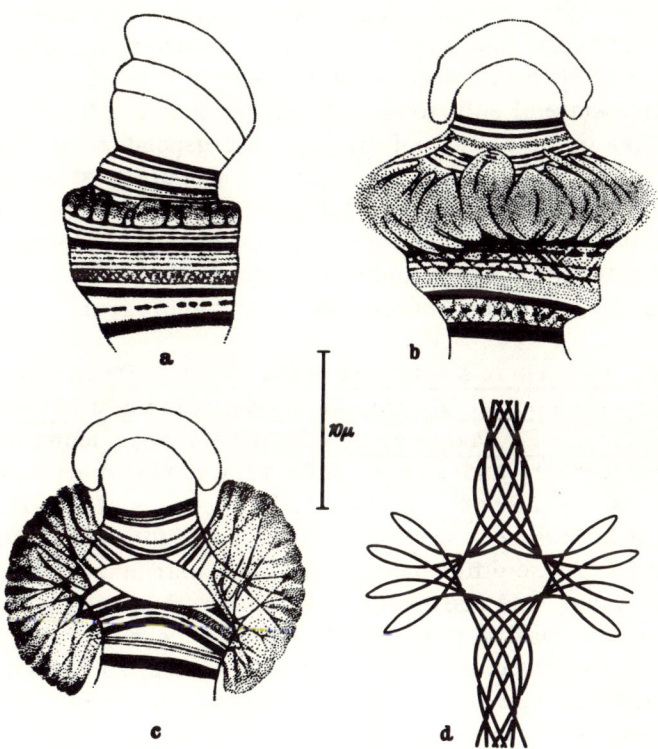

Fig. 55. a-c) Several stages of formation of a puff (Balbiani's ring) at locus BR_1 in chromosome IV of a salivary gland cell of *Chironomus*; d) interpretation (after Beermann) of mechanism of puff formation during change in length and shape of chromosomal fibrils (Clever, 1964b; Beermann, 1952).

Puff formation starts morphologically with the formation of a pale zone in the chromosome and disappearance of the precise outlines of the transverse strips and discs. The disc outlines then disappear and the chromosome starts to thicken in this area. There is a wide gradation of puffs in size: from those in which all that can be seen is a pale zone, through small puffs to very large ones (Balbiani's rings). Large puffs contain masses of products synthesized at that particular locus, namely RNA and protein, and these can easily be detected by specific chromosomal staining methods. Several models have been produced to illustrate the structure of puff formation, and one of them is illustrated in Fig. 55.

As a result of the intensive study of the dynamics and morphology of puffs, the following tentative conclusions have been drawn: a) puff formation is intimately connected with the morphogenesis and differentiation of the tissues; b) it is an expression of the active state of only some of the genes in specialized tissues; c) changes in the pattern of the puffs correspond to morphogenetic changes in the functions of organs; and d) the study of puffs reveals the external and morphological pattern of how the genetic system of the chromosome controls development and differentiation by determining RNA and protein synthesis only with certain genes and only in accordance with a certain program.

a) Dynamics of Formation and Disappearance of Chromosomal Puffs in Connection with Morphogenetic Processes. It was very soon found that the pattern, number, and size of puffs very during morphogenetic processes. In the early, middle, and late stages of development of insect larvae these characteristics of puff formation differ appreciably. These differences can be demonstrated graphically in several ways. The material now available in this field has been examined in detail in a series of surveys (Beermann, 1963; Clever, 1964a, b; Beermann and Clever, 1964; Berendes, 1965; Kiknadze, 1965; Kroeger and Lezzi, 1966), so that we need only discuss the main results obtained and some of the latest research.

Changes in puffs in salivary gland cells of larvae of *Chironomus dorsalis* at different stages of larval development (larvae aged 7-8 and 9-10 days, prepupae, young pupae, old pupae) are shown schematically in Fig. 56 (Kiknadze, 1965). It is clear that

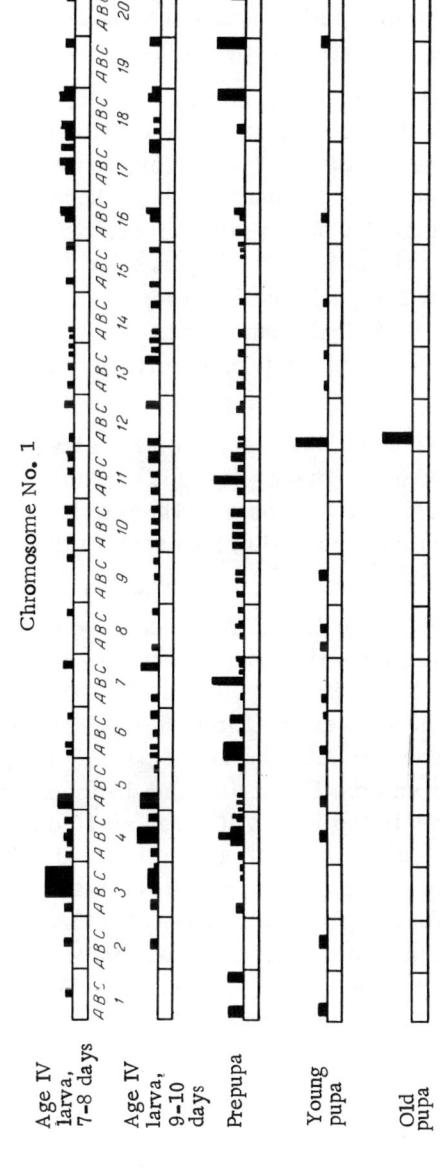

Fig. 56. Scheme showing distribution of activity of different classes of puffs in chromosome No. 1 of *Chironomus dorsalis* during larval development (Kiknadze, 1965). Five stages are shown: a larva of age IV 7-8 days, 8-10 days, prepupa, and young and old pupae.

some puffs disappear while others appear between the 7th and 10th days of morphogenesis. Many of the puffs disappear during pupation (the function of the gland also diminishes sharply), and by the end of the pupal stage all puffs except one have disappeared.

This pattern is shown in Fig. 56 for only one chromosome, but in the paper cited similar results were obtained for all four polytene chromosomes of this insect. Experiments by Kiknadze and Filatova (1960, 1963) and Kiknadze (1965) showed that during larval development of *C. dorsalis* a gradual increase in the number of actively functioning zones first takes place; at the beginning of age IV there are about 100 puffs on the complete set of chromosomes, while at the end of H^4 the number is about 160. In the late pupal stage the puffs disappear just before lysis of the salivary gland.

Besides the puffs which appear at certain stages there are others which are found at all stages (except the last). Kiknadze suggests that these puffs are connected with activity of genes responsible for fundamental metabolic processes in the salivary gland.

Clever (1961; 1962a, b; 1963; 1964a) carried out a series of particularly outstanding investigations of changes in the puffs in the salivary gland of *Chironomus* in relation to morphogenesis. He worked with a different species of *Chironomus* (*C. tentans*). He studied in detail the general distribution of puffs during the stages of larval development and the formation and disappearance of individual puffs, both small and large. Clever started from the assumption that dimensions of the puff characterize the degree of activity of the corresponding genes. His investigations also showed clearly that some puffs are found at nearly all stages, while others are typical of certain periods of development only. Some puffs apparently "work" on a principle of "pulsation." He used several methods of illustration to demonstrate the dynamics of puff activity.

The course of these changes for some of the puffs on the first chromosome is illustrated in Fig. 57. Clever made a particularly close study of puff formation accompanying the molting of larvae which takes place under the control of a special hormone, ecdysone.

It must be realized that puff formation is a very dynamic process. Patterns such as those illustrated in Fig. 57 are the

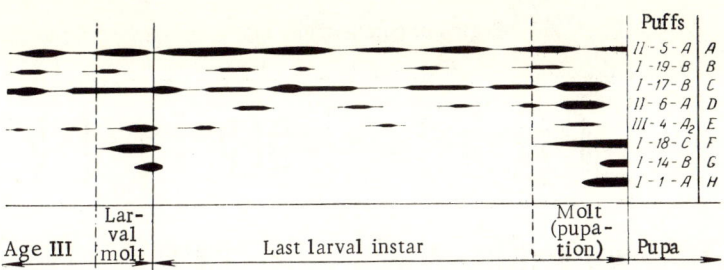

Fig. 57. Typical changes in the pattern of puffs in chromosomes of the salivary glands of *Chironomus tentans* during the 3rd and last larval age (Clever, 1964).

mean result of measurements of very many homonymous zones on chromosomes with a wide scatter of parallels. Generalized conclusions can be drawn only by statistical analysis of a large number of repeated measurements (Lychev and Medvedev, 1967). Another insect which has been used as the object of intensive study of the dynamics of puff formation in connection with morphogenesis in the larval stage is *Drosophila*. A series of outstanding investigations in this direction has recently been carried out by Berendes (1965a, b). He chose the chromosomes of the salivary glands of *Drosophila hydei* for his observations. A detailed chromosome-cytological map of all the chromosomes of this insect was first compiled (Berendes, 1963). In these investigations also a very clear correlation was observed between the dynamics of puff formation and disappearance in the chromosomes and the various phases of development. A scheme of one of the typical patterns of this dynamics in one zone of chromosome No. 5 between the 143rd and the 184th hours of development is shown in Fig. 58 (Berendes, 1965b).

However, variation of the puff pattern with the stages of development is only indirect evidence that puffs are connected with functional, morphological, and biochemical differentiation of cells during the processes of morphogenesis. More direct evidence in this respect was obtained in experiments to determine the tissue specificity of the puff map and correlation between a particular puff and a particular process.

Several results must be mentioned in this respect. Slight yet significant differences were found in *Drosophila* between the puff pattern in the distal (posterior) and proximal (anterior) parts

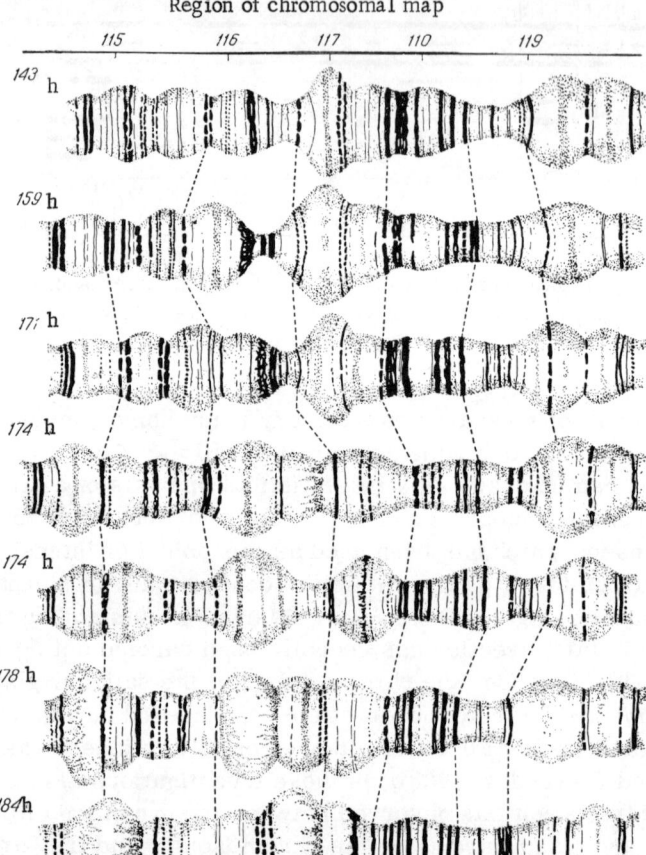

Fig. 58. Schematic drawing of region 115-119 (of chromosome No. 5) at various time intervals (in hours after oviposition), in periods of age III, prepupa, and young pupa. The pattern of the puffs in this part of the chromosome differs at different time intervals (Berendes, 1965b).

of the salivary gland (Berendes, 1965b). In many puffs, differences in size were found, but one puff (47B) was found only in the distal part. In his other investigations (Berendes, 1965a, 1966), the puff pattern was compared in cells of the salivary gland and the malpighian tubules of the stomach in larvae of *D. hydei*. As a whole, 116 puffs were counted in the period from 10 to 6 h before formation of the pupa. In the case of 61-69% of puffs, no appreciable differences were found in their activity between the two tissues, 6% of puffs showed active formation only in the cells of the malpighian tubules of the stomach, and 3.4% were typical of the salivary gland only. Differences in activity between the two tissues were found for the remaining puffs (29.4%).

A particularly interesting and successful attempt to associate the activity of certain puffs with particular functions was made by Beermann (1961). These experiments were carried out on *Chironomus*. In the salivary gland of larvae of this insect, in one species (*Chironomus pallidivittatus*) four cells located near the duct produced a granular secretion. In the closely related species *C. tentans*, these same cells produced a simple liquid without granules. In hybrids between the two species, this character obeys the simple mendelian laws of inheritance. Beermann succeeded in locating the difference in a group of fewer than ten bands in one of the four chromosomes. Cells producing granules (*C. pallidivittatus*) have a puff associated with this group of bands. The analogous puff is completely absent from this locus in *C. tentans*. In hybrids the puff appears only on the chromosome crossing from *C. pallidivittatus*, and the hybrid produces far fewer granules than this parent. The dimensions of this puff also were correlated with the number of granules produced. However, it has been shown (Laufer and Nakase, 1965) that one of the important functions of the salivary gland in Chironomus is the selective absorption of salivary enzymes, their concentration and secretion, rather than their synthesis, and the puffs are evidently connected functionally with these processes.

Another trend in the study of changes in puffs in connection with a change in certain functions is also extremely interesting: determination of the pattern and activity of puffs in mutations in animals. Kiknadze (1965) points out, in particular, that in the "yellow" mutation in her experiments changes were observed in the activity of several puffs; in some other mutations one or two new puffs appeared, although these had not functioned at all in the original forms. Changes in activity of the puffs in *Drosophila* in the course of morphogenetic mutations have also been described by Roodman and Kopac (1964).

Investigations to study the effect of somatic radiation-induced mutations in the salivary gland cells of *Drosophila* after irradiation of embryos and larvae in the early stages of development (before polytenization of the chromosomes) have now been commenced in our laboratory (Lychev and Medvedev, 1967; Lychev, 1967). Recently several new species of insects have begun to be used as objects for study of the dynamics of puffs in connection with morphogenesis and with particular (especially secretory) functions (Whitten, 1964), and it must be anticipated that this extension of the research must soon lead to the discovery of new underlying principles.

b) <u>Experimental Changes in the Pattern and Activity of Puffs Caused by Hormones, Chemical Agents, and Environmental Factors.</u> The manifestation of a particular level of activity of the puffs and their appearance and disappearance are not entirely a reflection of some form of development program built into the genetic system of a particular chromosome and specific for a given organ or given function. The genetic development program acts at the level of the genome as a whole and, consequently, its phenotypic manifestation must affect the whole organism and the biochemical, physiological, and other forms of interaction between the organs. Accordingly, particular acts of morphogenesis and differentiation must not be examined in isolation from genetic control over development, even if it is discovered that they are induced by a hormone, activator, or environmental factor. The hormone produced by another tissue is a product controlled by the genetic system, and its action on another tissue is often largely dependent on its action on the activity of the genes, on the presence of special receptor systems built into the genome. The phenomenon of puff formation thus serves as a wonderful model for studying the action of factors of this type on the manifestation of gene activity.

The first studies in this direction, which were started by Clever (1961), led to some very interesting results. Clever carried out experiments to study the effect of the hormone ecdysone, responsible in insects for induction of the metamorphosis from larva to pupa (a hormone of steroid type) on the pattern and activity of puffs. This hormone is produced by the prothoracic gland. It acts directly on the individual cells, inducing changes in them associated with subsequent formation of the pupa. Since, as has already been mentioned, the formation of certain puffs is clearly correlated with this morphogenetic process, the attempt to associate induction of these puffs with the action of ecdysone was perfectly rational. In fact, in cases when the secretion of ecdysone by the gland was stopped in some way or other, or the salivary gland was isolated from the action of ecdysone by ligation, the formation of those puffs which were associated with the molting process was halted (Clever, 1961, 1962a, b; Becker, 1962). However, ecdysone acts indirectly (through other systems) on some puffs and directly on others. Injection of ecdysone directly into young larvae caused the rapid formation of several puffs (15-60 min after injection). The puffs formed as a result of induction in this way (I-18-C and IV-2-B) are specific in normal morphogenesis for the period of development at which ecdysone production by the prothoracic gland begins and it appears in the hemolymph. The size and length of life of these puffs depended on the ecdysone concentration (Clever, 1961, 1963b, 1965a). Puffs can be caused to appear and disappear several times over in strict accordance with the ecdysone, the reactions (especially to concentration of the hormone) of each of two puffs to ecdysone being specific in a number of characteristics. Clever draws attention to the high sensitivity of the process: a reaction to ecdysone can be detected when there are only between 10 and 100 molecules of the hormone in the cells for a haploid set of chromatids. Clever (1964b) postulates that other puffs, connected with particular stages of morphogenesis, are induced by specific cytoplasmic factors, each responsible for controlling the activity of one or several chromosomal loci.

The results of one of these experiments are given in Fig. 59 (Clever, 1961). To determine the products whose synthesis in the chromosome is connected with the action of ecdysone or begins to take place under the influence of ecdysone, or whether ecdysone is essential for the activation reaction as in the case of regulation of genetic activity in bacteria, Clever carried out some interesting investigations to study the effect of antibiotics inhibiting the synthesis of proteins, RNA or DNA on the activating effect of ecdysone. In these experiments puromycin, chloramphenicol, actinomycin C, and mitomycin C were injected in different combinations with ecdysone (Clever, 1964b, c; 1965b; 1966). Actinomycin and mitomycin partially or completely suppressed the formation of

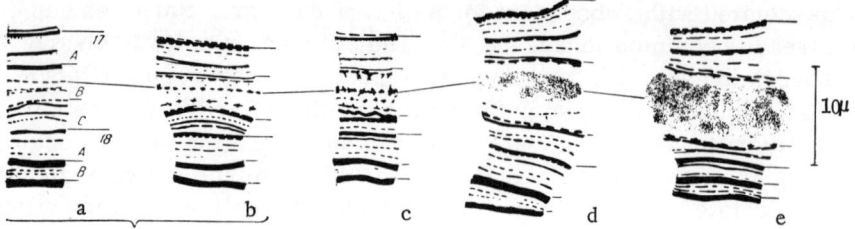

Fig. 59. Degrees of development of puffs at locus I-17-B in *Chironomus*. a-c) Untreated control animals; d, e) larvae receiving an injection of the hormone ecdysone (Clever, 1961).

ecdysone-specific puffs (I-18-C and IV-2-B), whereas puromycin and chloramphenicol had no effect.

Actinomycin, with the strongest inhibitory action, is a specific inhibitor of DNA-dependent RNA synthesis (it inhibits RNA-polymerases and is firmly bound to DNA). Mitomycin also acts at the DNA level. Chloramphenicol and puromycin inhibit protein synthesis at the level of polypeptide formation during interaction between sRNA and mRNA templates. However, two antibiotics (cyclohexamide and pactamycin) stimulated the formation of certain puffs (Clever, 1967).

Besides work to study the effect of ecdysone when injected into *Chironomus* larvae, I must also mention the interesting work of Becker (1962) on transplantation of the salivary gland of the young larva of *Drosophila melanogaster* into the body of a more adult insect. In this case, puffs characteristic of the later stage of development began to be formed in the chromosomes of the donated gland. Similar results with the gland of *Chironomus* were obtained by Kroeger (1964).

Besides endogenous factors of this type, many exogenous factors such as temperature, salt solutions and other chemical agents, and anaerobiosis also have a definite and specific action on the pattern and activity of puffs (Ritossa, 1964; Laufer et al., 1964; Kroeger, 1964; Berendes et al., 1965; Kiknadze, 1965). Some simple chemical compounds (zinc chloride, urethane, chloroform, etc.) simulate the action of ecdysone. It is interesting to note that prolonged inbreeding in *Drosophila* also influences (mainly depresses) the formation of several puffs (Lychev, 1965).

Information on other investigations in this direction is summarized in the survey by Kroeger and Lezzi (1966).

Work has now begun in our laboratory to examine the effect of irradiation of *Drosophila* embryos and larvae with various doses on puff formation and RNA synthesis by the chromosomes of salivary gland cells.

c) Biochemical Specificity of Chromosomal Puffs and Characteristics of Forms of RNA Synthesized by Puffs. Pelling (1964), in a microautoradiographic investigation using H^3-uridine and a ribonuclease control, clearly demonstrated that RNA is the main product of synthesis in the puffs of salivary gland chromosomes.

One of the many autoradiographs obtained in this investigation is illustrated in Fig. 60. Labeled RNA is synthesized in the puffs within a few minutes of injection of H^3-uridine which is evidence of primary synthesis *in situ*. The dimensions of the puffs correlated precisely with the rate and volume of RNA synthesis. At the same time, the proteins contained in the puffs are evidently synthesized in the cytoplasm.

RNA formed in the puffs bears every mark of messenger RNA. This was shown by the use of an ultramicromethod of determination of the nucleotide composition of RNA developed by Edström (1960). This method is based on microelectrophoresis of a hydrolyzate of RNA microsamples on thin viscose silk threads followed by microultraspectrophotometry. The use of this method to determine the nucleotide composition of puffs and chromosomes (Edström and Beermann, 1962), in which the composition of RNA from the cytoplasm, nucleolus, the entire No. 1 chromosome, and three large puffs (Balbiani's rings) of chromosome No. 4 was compared, showed that each of these nucleic acids possessed its own specific nucleotide composition. The differences between each of the three puffs of chromosome No. 4 were definitely significant. One puff (the middle one) contained RNA with an exceptionally high content of adenine (38%), twice as high as the uracil content (17.1%). This difference indicates that during synthesis of this RNA, only one polynucleotide chain of DNA is copied, because in the complete double strand of DNA the ratio of thymine to adenine is 1:1 in accordance with the principle of complementarity. In the total

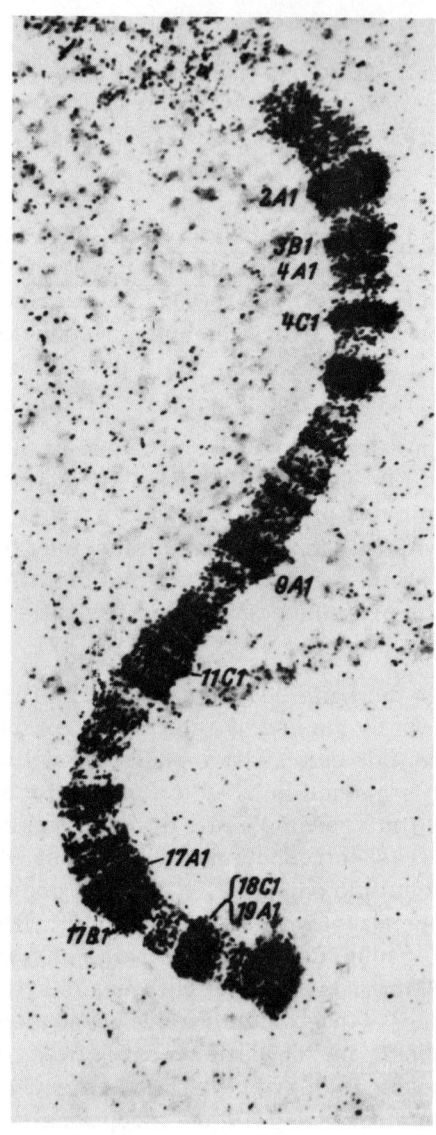

Fig. 60. Microautoradiograph of chromosome No. 1 of *Chironomus tentans* after incubation with H^3-uridine. Zones of RNA synthesis in the puffs can be distinguished (840×)(Pelling, 1964).

cytoplasmic RNA of the salivary gland of *Chironomus* the ratio between the adenine and uracil contents likewise is close to 1 (29.4: 25.7). Several other investigators have studied RNA synthesis by puffs of giant chromosomes (Ritossa, 1963; Clever, 1964c; Sirlin and Schor, 1962).

It is interesting to note that ribonuclease, when applied experimentally to living salivary gland cells, destroys RNA in the nucleolus but has the opposite action on the chromosomes: it stimulates the intensive formation of new puffs and rapid synthesis of RNA in them (Ritossa and Borstel, 1964; Ritossa et al., 1965). The mechanism of this phenomenon is not yet clearly understood, it may be that stimulation by RNA breakdown products takes place.

§ 5. Study of Differentiation of RNA Synthesis by Loci and in Time in Lampbrush Chromosomes of Vertebrate Oocytes

Chromosomes of lampbrush type are another form of giant chromosomes which, because of their structural features, are suitable for genetic and functional investigations of chromosomes. They were discovered in 1882 in the oocytes of fish, amphibians, birds, and other vertebrates and also in some invertebrates. Growth of chromosomes of this type is associated with generalized hypertrophic growth of the cell, in which the parallel chromatids are not only increased in number, but are also packed in a special way to produce numerous processes giving the outward appearance of lampbrushes (Fig. 61, Gall, 1958). Most

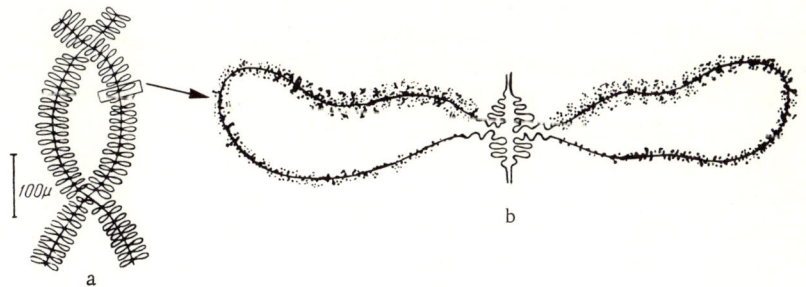

Fig. 61. Scheme of a lampbrush chromosome from triton oocyte (Gall, 1958): a) Low magnification; b) high magnification.

modern geneticists and I myself have never seen a lamp brush, but the name has persisted since the days when lamp brushes were in use; they bear some resemblance to test tube brushes. Chromosomes of this type are very long, for example, 700-900 μ in the triton, and if the loops are stretched out to their full theoretical extent the length of the chromosome would be measured in centimeters or even in meters. The loops of polytene chromosomes are often compared with puffs. The structure, biochemistry, and functional activity of lampbrush chromosomes have been fully surveyed by Callan and Lloyd (1960), Callan (1963), Gall (1963a), and Miller (1965), and after considering some of the more important matters concerned with analysis of the role of chromosomes as a regulatory system, we shall therefore scrutinize only the most important of the recent findings.

Chromosomes of lamp brush type possess high elasticity, so that they can be stretched to 2.5 times their length. The loops are usually in pairs to the right and left of the chromomeres and main axis. During stretching of these chromosomes, rupture always takes place in the region of the chromomeres, through the chromomere (Callan, 1955, 1963) (Fig. 62). In the haploid set of chromosomes of the triton (*Triturus cristatus*) there are about 5000 chromomeres. Toward the end of growth of the oocyte, and especially toward the period of ovulation, the loops begin to reduce in size and to disappear. Their presence is undoubtedly an indication of an active synthetic function of the chromosomes. A positive Feulgen reaction (for DNA) is given by the principal axis of the chromosomes, the centromeres, chromomeres, and telomeres. The lateral loops diverging from the chromomeres do not give this reaction. Most of the so-called normal loops have an axis and numerous thin fibers diverging from it (Callan, 1955, 1963). In salt solutions this axis oscillates (Brownian movement, and each fibril oscillates as if it were attached to the axis only by one end. The loops of the chromosomes are highly heterogeneous in thickness, size, and other properties. On this basis, maps of these chromosomes can be drawn using the most specific loops as markers. Some loops have a special granular structure, the granules being apparently secreted into the nuclear juice.

A diagram of the structure of various types of lateral loops is given in Fig. 63 (Callan, 1963). Asymmetry of the granules can be seen in the granular loops, possibly connected with their

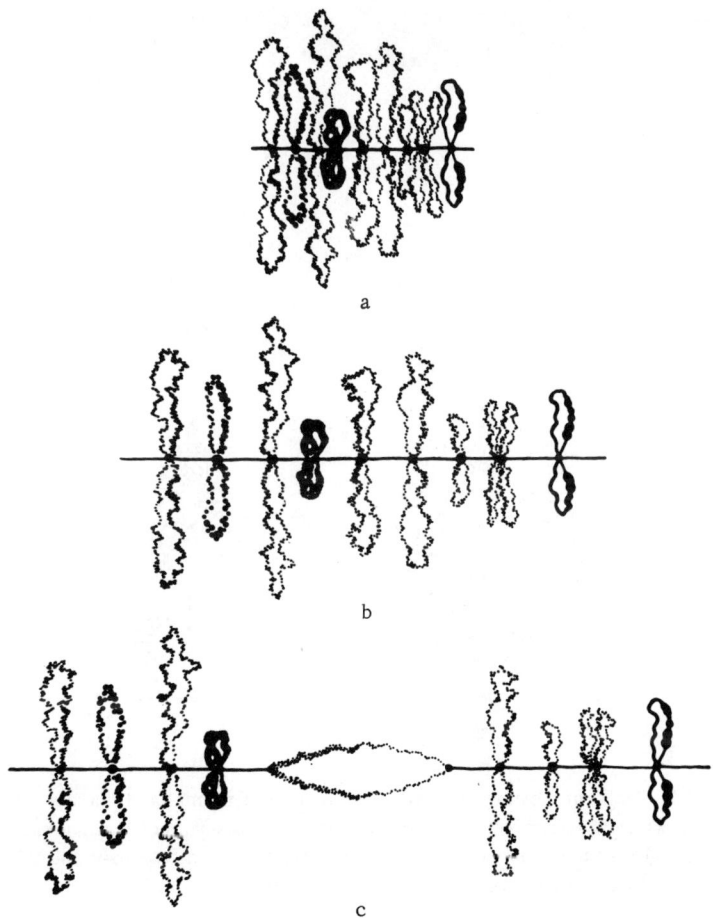

Fig. 62. Scheme showing the result of stretching a segment of a lampbrush chromosome. a) Normal structure; b) stretched to the limit of elasticity; c) stretched beyond the threshold of elasticity: one chromomere has broken and the two lateral loops are stretched across the gap (Callan, 1963).

consecutive development. Some loops of this type are very large in size. The substance forming the granules is also present in the nongranular loops, although evidently merged into a special gel.

The loops mainly contain RNA (Macgregor and Callan, 1962) which disappears after ribonuclease treatment. Trypsin and pepsin cut off the fibers from the loops and the gel around the loops,

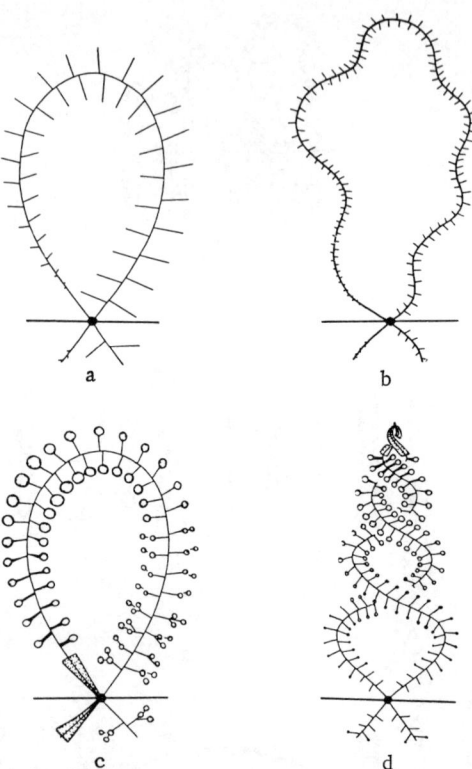

Fig. 63. Diagram showing approximate structure of normal and granular lateral loops. a) Highly asymmetrical normal loops; b) less asymmetrical normal loops; c) typical granular loop with unusual symmetry (in chromosomes V and VIII of the triton) (Callan, 1963).

but have no effect on the structure of the chromosome axis. These same fibers and gel (matrix) are cut off by ribonuclease treatment. However, the chromomeres and chromosome axis are resistant to the action of ribonuclease. Desoxyribonuclease acts on the chromomeres and axis, causing considerable changes in the loops and fragmenting them into small segments (Macgregor and Callan, 1962). These workers suggest on the basis of their enzyme tests that a continuous fiber or fibers of DNA may be present which join the neighboring chromomeres and pass along the whole length of each lateral loop to form its axis. They suggest that the DNA fiber may pass without interruption from one end of the chromosome to the other, forming more compact zones in the chromomeres.

Absence of a Feulgen reaction in the lateral loops is not evidence that they do not contain DNA, because this reaction is insufficiently sensitive in cases when there are no large concentrated masses of DNA, and in this case proteins and RNA mask the axial DNA fiber. Passage of double-helical DNA (or a series of DNA molecules) through all the loops is also admissible on the basis of experiments to study the kinetics of disintegration of the loops by desoxyribonuclease (Gall, 1963b; Nebel, 1962). Izawa and coworkers (1963a) also demonstrated the presence of DNA in the loops. The polarity of the chromosomal loops must also be noted, one of the chromomere ends of the loop usually being thinner (containing shorter fibers) than the other (Fig. 64). This feature of the loop structure led Callan and Lloyd (1960) to postulate that the thin end of the lateral loop corresponds to a part of the axis which recently unwound from the chromomere and has not yet formed many products of synthesis. The thick end of the loop, according to their hypothesis, is being wound up again into the chromomere, and in this way the DNA-axis of the loop is in apparent movement (in the direction of the arrows in Fig. 64), carrying ever-changing masses of DNA into the zone of synthesis (the hypothesis of polarized movement).

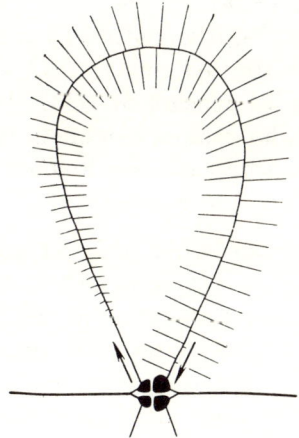

Fig. 64. Diagram illustrating the theory of polarized lateral unwinding and winding of the loop in part of the chromomere (direction of movement indicated by arrows), causing asymmetry of the loops (Callan, 1963).

This hypothesis served as a basis for later and very interesting autoradiographic experiments to study synthesis of RNA and proteins in the loops of chromosomes of this type (in the triton), and these led to results of fundamental importance (Gall and Callan, 1962; Callan, 1963). After subcutaneous injection of H^3-uridine, the label appeared initially in the nuclei of the oocytes (after a few hours) and not until two or three days later in the cytoplasm. The H^3-uridine concentration in the oocyte nuclei reached a maximum after 4 days, but in the cytoplasm not before one month after injection. No polarization of H^3-uridine incorporation was found in the small loops, because of the slow rate of its detectable incorporation, but when the giant loops of chromosome No. 12 were studied, the label distinctly moved in the course of time along the loop from one chromomere end to the other (Gall and Callan, 1962).

In preparations and autoradiographs made on the day after injection of H^3-uridine, only a small zone near the thin end of the loop was labeled, two days after injection the label had spread a little further and after four days it had reached the middle of the loop. After 14 days all the granules of the loop were labeled (Fig. 65) (Callan, 1963). The speed of movement of the label was almost constant, and only about 10 days were needed for it to pass along the whole loop. Incorporation of H^3- uridine is associated with RNA synthesis. Gall and Callan put forward the suggestion that this loop may contain a limited zone of RNA synthesis which moves along the loop (or else its product moves). The hypothesis of the wavelike movement of the zone of synthesis they considered to be less probable, because small loops are known to be capable of synthesizing RNA along their whole length.

The minimal duration of the stage of existence of lampbrush chromosomes in triton oocytes is about 200 days, if it can be taken as equal to the duration of maturation of these animals. Calculations (Callan, 1963) show that during this time, if the chromatid DNA does in fact move, unwinding from one side of the loop and winding up again into the other side, given that the length of the loop is 50-100 μ, it passes through a zone of synthesis occupying from 1 to 2 mm of the chromatid DNA. It is interesting to note that gonadotropic hormone accelerates ovulation and at the same time it doubles the rate of linear incorporation of H^3-uridine into the giant loop (Gall and Callan, 1962).

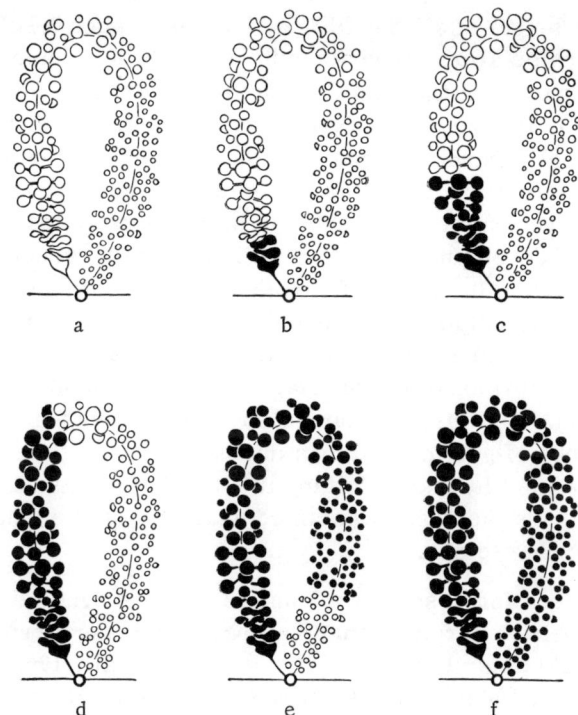

Fig. 65. Diagram showing progressive incorporation of H^3-uridine into RNA on the giant loop of chromosome No. 12 of *Triturus cristatus*. H^3-labeled granules are stained black. a) Before injection of triton with H^3-uridine; b) 1 day after injection; c) 2 days; d) 4 days; e) 7 days; f) 14 days after injection (Callan, 1963).

It has not yet proved possible to demonstrate the linear sequence of uptake of the label into other loops of the oocyte chromosomes. This is presumably because of the impossibility of pulse labeling because of the high concentration of products of nucleic acid metabolism in the oocytes and rapid removal of synthesized RNA from the zones of the chromosomal loops. It is also important to note that when protein synthesis in the giant granular loop of the triton chromosome was studied by a method based on incorporation of H^3-phenylalanine, incorporation was uniform throughout the length of the loop and did not follow a linear sequence like that of uridine incorporation (Gall and Callan, 1962). According to Gall (1956), the total sum of the lengths of all loops in the haploid set of lampbrush chromosomes in the triton is 50 cm. This suggests that the total length of all the DNA molecules in such a set is about 10 m. This calculation agrees completely with that made in the same investigation on the basis of the photometrically determined DNA content in the nucleus (3×10^{-5} μg) on linear parameters of the DNA double helix in accordance with the Watson and Crick model. The wide variation in size of the loops in different species of amphibians, or even in the same species in different states and different periods of development, is interesting. In this connection work has now started on the study of chromosomal loops along the same lines as the long-established studies of giant chromosomes of insect glands, i.e., the study of the character of the loops and its dependence on hormonal action, stages of development, and external conditions. Experiments have also been undertaken to study the inheritance of particular loops during interspecific crossing and segregation of hybrids (Callan, 1963) and attempts have been made to compare the results obtained with data on the inheritance of particular characters.

Some very interesting work on the study of synthetic functions of lampbrush chromosomes has recently been carried out by Allfrey, Mirsky, and their co-workers (Izawa et al., 1963a, b; Davidson et al., 1964). They showed (Izawa et al., 1963a, b) that the RNA/DNA ratio in lampbrush chromosomes of amphibians is unusually high, at least 100 times greater than in the chromatin substance of typical differentiated cells, such as liver cells. The most abundant products of synthesis of the chromosomes during oogenesis in *Xenopus laevis* are ribosomal RNA molecules (about 97-98%). Synthesis of messenger RNA accounts for only 2-3%. The method of DNA/RNA hybrid analysis showed that in this period only a small part of the genome, roughly 2.7%, is active in the nuclei (Davidson et al., 1966). However, it has been shown that RNA synthesized in the chromosomes and nucleolus of amphibian (*X. laevis*) oocytes is "conserved" in the nuclei and accumulates in them until the ovulation stage without passing over into the cytoplasm. Investigations have shown (Brown and Gurdon [1964]) that synthesis of ribosomes and ribosomal RNA in the fertilized oocytes of *X. laevis* does not take place until late gastrulation: the whole process of primary development takes place on the basis of ribosomes which have accumulated during oogenesis. An enormous excess of RNA over DNA, especially in the loop zones, was also demonstrated by Edström and Gall (1963) who studied the composition of the RNA

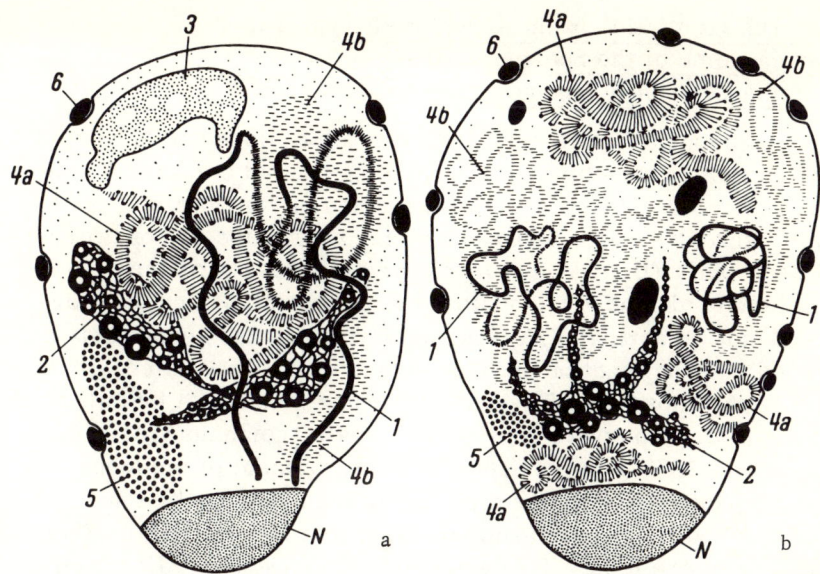

Fig. 66. Scheme of formation of chromosomes with loops in nuclei of spermatocytes of *Drosophila hydei* (a) and *D. neohydei* (b): N denotes nucleolus; 1) fibrils; 2) glomeruli; 3) pseudonucleolus; 4a) ribbons; 4b) tubular material associated with fibrils; 5) accumulation of granules; 6) granules on nuclear membrane (Hess and Meyer, 1963).

fraction in triton oocytes by ultramicroelectrophoresis. It has also been shown (Izawa et al., 1963a) the RNA synthesis in the chromosomes of triton oocytes is DNA-dependent. It is suppressed by inhibitors of DNA-dependent RNA synthesis (actinomycin D, arginine-rich histone, polylysine). Characteristically, under these circumstances, inhibition of DNA-dependent RNA synthesis is accompanied by reduction and disappearance of the loops in such chromosomes. In the opinion of these workers, this shows that the specific morphology of active chromosomal zones of this type is not only intimately connected with their functional role in RNA synthesis, but is itself also directly dependent on this synthesis.

§6. Differentiation of the Lampbrush Type Y Chromosome of the Spermatocytes of Some *Drosophila* Species

Comparatively recently (Meyer, 1963; Hess and Meyer, 1963) chromosomal structures also containing paired chromatid loops and resembling lampbrush chromosomes of vertebrate oocytes in their structure, were found in growing spermatocytes of two species of drosophila (*D. hydei* and *D. neohydei*). These structures were completely absent in the nuclei of X/O-spermatocytes and were found only in normal X/Y-spermatocytes. It was accordingly concluded that the function of these structures is associated with the sex Y-chromosome. This was confirmed by the fact that all structures of this type were duplicated in male spermatocytes with an anomalous XYY-complement. The arrangement of the loops was species-specific, and was different in the closely allied species *D. hvdei* and *D. neohydei* (Fig. 66) (Hess and Meyer, 1963). These structures are feebly developed in *D. melanogaster*. After interspecific crosses these structures were of the type characteristic of the species from which the hybrid obtained its Y-chromosome. The discovery of this new type of functionally differentiated chromosomes naturally laid the foundations of a series of investigations (Meyer and Hess, 1965; Hess, 1965, 1966) in which structural changes of these chromosomal "loops" were studied during the action of factors influencing RNA synthesis (actinomycin) and of mutations in *Drosophila*. These investigations showed that lampbrush differentiation of the Y-chromosome (the formation of loops) corresponds to an active state of the genes.

Literature Cited

Atkins, L., Book, J. A., Gustavson, K. H., Hasson, O., and Hjelm., 1963, Cytogenetics, 2:208.
Bauer, H. Z., 1935, Zellforsch., 23:280.
Becker, H. J., 1962, Chromosoma, 13:341.
Beermann, W., 1952, Chromosoma, 5:139.
Beermann, W., 1956, Cold Spring Harbor Symp. Quant. Biol., 21:217.
Beermann, W., 1961, Chromosoma, 12:1.
Beermann, W., 1963, Am. Zoologist, 3:23.
Beermann, W., and Clever, U., 1964, Sci. Am., 210:50.
Bendich, A., and Rozenkranz, H. S., 1962, In: Progress in Nucleic Acid Research, ed. W. E. Cohn and J. N. Davidson, Vol. I, Academic Press, New York.
Berendes, H. D., 1963, Chromosoma, 14:195.

LITERATURE CITED

Berendes, H. D., 1965a, Develop. Biol., 11:371.
Berendes, H. D., 1965b, Chromosoma, 17:35.
Berendes, H. D., 1966, J. Exp. Zool., 162:209.
Berendes, H. D., Breugel van F. M. A., and Holt, T. K. H., 1965, Chromosoma, 16:35.
Beutler, E., Seh, M., and Fairbanks, V. F., 1962, Proc. Nat. Acad. Sci., USA, 48:9.
Bianchi, N. O., and De Bianchi, M. S. A., 1965, Chromosoma, 17:273.
Bogdanov, Yu. F., Iordanskii, A. B., and Gindilis, V. M., 1965, Genetika, No. 5:82.
Bridges, C. B., 1935, Am. Naturalist, 69:59.
Brodskii, V. Ya., 1965, In: Textbook of Cytology, Izd. Nauka, Moscow, p. 269.
Brown, D. D., and Gurdon, J. B., 1964, Proc. Nat. Acad. Sci., USA, 51:139.
Callan, H. G., 1955, In: Symposium on Fine Structure of Cells, Leiden, Holland, 1954, Intern. Union Biol. Sci. Publ. Ser. B., 21, 89, P. Hoodhoff, Groningen, The Netherlands.
Callan, H. G., 1963, Intern. Rev. Cytol., 15:1.
Callan, H. G., and Macgregor, H. C., 1958, Nature, 181:1479.
Callan, H. G., and Lloyd, L., 1960, Phil. Trans. Roy. Soc. Lon., Ser. B, 243:702, 135.
Clever, U., 1961, Chromosoma, 12:607.
Clever, U., 1962a, Chromosoma, 13:385.
Clever, U., 1962b, J. Insect. Physiol., 8:357.
Clever, U., 1963a, Chromosoma, 14:651.
Clever, U., 1963b, Develop. Biol., 6:73.
Clever, U., 1964a, Naturwissenschaften, 19:449.
Clever, U., 1964b, In: The Nucleohistones, ed. J. Bonner and P. Ts'o, Holden-Day, San Francisco, p. 317.
Clever, U., 1964c, Science, 146:794.
Clever, U., 1965a, Chromosoma, 17:309.
Clever, U., 1965b, Brookhaven Symp. in Biol., No. 18, Genetic Control of Differentiation, p. 242.
Clever, U., 1966, Am. Zoologist, 6:33.
Clever, U., 1967, In: The Control of Nuclear Activity, ed. L. Goldstein, Prentice-Hall Inc., New York, p. 161.
Cole, A., 1962, Nature, 196:211.
Davidson, E. H., Allfrey, V. G., and Mirsky, A. E., 1964, Proc. Nat. Acad. Sci., USA, 52:501.
Davidson, E. H., Crippa, M., Kramer, F. R., and Mirsky, A. E., 1966, Proc. Nat. Acad. Sci., USA, 56:56.
Deepesh, N. De, 1964, Nature, 203:343.
Defendi, V., and Manson, L. A., 1963, Nature, 198:359.
Dodge, J., 1963, Arch. Mikrobiol., 45:46.
Du Praw, E. J., 1965a, Proc. Nat. Acad. Sci., USA, 53:161.
Du Praw, E. J., 1965b, Nature, 206:338.
Du Praw, E. J., 1966, Nature, 209:577.
Edström, J. E., 1960, J. Biophys. Biochem. Cytol., 8:39, 47.
Edström, J. E., and Beermann, W., 1962, J. Cell Biol., 14:371.
Edström, J. E., and Gall, J. G., 1963, J. Cell Biol., 19:279.

Freese, E., 1958, Cold Spring Harbor Symp. Quant. Biol., 23:13.
Gall, J. G., 1958, In: The Chemical Basis of Development, ed. W. D. McElroy and B. Glass, Johns Hopkins Press, Baltimore, pp. 103-135.
Gall, J., 1963a, Science, 139:120.
Gall, J. G., 1963b, Nature, 198:36.
Gall, J. G., 1963c, In: Cytodifferential and Macromolecular Synthesis, ed. M. Locke, Academic Press, New York, p. 119.
Gall, J. G., and Callan, H. G., 1962, Proc. Nat. Acad. Sci., USA, 48:562.
Galton, M., and Holt, S. F., 1964, Cytogenetics, 3:97.
Galton, M., and Holt, S. F., 1965, Exp. Cell Res., 37:111.
Georgiev, G. P., and Chentsov, Yu. S., 1963, Biofizika, 8:50.
German, J. L., 1962, Trans. N.Y. Acad. Sci., 24:395.
German, J. L., 1964a, J. Cell Biol., 20:37.
German, J. L., 1964b, In: Symposia of the Internal Soc. for Cell Biology, Vol. 3, Cytogenetics of Cells in Culture, ed. R. J. C. Harris, Academic Press, New York, p. 191.
Gilbert, C. W., Muldal, S., Lajtha, L. G., and Rowley, J., 1962, Nature, 195:869.
Goldstein, A., and Brown, B., 1961, Biochim. Biophys. Acta, 53:19.
Grunbach, M. M., Morishima, A., and Taylor, J. H., 1963, Proc. Nat. Acad. Sci., USA, 49:581.
Hess, O., and Meyer, G. F., 1963, J. Cell Biol., 16:527.
Hess, O., 1965, Chromosoma, 16:222.
Hess, O., 1966, In: Chromosomes Today, ed. C. D. Darlington and K. R. Lewis, Vol. I, Oliver and Boyd, Edinburgh, p. 167.
Hsu, T. C., 1962, Exp. Cell Res., 27:332.
Hsu, T. C., 1964a, In: Biology of Cells and Tissues in Culture, ed. E. N. Willmer, Academic Press, New York, p. 397.
Hsu, T. C., 1964b, J. Gell Biol., 23:53.
Hsu, T. C., Schmid, W., and Stubblefield, E., 1964, In: The Role of Chromosomes in Development, ed. M. Locke, Academic Press, New York, p. 83.
Izawa, M., Allfrey, V. G., and Mirsky, A. E. 1963a, Proc. Nat. Acad. Sci., USA, 49:544.
Izawa, M., Allfrey, V. G., and Mirsky, A. E., 1963b, Proc. Nat. Acad. Sci., USA, 50:811.
Jacob, J., and Sirlin, J. L., 1963, J. Cell Biol., 17:153.
Kiknadze, I. I., 1965, In: Cell Differentiation and Mechanisms of Induction, Izd. Nauka, Moscow, p. 78.
Kiknadze, I. I., and Filatova, I. T., 1960, Izv. Sibirsk. Otdel. Akad. Nauk. SSSR, 12:131.
Kiknadze, I. I., and Filatova, I. T., 1963, Dokl. Akad. Nauk SSSR, 152:450.
Kikuchi, Y., and Sandberg, A. A., 1964, J. Nat. Cancer Inst., 32:1109.
Kimball, R. F., and Prescott, D. M., 1962, J. Protozool., 9:88.
Koltzoff, N. K., 1934, Science, 80:312.
Kroeger, H., 1964, Chromosoma, 15:36.
Kroeger, H., and Lezzi, M., 1966, Ann. Rev. Entomology, 11:1.
Laufer, H., and Nakase, Y., 1965, Proc. Nat. Acad. Sci., USA, 53:511.

LITERATURE CITED

Laufer, H., and Nakase, Y., and Vanderberg, J., 1964, Develop. Biol., 9:367.
Lobashev, M. E., 1963, Genetics, Izd. LGU, Leningrad.
Lychev, V. A., 1965, Tsitologiya, 7:325.
Lychev, V. A., 1967, In: Proceedings of a Scientific Conference of Junior Research Workers at the Institute of Medical Radiology, Academy of Medical Sciences of the USSR, Obninsk, p. 69.
Lychev, V. A., and Medvedev, Zh. A., 1967, Genetika, No. 8, 53.
Macgregor, H. C., and Callan, H. G., 1962, Quart. J. Microscop. Sci., 103:173.
Mazia, D., 1961, Mitosis and the Physiology of Cell Division: The Cell, Vol. III, Academic Press, New York and London.
Meyer, G., 1963, Chromosoma, 14:297.
Meyer, G. F., and Hess, O., 1965, Chromosoma, 16:249.
Miller, O. L., 1965, In Symp: Genes and Chromosomes - Structure and Functions, Nat. Cancer Inst. Monograph, No. 18, p. 79, Bethesda.
Mirsky, A., and Osawa, S., 1961, In: The Cell, ed. J. Brachet and A. Mirsky, Vol. II, Academic Press, New York and London, p. 677.
Moorhead, P. S., and Defendi, V., 1963, J. Cell. Biol., 16:202.
Morishima, A., Grumbach, M. M., and Taylor, J. H., 1962, Proc. Nat. Acad. Sci., USA, 48:756.
Moses, M. J., and Coleman, J. B., 1964, In Symp: Role of Chromosomes in Development, ed. M. Locke, Academic Press, New York, p. 11.
Mueller, G. C., and Kajiwara, K., 1966, Biochim. Biophys. Acta, 114:108.
Nebel, B. R., 1957, J. Heredity, 47:51.
Nebel, B. R., 1962, In: Electron Microscopy, Vol. 2, Academic Press, New York and London, p. 28.
Nørrevang, A., 1963, Exp. Cell Res., 31:603.
Novick, A., McCoy, J. M., and Sadler, J. R., 1965, J. Mol. Biol., 12:328.
Osgood, E. E., Jenkins, D. P., Brooks, R., and Lowson, R. K., 1964, Ann. N.Y. Acad. Sci., 113, Art. 2:717.
Painter, T. S., 1964, Proc. Nat. Acad. Sci., USA, 51:1282.
Pavan, C., and Breuer, M. E., 1955, In: Symp. on Cell Secretion, Belo Horizonte, Brazil, p. 14.
Peacock, W. J., 1963, Proc. Nat. Acad. Sci., USA, 49:793.
Pelling, C., 1964, Chromosoma, 15:71.
Plaut, W., 1963, J. Mol. Biol., 7:632.
Plaut, W., and Nash, D., 1964, In: Role of Chromosomes in Development, ed. M. Locke, Academic Press, New York, p. 113.
Prescott, D. M., 1962, J. Histochem. Cytochem., 10:145.
Prescott, D. M., Kimball, R. F., and Carrier, R. F., 1962, J. Cell Biol., 13:175.
Prokof'eva-Bel'govskaya, A. A., and Bogdanov, Yu. S., 1963, Zh. Vses. Khim. Obshchestva im. D. I. Mendeleeva, 8:33.
Prescott, D. M., and Bender, M. A., 1963, Exp. Cell Res., 29:430.
Ris, H., 1961, Canad. J. Genetics and Cytology, 3:95.
Ritossa, F. M., 1963, Exp. Cell. Res., 35:515.
Ritossa, F. M., 1964, Exp. Cell. Res., 36:601.
Ritossa, F. M., and Borstel, R. C. von, 1964, Science, 145:513.

Ritossa, F. M., Pulitzer, J. F., Swift, H., and Borstel, R. C. von, 1965, Chromosoma, 16:144.
de Robertis, E., Novinski, W., and Saez, F., 1960, General Cytology, Saunders, Philadelphia.
de Robertis, E., Novinski, W., and Saez, F., 1965, Cell Biology, Saunders, London.
Roodman, T. C., and Kopac, M. J., 1964, Nature, 202:876.
Rowley, J., Muldal, S., Gilbert, C. W., Lajtha, L. G., Linsten, J., Fraccaro, M., and Kajser, K., 1963, Nature, 197:251.
Sadron, C., Pouyrt, J., and Vendrely, R., 1957, Nature, 179:263.
Sager, R., and Ryan, F., 1961, Cell Heredity, Wiley, New York.
Sirlin, J. L., and Schor, N. A., 1962, Exp. Cell Res., 27:363.
Stockdale, F. E., and Holtzer, H., 1961, Exp. Cell Res., 24:508.
Stubblefield, E., 1964, Federated Proc., 23:332.
Stubblefield, E., and Mueller, G. C., 1962, Cancer Res., 22:1091.
Taylor, J. H., 1958, Exp. Cell Res., 15:350.
Taylor, J. H., 1960, J. Biophys. Biochem. Cytol., 7:455.
Taylor, J. H., 1962, Internat. Rev. Cytol., 13:39.
Taylor, J. H., 1963, In: Molecular Genetics, ed. J. H. Taylor, Academic Press, New York, pp. 65-111.
Taylor, J. H., 1964, In: Cytogenetics of Cells in Culture, Including Radiation Studies, ed. R. J. C. Harris, Academic Press, New York.
Tsanev, R. G., and Markov, G. G., 1964, The Biochemistry of Cell Division, Izd. Meditsina, Moscow.
Vogel, F., 1964, Nature, 201:847.
Wimber, D. E., 1961, Exp. Cell Res., 23:402.
Whitten, J. M., 1964, Science, 143:1437.
Wolfe, S. L., 1965, J. Ultrastruct. Res., 12:104.
Woodard, J., Rasch, E., and Swift, H., 1961, J. Biophys. Biochem. Cytol., 9:445.
Zbarskii, I. B., and Georgiev, G. P., 1962, Tsitologiya, 4:605.

Chapter 5

Biochemical Realization of the Morphogenetic Program. Changes in Proteins and Nucleic Acids During Processes of Differentiation and Development

Introduction

    In the preceding chapter we examined a number of different indices of morphogenetic differentiation of the nucleus and chromosomes controlling genetic systems of devolopment. This differentiation must provide for the functional transition from total genetic potency of the nucleus to selective activation of strictly definite loci and equally selective repression of other loci. This genetic intrachromosomal differentiation results in the biochemical, physiological, and morphological differentiation of the cells, in their spocialization in many different directions, in those differences which can be observed when we compare the cells of the liver, kidneys, brain, blood, skin, connective tissue, and so on. These changes, moreover, appear at the necessary time and in a predetermined zone of the developing system. A particular type of biochemical specialization usually does not arise at once, but through a series of transitions in several cell generations. Differentiation of the nucleus is followed by differentiation in the cytoplasm, and the new cytoplasmic products, in turn, have a further morphogenetic differentiating influence on the genetic system in the cells, sometimes not in the cells in which they were formed (embryonic and hormonal induction). They switch on new components of the genetic program of development. Naturally, therefore, before turning to the analysis of molecular-genetic mechanisms of development, to the analysis of the material nature of individual ontogenetic programs insuring the usefulness and synchronization of all acts of development in space and time, we must examine the biochemical aspects of cytoplasmic differentiation during morphogenesis, the basic principles governing changes in proteins and RNA associated with morphogenesis, which are the outward manifestation of the action of this program and which, in turn, bring about all other external manifestations of specialization (morphological, functional, and so on).

I have already examined these problems in a rather compressed manner in earlier publications (Medvedev, 1963, 1965). More recently certain individual aspects have been dealt with in a number of theoretical surveys, but the progress now being made in this field is so rapid that many aspects of the problem demand a thorough reappraisal.

1. Morphogenetic Changes in the Protein Spectrum of Cells and Tissues

As I stated in the introduction, morphogenetic differentiation above all gives rise to considerable modifications of the protein spectrum of cells and tissues, both quantitatively (changes in the relative proportions of different proteins) and qualitatively (the appearance of "new" proteins). Since the synthesis of each protein is under genetic control and is determined by messenger RNA molecules formed at particular loci of the genome, it can naturally be assumed that changes in the protein pattern are a reflection of changes in the spectrum of active loci of the chromosomes. Interaction between active loci of the genome and the pattern of protein synthesis may be either synchronous or asynchronous (retrospective) during the formation of loci of the stable RNA templates which function for long periods of time after the activity of the locus has ceased. Cases of this type can be observed, for example, during synthesis of hemoglobin and other proteins by mammalian reticulocytes. These cells destroy their own nucleus, but even in the anuclear form they continue to synthesize hemoglobin and other proteins for several days on account of stable forms of messenger RNA formed previously.

a) Genetic Control of Synthesis of Individual Proteins in Cells and Tissues. The existence of direct genetic control of the synthesis of individual proteins of the gene → protein type, as we have seen, has been studied in particular detail in viruses and bacteria. The protein composition of multicellular organisms is much more complex, and we see in them a great variety of proteins whose functional forms possess a very high molecular weight and consist of a series of subunits differing in their amino acid composition, i.e., they are aggregated in character. In such cases, as investigations have shown, one gene determines the structure of a polypeptide subunit with a continuous chain, while other subunits are determined by other genes, not necessarily by neighboring genes.

The literature on genetic determination of the structure of protein subunits is very extensive, and I do not propose to give a comprehensive survey of this subject. I shall dwell only on some typical examples which will serve to bind together the material concerned with changes in protein patterns and changes in the patterns of active and repressed genes.

1) Genetic control of the structure and synthesis of hemoglobins. The genetics of synthesis and structure of the hemoglobins, primarily human hemoglobins, has been studied more completely than the genetics of synthesis of other proteins because of its practical importance to the understanding of the many inborn abnormalities of hemoglobin. The hemoglobins of man and various animals are classical objects for the study of strict genetic control of the synthesis of protein subunits in animals. After the first work (Pauling et al., 1949), in which differences in electrophoretic mobility were found between normal hemoglobin and hemoglobin isolated from the erythrocytes of persons with congenital sickle-cell anemia, no fewer than 2000 investigations of genetic structural anomalies of hemoglobin have been published, providing fundamental evidence for the concept of genetic determination of protein structure. As a result of these investigations, scores of inherited hemoglobin anomalies have been discovered in homozygotic and heterozygotic states (together with scores of forms of abnormal hemoglobins), and the genealogical analysis of inheritance of these anomalies has revealed the number of genetic hemoglobin loci, demonstrated that inheritance of these

anomalies obeys Mendel's laws, and has enabled progress to be made with the study of other problems in the genetics of hemoglobin. The genetics of abnormal and normal human hemoglobins has become an extensive and independent field of genetics and biochemistry, and many outstanding surveys and monographs on this subject have been published (Itano, 1957; Lehmann, 1960; Efroimson, 1961; Rucknagel and Neel, 1961; Ingram, 1961, 1963; Chernoff, 1961; Baglioni, 1963; Zuckerkandl and Pauling, 1962; Jonxis, 1963; Korzhuev, 1964), so that we are able to take a more general look at this problem without describing its history or going into the genetic details of the whole range of hemoglobin anomalies.

Advances in the genetics of human hemoglobins has stimulated interest in similar problems from the standpoint of evolution, and extensive investigations have recently been undertaken to study hemoglobins of various animals. Surveys of their findings (Gratzer and Allison, 1960; Manwell, 1964; Korzhuev, 1964; Popp, 1965; Braunitzer, 1966; Alekseenko, 1966) provide a clear picture of the wide scale of research in this field.

Normal human hemoglobin consists of three fractions. About 90% is hemoglobin A (HbA), 2-3% consists of hemoglobin A_2 (HbA$_2$), moving more slowly during electrophoresis, and 4-10% of the fast moving hemoglobin A_3 (HbA$_3$). In many cases in adults a very small proportion, about 1%, of embryonic (fetal) hemoglobin (HbF) still persists. Hemoglobin A_3 is not a stable individual fraction, but a product of changes in hemoglobin A as a result of aging of the erythrocytes.

Hemoglobin A is a tetramer, consisting of two pairs of identical polypeptide chains known as α and β chains. Each of these chains contains about 140 (α 141 and β 146) amino acid residues, and one of the surprising features of their structure, discovered by decoding the amino acid sequence of both chains, is that, if the amino acid sequences of the chains are compared, it is found that both chains contain many identical polypeptides in identical positions. This suggests the common origin in evolution of both these chains and their phylogenetic differentiation as a result of mutations affecting different parts of the molecules.

A diagram of the arrangement of the two α and the two β chains in the hemoglobin molecule is given in Fig. 67 (Ingram and

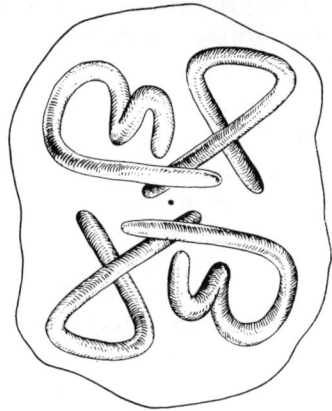

Fig. 67. Two-dimensional drawing of a model of the arrangement of the α and β chains of hemoglobin (Ingram and Stretton, 1959).

Stretton, 1959). Hemoglobin A_2 is also a tetramer, two of its polypeptide chains being identical with the α chains of HbA and the other chains differing from the β chains in the positions of certain amino acid residues. They have been described as delta-chains. Fetal hemoglobin (HbF) like hemoglobin A_2, contains two α chains identical with the α chains of HbA. The two other polypeptide chains of this hemoglobin differ from the β chains in the position of about one-quarter of the amino acid residues; they are described as γ chains. The order of the amino acids in each polypeptide monomer (α, β, γ, δ,) is controlled by an independent gene, mutations of which affect only the corresponding polypeptide.

Ingram (1959) suggested the "one gene—one polypeptide" hypothesis to explain the genetic determination of synthesis of the peptide chains of the hemoglobin. Genes controlling the synthesis of polypeptides are usually designated by the same symbols as are used for the peptides. Gene α^A relates to the gene carrying information for synthesis of the α chain of hemoglobin A, gene β^A is the gene controlling synthesis of the β chain of hemoglobin A, gene β^S controls synthesis of the β chain in abnormal hemoglobin S (sickle-cell anemia), and so on.

Investigations have clearly shown that both the genes determining synthesis of normal α and β chains in man and the genes controlling synthesis of abnormal chains are inherited by Mendelian principles. A considerable amount of statistically significant evidence for this has been obtained by studying the offspring of marriages in which one parent possesses two different abnormal hemoglobins (HbC and HbS at the same time, for example), while the other parent has normal hemoglobin (Rucknagel and Neel, 1961; Ranney, 1954; Neel, 1956; Hunt and Ingram, 1958; Askoy and Lehmann, 1957; Atwater et al., 1960).

Hemoglobin A is completely absent from most persons with two abnormal hemoglobins (for example, HbS and HbD punjab or HbS and HbN). This is evidence of allelism of the abnormal hemoglobins. The degree of abnormality reaches a maximum in the persons described by Atwater and co-workers (1960), with four abnormal hemoglobins and carrying different abnormalities in one of two genes in different alleles (double heterozygotes): α^A/α^C, $\beta A/\beta C$, or $\alpha A/\alpha$Hopkins-2, $\beta A/\beta S$ (Itano and Robinson, 1960). Meanwhile, no molecules have been found which contained two differently abnormal α or two abnormal β chains. This suggests either that such variants are lethal or that the genes also control the formation of dimers, or most probably of all, that the formation of a tetramer is preceded by the formation of a dimer, directly at the sites of synthesis of these proteins, for example.

Many very interesting investigations have been made of the genetics of abnormal hemoglobins and detailed tables of all possible types of heterozygosis and of their incidence have been published (Walter, 1963; Chernoff, 1961). The possible nature of the mutations has been analyzed, and attempts have even been made to solve the problem of the genetic code on the basis of variation among the hemoglobins.

Investigations have revealed genetic determination and Mendelian relationships in the progeny when studying inheritance of the structure of hemoglobin subunits in various animals (from the results of crossing): in monkeys (Buettner-Janusch, 1962), in mice of various lines (Hutton et al., 1962; Popp, 1962a, b, 1963, 1965), in cattle (Crockett et al., 1963), and in invertebrates (Manwell, 1964; Manwell et al., 1963). Genetic determination of variations in composition of hemoglobins in 12 different inbred lines of mice has been studied in our own laboratory since 1965 (Strekalov, 1967). However, I do not propose to give a detailed description of all these investigations here. All that is necessary at this stage is to show that the structure of each subunit of hemoglobin is under the direct control of one gene and, consequently, that changes in the hemoglobin pattern typical of changes from one period of morphogenesis to another reflect changes in the work of particular genetic loci.

2) Genetics of lactate dehydrogenase and some other enzymes. Lactate dehydrogenase (LDH) is the protein for which genetic control of synthesis of its subunits has been demonstrated most fully and in the greatest detail in mammals in connection with the existence of numerous variants in structure of this enzyme resulting from different tetramer combinations of two subunits A and B. This enzyme is usually present in animals in several forms or isozymes (five types), depending on the combinations of the subunits. The problem of of variations of isozymes of this enzyme in relation to their precise ontogenetic variations will be examined in the next section. Here I shall discuss only the question of genetic control, the study of which is made easier by the existence of variations in structure (and also in electrophoretic mobility and immunologic behavior) of lactate dehydrogenase molecules associated with different types of tetramer combinations of subunits.

(LDH-1; LDH-2; LDH-3; LDH-4; LDH-5)
BBBB   BBBA   BBAA   BAAA   AAAA

It is naturally assumed that the A and B fragments are under the control of separate genes. To verify this assumption, variants of animals with different types of combinations must be selected for crossing and the type of inheritance of this characteristic must be examined. Investigations along these lines has been undertaken by several workers with various organisms. In one investigation (Shaw and Barto, 1963) mice of the species *Peromyscus municulate bairdi* were crossed with a heterozygous hybrid mouse from two lines. In the study of three hybrid combinations (normal × normal, normal × heterozygous, and heterozygous × heterozygous) the resulting variations of isozyme combinations were present in the Mendelian ratio. Inheritance was autosomal in character. Distribution of the A-B combinations in the progeny showed that each subunit was determined by an independent genetic locus. The composition of the enzyme (for example, AAAB or BBBA) in this case was determined by the relative activity of the two loci. In some animals a sixth and unusual form of lactate dehydrogenase (X) has been discovered. The study of this spectrum of lactate dehydrogenase in various crosses in pigeons (Zinkham et al., 1964; Blanco et al., 1964) led to the conclusion that lactate dehydrogenase synthesis in this case is controlled by three loci, A, B, and C. All the pigeon variants (6) which were studied were homozygous relative to loci A and B, and some were heterozygous relative to the C locus. Characteristically the X fraction of the enzyme and, consequently, locus C were active only in one organ of the animal, namely the testis. The remaining organs, (lungs, liver, muscles, etc.) contained forms of the enzyme consisting only of A and B subunits. Separate genetic control of the A and B subunits of lactate dehydrogenase was also demonstrated by the study of this enzyme in human erythrocytes. Analysis of several generations in families possessing LDH variants (Kraus and Neely, 1964) showed that the loci of LDH-A and LDH-B are located in autosomes.

Similar results were obtained for two other enzymes. Genetic determination of the structure and heterogeneity of salivary amylase of mice was discovered by studying the inheritance of isozyme forms of amylase in the progeny of mouse lines differing in this characteristic (lines with homogeneous and heterogeneous amylase composition) (Sick and Nielson, 1964). Three amylase phenotypes (A, B, AB) were discovered in the saliva; these were determined by two alleles. It is interesting to note that pancreatic amylase differed from salivary, and, although it was determined by other alleles, the character of its variations was the same (two alleles and three phenotypes). In cows and bulls, as large numbers of crosses showed (Ashton, 1965), polymorphism of amylase is controlled by three autosomal codominant alleles.

A series of interesting investigations has been carried out to study genetic determination of isozyme heterogeneity of phosphatases in man (Hopkinson et al., 1963; Lai et al., 1964; Arfors et al., 1963). The first of these studies showed that the acid phosphatase of erythrocytes is determined by three alleles P^a, P^b, and P^c in an autosomal locus. By mass examination of families they discovered five phenotypes, two homozygous (A and B) and three heterozygous (B^a, C^a, and C^b) and they postulated the existence of yet another heterozygote (C). This predicted genotype was soon found by another group of workers (Lai et al., 1964) as a result of the examinations of 80 families and 369 unrelated individuals of different nationalities.

Differences in the electrophoretic mobility of a series of phenotypes are shown in Fig. 68, clearly demonstrating the difference between homozygous and heterozygous variations. It is very interesting to note that the genetically determined polymorphism of alkaline phosphatase in the human placenta was determined not by the maternal genes, but by the fetal genes (Robson and Harris, 1965). This confirms findings indicating that a considerable proportion of the placenta is derived from embryonic tissue. To demonstrate genetic control of synthesis of particular types of enzymes, the classical method of analysis of twins can be used. This method was used to establish genetic variants of phosphatases by Arfors and co-workers (1963), who studied the character of phosphatase heterogeneity in 89 monozygous and 111 dizygous twin pairs.

Other work (Boyer et al., 1962; Escobar et al., 1964; Mohler and Crockett, 1964) has revealed a clear pattern of the genetic control of glucose-6-phosphate dehydrogenase. The existence of two different types of enzymes in erythrocytes of heterozygotes was

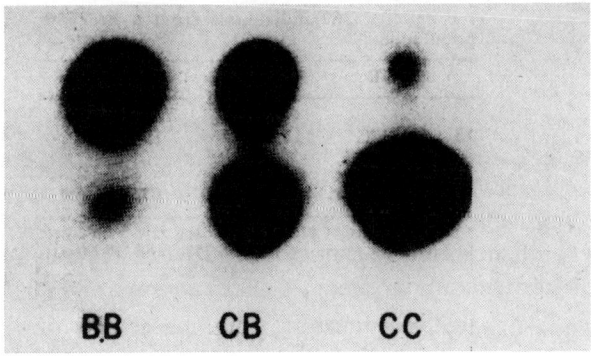

Fig. 68. Pattern of alkaline phosphatase of erythrocytes during starch-gel electrophoresis. Three phenotypes BB, CB, CC. Point of application was in the lower part of the picture and migration was toward the anode (above) (Lai et al., 1964).

shown to be associated with two different cell populations. Synthesis of this enzyme was determined by a locus situated in the X sex chromosomes (Davidson et al., 1963; Ohno et al., 1965). The same enzyme, not in the erythrocytes but in several other mouse tissues, is determined not by the X-chromosome but by an autosomal locus (Shaw and Barto, 1965).

This description of cases of genetic determination of variations in structure and composition of enzymes and other proteins, clearly established by hybrid analysis and other methods, could be continued. Genetic determination has been studied for β-glucuronidase (Paigen, 1959, 1961), for variants of the transferrins (Beckman and Holmgren, 1963), for haptoglobins (Javid, 1964), for carbonic anhydrase (Tashian et al., 1963), for DOPA-oxidase in *Drosophila* (Lewis and Lewis, 1963), and several others. Investigations of genetic variations in enzymes in *Drosophila* mutants have been particularly numerous in recent years (see the survey by Shaw, 1965). There seems, therefore, to be little doubt that this determination is universal for all enzymes and proteins of the organism, and it follows naturally from the evidence that the mechanism of protein synthesis corresponds to information carried in DNA. Consequently, ontogenetic changes in the pattern, composition, and type of heterogeneity of the enzymes, which I now propose to discuss, are also connected with definite processes in the genetic systems of cells.

b) <u>Quantitative and Qualitative Changes in the Pattern of the Structural Proteins of Cells during Morphogenesis and Differentiation.</u> Changes in the pattern of structural nonenzyme proteins associated with morphogenesis and differentiation in the embryonic and postembryonic states of development have been studied for a long time, although until recently the information obtained was only fragmentary and concerned mainly with a few proteins only. Most of the results were descriptive in character and were not connected with the attempt to discover the genetic mechanisms of morphogenesis.

For this reason, although the existing material in this field was of undoubted interest as an illustration of the general biochemical pattern of development, it did not provide a serious basis for analysis of the molecular-genetic mechanisms of development. Since the differentiation of structural proteins has repeatedly been examined from the descriptive point of view (Dorfman, 1958; Brachet, 1960; Medvedev, 1963; Herrmann, 1963; Solomon, 1965), I shall discuss only its general principles insofar as this is necessary for a more complete understanding of the general pattern of development from the biochemical standpoint.

Muscle proteins. Synthesis of proteins responsible for the muscular type of specialization (actin, myosin, myogen, etc.) has been most fully studied from the aspects of embryogenesis and morphogenesis. These proteins have not been found in the oocyte or in the first stages of embryogenesis either by chemical analysis or by use of immunologic methods. Their appearance in particular groups of cells at particular periods of morphogenetic differentiation is therefore the result of activation of a hitherto inactive group of genes. From the theoretical point of view it would be very interesting to know whether this "switching on" of the systems concerned with the synthesis of the different muscle proteins takes place simultaneously and synchronously, or whether each protein starts to appear independently of the rest, asynchronously. As we shall see later, the solution of this problem is essential for the theoretical analysis of mechanisms of differentiation.

Evidence concerning the ontogenesis of muscle proteins does not yet provide a sufficiently precise answer to this question, and the character of morphogenesis of muscles in different animals differs in many respects. The basic structural proteins of muscles are myosin and actin, forming the actomyosin complex in muscle fibers. Structurally speaking, myosin is a very complex protein, whose molecule consists of many subunits with very high molecular weight, linked together by secondary bonds. The results of many investigations show that the molecular weight of myosin is between 420,000 and 500,000 (Poglazov, 1965).

Recent investigations have shown that the myosin molecule is built up on the basis of two or three molecules of meromyosin, differing in their molecular weight, each of which breaks up, if the secondary bonds are severed, into polypeptide subunits or protomyosins, with a molecular weight of about 10,000. One group of subunits of this protein possesses adenosine-triphosphatase (ATPase) enzyme activity, connected with the provision of energy for the contractile function of the myofibrils. All studies of the ontogenesis of myosin and actomyosin must therefore be accompanied by determination of the properties of this protein, not only its solubility, contractility, antigenicity, and other properties, but also its ATPase activity. Many individual differences in these properties are found in different animals, because they acquire the function of active movement at different periods. Myosin is not synthesized at once in the form of finished myofibrils.

The available data (Winnick and Goldwasser, 1961) indicates that protomyosin, which is immunologically identical with myosin, is formed in the ribosome and mitochondria of chick muscle fibers, and after leaving the place where it is synthesized, it is aggregated into specific myofibrillary structures. The existence of the process of aggregation of typical myosin molecules from protomyosin molecules in embryogenesis of the chick was also clearly demonstrated by the work of Robinson (1952), who compared the appearance of ATPase activity in myosin of the chick embryo with the dynamics of this activity in the muscle sarcoplasm. He found that the accumulation of ATPase activity in myosin fibers is accompanied by a strictly corresponding decrease in the ATPase content in the sarcoplasm. Robinson accordingly postulates that embryonic muscle contains a water-soluble prototype of myosin, which can subsequently undergo aggregation into an elementary protofibrillary structure. The mechanism of this phenomenon possibly differs from that of simple aggregation, and it may be a more complex process consisting of the formation of a complex of a protein with ATPase activity together with other proteins.

Comparison of the dynamics of appearance of myosin fibrils in embryogenesis with the dynamics of ATPase activity in chick embryos (Kasavina, 1950) supports this hypothesis, i.e., that sarcoplasmic ATPase is incorporated as a subunit of typical myosin only at a certain stage of development, before which ATPase and protomyosin were synthesized separately. Kasavina found that fibrils of proteins from muscle extracts of chick embryos before the 16th day of development did not possess appreciable ATPase activity and did not contract in the presence of ATP. These properties of the fibrils began to appear on the 17th day of embryonic development. In later investigations, Kasavina (1954) studied the appearance of ATPase activity during ontogenesis of muscles of several animal species. She found that guinea pigs and hens possess a marked actomyosin reaction to ATP while still in the embryonic period of development, and that it appears all at once, in one stage. Other animals (mice, rats, rooks) are characterized by a gradual increase in the reaction between actinomyosin and ATP, becoming clearly defined on the 10th and 13th day of postembryonic development. In man it begins to appear at the 8th month of intrauterine life. Myosin from mice and rats at the age of 11-13 days was able to bind itself to actin from adult animals, whereas actin from rats and mice aged 11-13 days did not react with myosin from the adult animal. Many qualitative indices of embryonic muscle ATPase and of embryonic myosin differ from those of the corresponding proteins in the adult. It has therefore been suggested that these proteins possess embryonic specificity, and that proteins also undergo morphogenetic changes from some forms to others, associated with definite changes in their molecules. Studies of ATPase activity of myosin isolated from muscles at different periods of morphogenesis and from two different animals, rats (Villafranca, 1954) and frogs (Nass, 1962), clearly demonstrated that myosin ATPase in early embryos differs considerably in its properties (activation by Ca and Mg, pH optimum, effect of substrate concentration, and so on) from the corresponding ATPase of late embryos and adult animals. Actomyosin appeared in frog embryos at stage 17 (tail bud), and ATPase activity appeared during the same period, rising parallel with the accumulation of actomyosin. ATPase of tadpoles differed from ATPase of adult frogs not only in physico-chemical properties but also in specificity. Embryonic ATPase detaches both labile phosphate groups from the ATP molecule, and in the character of its action it is therefore an apyrase, in contrast to the true ATPase of the adult frog, whose action is directed only at the terminal phosphate group in the ATP molecule.

I should mention that not only the ATPase activity, but other properties of the actomyosin complex in different animals also undergo certain changes during morphogenesis, and this led Ivanov and co-workers to postulate the existence of proteins of a proactomyosin type in embryos (Ivanov et al., 1956a, b). Differences between the actomyosins of embryos and adult animals in their physico-chemical properties have recently been demonstrated for rabbit muscles (Pinaev, 1965). As regards the time of appearance of the proteins of actomyosin complex, the only relevant work is that of Osawa (1958), who showed that actin appears one day earlier than myosin in frog embryos.

Clearly defined morphogenetic changes in the sarcoplasm of the muscles, which has a complex fractional composition and contains many different enzymes, have also been found in the embryonic and postembryonic periods (Kasavina and Umanskaya, 1958; Kadykov, 1963; Hartshorne and Perry, 1962).

Connective-tissue proteins. The chief protein of connective tissue, collagen, is also of a highly complex character in its ultimate functional form of collagen fibrils. It consists of a series of substructures which, besides protein, include mucopolysaccharides. During embryonic development, collagen appears in various organs as a new protein at strictly definite periods of development (Edds, 1958; Prockop et al., 1962; Coleman et al., 1965). Just as with myosin, definite morphogenetic changes are found in collagen; in the course of development the composition of

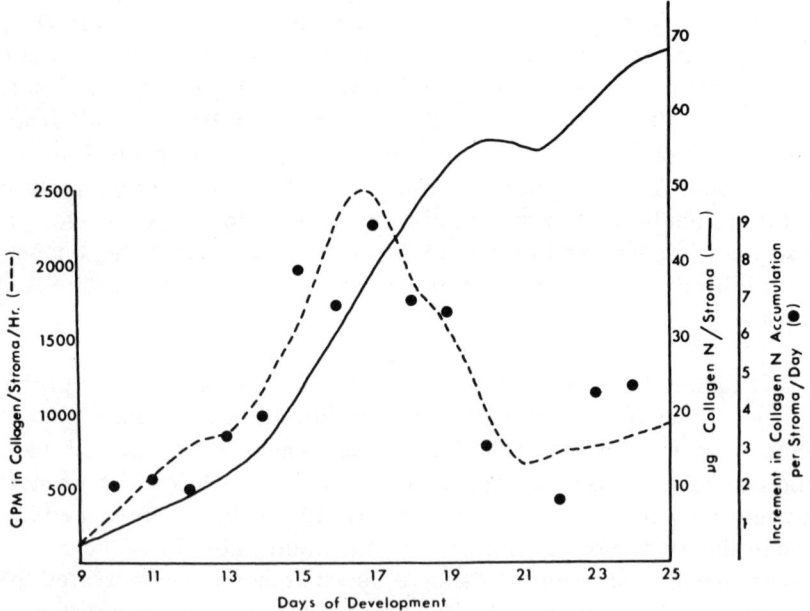

Fig. 69. Comparison of accumulation of collagen (collagen N) and rate of incorporation of glycine-1-C^{14} into collagen of the corneal stroma of the chick embryo (Coleman et al., 1965).

the protein fractions of collagen changed (a decrease in the soluble and increase in the insoluble fractions, changes in hydroxyproline content) (Edds, 1958). The change in the properties of collagen during development of the corneal stroma of the chick embryo is illustrated in Fig. 69 (Coleman et al., 1965). It is clear that at a certain period the metabolic activity of the collagen changes sharply, despite its progressive accumulation.

The initial appearance and morphogenetic changes of other structural proteins — elastin (Keech and Reed, 1957) and keratin (Pels, 1959; Malt and Bell, 1965) — have common features with the case of collagen just examined. During embryogenesis of the chick, typical keratin appears at the end of the 13th or beginning of the 14th day of development. However, a number of nonkeratinized subfractions of this protein can be found as early as the 6th day of development, and the formation of a particular terminal fragment evidently triggers the process of keratinization. The relative content of sulfur-containing amino acids in embryonic keratin characteristically increases in the course of development, indicating morphogenetic changes in this protein. Solubility of keratin also changes.

c) Morphogenesis of the Plasma Proteins. The many and varied plasma proteins are products of syntheses taking place in numerous organs and tissues, so that their ontogenetic changes reflect highly diverse morphogenetic processes. Serum albumin, the principal plasma protein of the adult, is formed mainly in the liver, so that in the earliest stages of development this protein naturally is absent. Production of immune γ-globulins is functional in nature, the plasma enzymes represent the sum of products formed by practically all the organs, and so on.

Changes in the composition of the plasma proteins during embryonic development have been the subject of much research undertaken on various organisms ranging from insects to man. I shall discuss here, in a simple and descriptive manner, only a few of these investigations to illustrate the typical changes taking place, because from the point of view of genetic mechanisms, the available data do not yet reveal any precise pattern. The system of morphogenesis of the hemoglobin, which we shall examine in the next section, is much more useful from the genetic point of view.

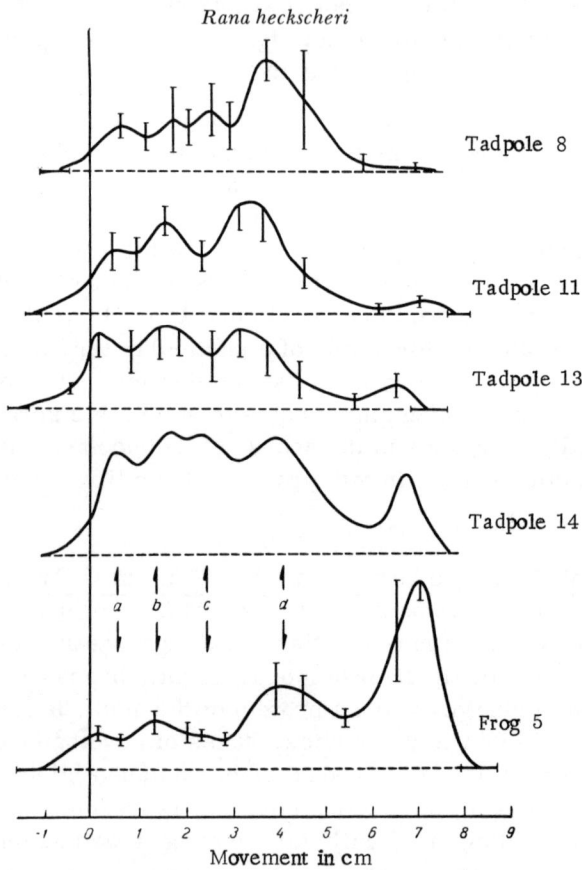

Fig. 70. Densitometric curves of electrophoretically separated serum proteins of tadpoles of the frog (*Rana heckscheri*) at various stages of metamorphosis. Globulin fractions designated a, b, c, d. Fraction c, with the fastest speed of migration, is albumin. From top to bottom: tadpole 8 in early metamorphosis; tadpoles 11 and 13, 3 and 6 days after injection of thyroxin; tadpole 14, middle of metamorphosis; frog 5, adult (Herner and Frieden, 1960).

Siakotos (1960) studied changes in the composition of the plasma proteins of the cockroach. Changes in the fractions were particularly marked before and during molting. Some proteins appeared distinctly in the period of sexual maturation. The lipoprotein fraction was particularly variable. Characteristics of the changes in the various plasma protein fractions during the stages of metamorphosis of the silkworm have been described by Japanese workers (Izawa et al., 1960). Composition of the protein fractions of the plasma and hemolymph of several species of insects (*Haylophora cecropia*, *Samia cynthia*) throughout the stages of morphogenesis have been studied by Laufer (1960, 1964).

Besides distinct species differences in the spectrum of the plasma proteins, characteristic changes have also been found in different phases of development, notably between the larva and adult insect. Some functions were typical of females only. The appearance of a "new," fast-moving albumin during metamorphosis, and its persistence as the principal plasma protein of the adult organism, are found in the lamprey *Petromyzon marinus dosatus* (Rall et al., 1961). Changes in the serum proteins have been closely studied during morphogenesis of frogs of various species (Herner and Frieden, 1960). In this case also, albumins typically do not appear until a certain stage of development (Fig. 70). In other species of frogs studied by these workers, the character of the changes in plasma proteins during development was similar in type.

Metamorphosis of the plasma protein of the chicken throughout the phases of embryogenesis and postembryonic development has been investigated very thoroughly (Weller and Schechtman, 1962).

A typical picture of these changes, which also reveals most distinctly the behavior of the albumins, is shown in Fig. 71. In mammals the morphogenetic spectrum of the plasma proteins has been studied most fully in rodents (rats, mice) (Shmerling and Uspenskaya, 1955; Pantelouris and Hall, 1962) and in man (Tatarinov et al., 1963; Woods et al., 1961). In this case also, albumins did not appear until about halfway through embryogenesis, whereas certain embryonic proteins were absent from adults.

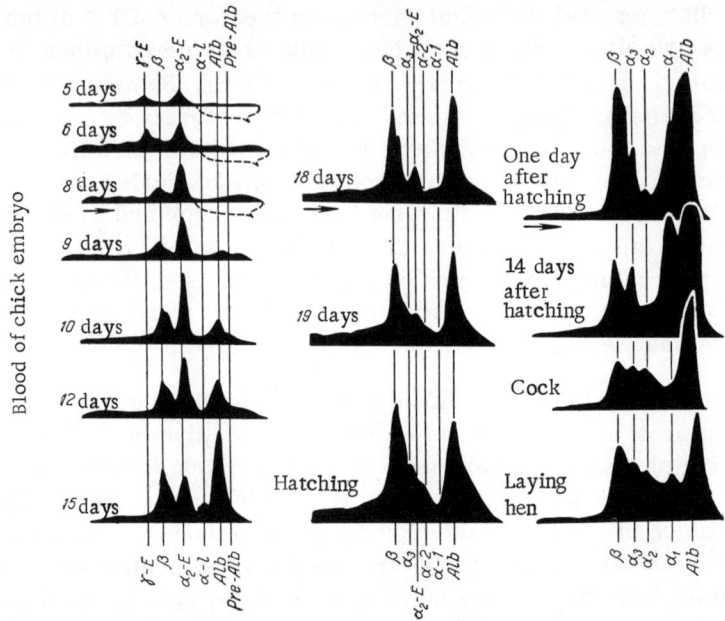

Fig. 71. Densitometric curves of electrophoretically separated serum proteins. Vertical lines drawn through protein components with equal electrophoretic mobility. Direction of movement of proteins from left to right (Weller and Schechtman, 1962).

d) Biochemical Morphogenesis of the Hemoglobins. Genetic Mechanisms of the Change from Embryonic Hemoglobin to Adult Forms. The change from the synthesis of the distinctive embryonic, or fetal type of hemoglobin, formed in the early stages of development, to adult forms of hemoglobin is one of the most thoroughly studied and most interesting phenomena of biochemical morphogenesis. We shall examine this phenomenon from the molecular-genetic aspect in rather more detail than the ontogenetic changes in structural and enzyme proteins. The closer attention which we shall pay to this biochemical morphogenetic system is justified for several reasons.

First, this system of protein morphogenesis, although incompletely studied, has nevertheless been investigated most systematically, whereas the data for most other body proteins are still very fragmentary. Second, I chose the system of morphogenetic transition HbF → HbA as the subject of an investigation to study the mutagenic action of irradiation on regulatory mechanisms of protein synthesis, and this naturally required a more intensive study of the facts relating to systems of "switching on and off" the synthesis of the different forms of hemoglobins.

In addition, until very recently no general biochemical surveys had been published of ontogenesis of the hemoglobins. This gap has recently been filled to some extent by two papers (Baglioni, 1966; Marks, 1966).

1) **Structure and functions of fetal hemoglobin.** Hemoglobin is responsible for the transport of oxygen and carbon dioxide and thus participates in cell and tissue respiration. The amount of energy required for all biochemical, chemical, or physico-chemical reactions is strictly definite. Any conversion, such as synthesis of a peptide bond, phosphorylation, or passage of ions through a membrane requires a definite quantity of energy (measured in calories), and this is practically independent of the species of the animal. However, conditions under which animals are supplied with oxygen vary greatly, depending on the environment in which they live (soil, water, air), the volume of their lungs and gills, the velocity of the blood flow, the hemoglobin concentration, the erythrocyte volume, and many other factors. Accordingly, the affinity of the hemoglobins for oxygen also varies. If external conditions are unfavorable for supplying oxygen, the ability of the hemoglobins to fix oxygen is hereditarily increased. This affinity for oxygen is determined by the structure of the polypeptide chains of the hemoglobin and the configuration of the Hb molecule. Details of variations in the affinity of different hemoglobins for oxygen in a phylogenetic series are described by Korzhuev (1964) in his monograph.

However, sharp differences in the conditions of the oxygen supply are also found between the embryonic and postembryonic periods. The mammalian embryo receives its oxygen from maternal blood, which does not come into direct contact with the fetal tissues. Under these conditions fetal hemoglobin must have a higher affinity for oxygen than maternal hemoglobin in order to remove oxygen from the maternal hemoglobin as they circulate together in the placenta. If this difference in the ability of fetal and maternal hemoglobins to bind oxygen did not exist, it would be impossible for the fetus to receive its supply of oxygen.

In birds and other egg-laying animals, the differences in the oxygen supply are determined by absence of active lung function in the embryo and by the entry of air into the egg through its membrane; in the case of amphibians the larvae develop in water, where these conditions of oxygen supply differ sharply from those in air, and so on. This all means that at these stages of development other forms of hemoglobin than those found in the adult must be present, subsequently to be replaced by adult forms of hemoglobin after birth or at a certain critical phase of development.

The fetal hemoglobin (HbF) of man and most other mammals, like HbA, is a tetramer composed of two subunits. The two α polypeptide chains which it incorporates are identical with the α polypeptide chains of HbA and they are controlled by the same gene, which is active throughout life (Baglioni et al., 1961; Minnich et al., 1962). The two other polypeptide chains differ from the β chains of HbA and have been called γ chains.

The difference between the amino acid composition of HbF and HbA was established several years ago (Huisman, 1954; Dustin et al., 1954). Since the identity of the α chains of HbF and HbA was subsequently demonstrated (Hunt, 1959; Schroeder et al., 1961, 1963a), the differences between them are localized entirely in the γ and β chains. In work which continued over many years, Schroeder and co-workers (1961, 1963b) established the sequence of all 146 amino acid residues of the γ polypeptide of hemoglobin and compared it with the polypeptide sequence of the β chain, also containing 146 residues. They found differences in the position of 17 amino acid residues in the α and β chains, giving rise to 39 different sequences. The β and γ chains differed in the overall balance between acid and basic amino acids, and finally, the γ chain has two supplementary histidine residues.

2) **Morphogenetic switchover from the system of synthesis of fetal hemoglobin to adult hemoglobin.** Hemoglobin synthesis has very complex systems of regulation. In this section we shall discuss only one of them, the morphogenetic system responsible for the switchover from synthesis of the γ chain to synthesis of the β chain. There are, in

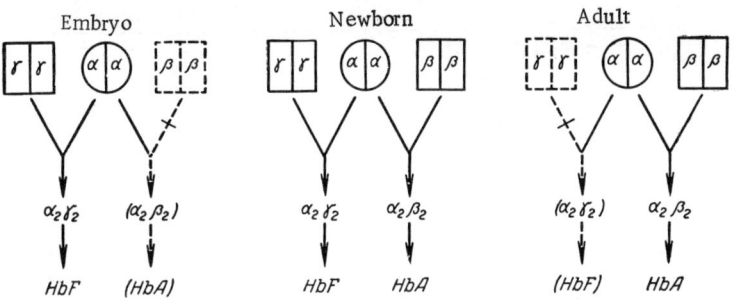

Fig. 72. Hypothetical scheme of genetic control of synthesis of hemoglobin tetramers in normal individuals (Chernov, 1961).

addition, other regulatory systems with very interesting action, responsible for example for the equivalent synthesis of all components of the hemoglobin structure, such as the heme of the α and β chains. Heme, for instance, induces the synthesis of protein subunits. Synthesis of α and β chains is interconnected at the level of liberation from the ribosomes, and so on. These forms of regulation are the ordinary functional, biochemical forms, and there is no need to analyze their mechanisms here.

The genetic scheme of the transition from HbF synthesis to HbA synthesis at the end of the embryonic period is illustrated in Fig. 72 (Chernov, 1961).

In Fig. 73 this scheme is shown as the results of actual biochemical determinations of concentrations of the γ and β chains in the last weeks of human embryonic development and during the first months after birth (Baglioni, 1963). HbF synthesis in man and animals falls abruptly and rapidly after birth, but usually it does not reach zero but stays at the level of about 0.5-1%, so that repression of the γ gene is not absolute. Recently, biochemical methods of HbF determination (the alkali-resistant form) have been supplemented by cytochemical (based on differences in solubility of HbF and HbA in an acid medium) and immunofluorescence methods. The last method is particularly demonstrative and accurate. It is based on the principle that antibodies obtained by immunization with HbF can be "loaded" with fluorescent dyes (fluorochromes). Fluorescent antibodies of this type are selectively bound with erythrocytes containing increased concentrations of

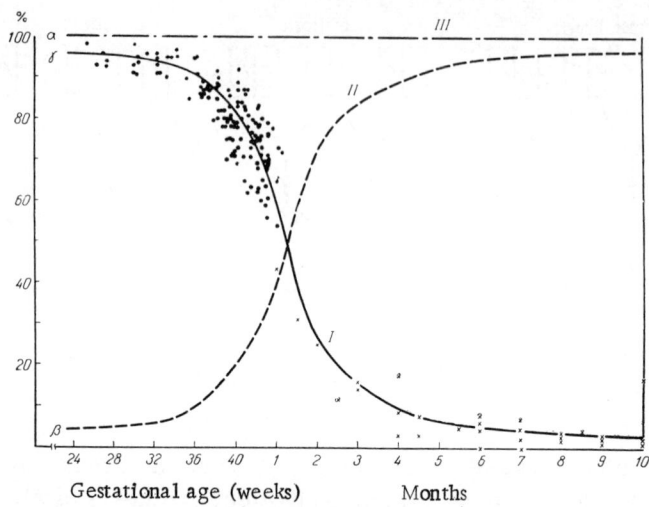

Fig. 73. The switch from HbF synthesis to HbA synthesis. I) Mean relative number of γ chains during last few months of intrauterine life and first 10 months after birth; II) mean relative number of β chains; III) relative number of α chains during the same period (number of α-peptide chains remains constantly 100% because this chain is common to both HbF and HbA); black dots represent HbF content (in percent) in relation to gestational age, crosses show HbF content (in %) in infants aged from 1 to 10 months (Baglioni, 1963).

fetal hemoglobin, so that the erythrocytes themselves give a distinct fluorescence under the luminescence microscope.

The presence of HbF-containing erythrocytes in adult human blood is demonstrated in Fig. 74 (Hosoi, 1965). As a result of chemical tests the blood was found to contain about 1% HbF. Besides erythrocytes (in adults) containing a high HbF concentration, there are other erythrocytes with weak and very weak fluorescence. Cells with strong fluorescence account on the average for 5% of the total number of erythrocytes. Such cells evidently contain a mixture of HbF and HbA, and it can thus be concluded that no cells are present which contain HbF only. HbF and HbA are synthesized together in these cells, and the essence of the matter is the presence of cells with different relative concentrations of HbF and HbA. The F-erythrocytes have an increased concentration of HbF, but do not contain HbF only.

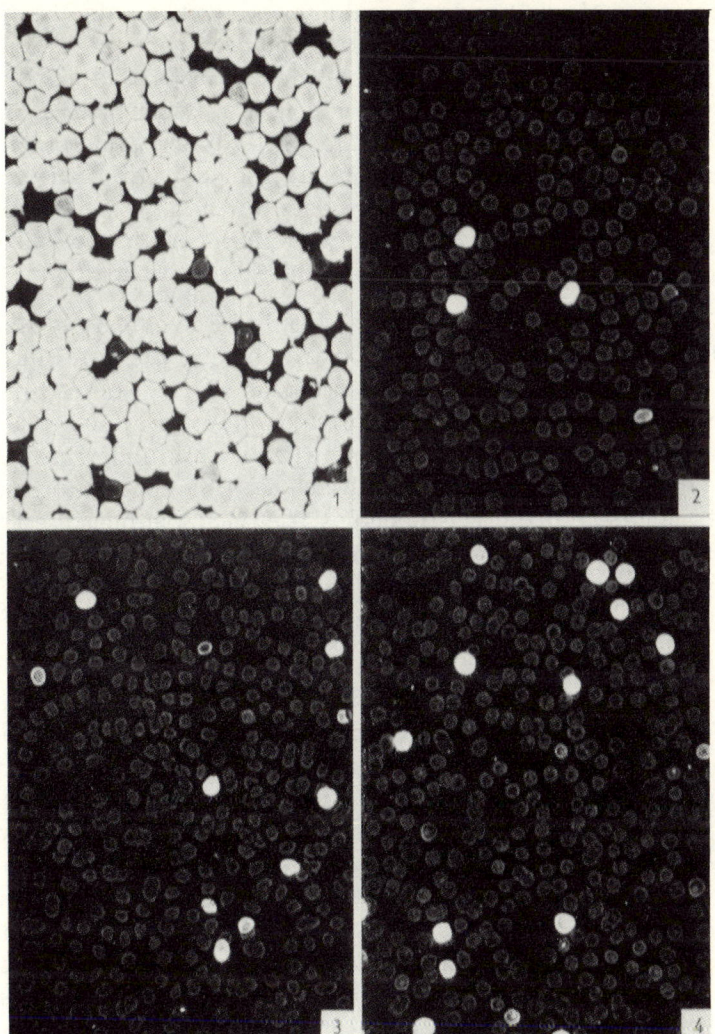

Fig. 74. Immunofluorescence of erythrocytes containing fetal hemoglobin. 1) Umbilical blood, 96% of cells contain HbF, alkali-resistant Hb 67.8%; 2) blood of a 21-year-old woman, 2.4% of cells contain HbF, alkali-resistant Hb 0.54%; 3) blood of a 21-year-old man, 3.2% of cells contain HbF, alkali-resistant Hb 0.48%; 4) blood of a 21-year-old man, 7.4% of cells contain HbF, alkali-resistant Hb 0.79% (Hosoi, 1965).

The possibility that HbF and HbA can coexist in individual erythrocytes of adult and neonatal human blood has also been demonstrated by spectrophotometric and cytochemical methods (Matioli et al., 1962; Zipursky et al., 1962; Shepard et al., 1962). The switch from HbF and HbA thus consists evidently of a change in the direction of work of what is essentially the same single erythropoietic system, and not a complete replacement of one erythropoietic system by another new one, which might be expected from data showing changes in the localization of erythropoiesis in different phases of embryonic development (Maximov and Bloom, 1957).

The time at which the switch from HbF synthesis to HbA synthesis begins and its total duration are specific for different species of animals. In man the switch begins shortly before birth and extends over several months. At birth on the average about 60-70% of HbF and 30-40% of HbA are synthesized, while in bovine embryos the HbF content begins to fall and the HbA content to rise three months before birth (Roche et al., 1953). In mouse embryos the HbF content begins to fall sharply at about the middle of pregnancy (Craig and Russell, 1964), and so on.

3) Phylogenetic distribution of the change in hemoglobin forms with morphogenesis. This phenomenon of a change from one form of hemoglobin to another in connection with morphogenesis is apparently very widespread, for it is found in most animals possessing erythrocytes and hemoglobin (see the survey by Korzhuev, 1966). In the simplest aquatic vertebrate, the hagfish (*Myxine glutinosa*), the switch from fetal to adult hemoglobin has been studied in detail by Manwell (1963a). The fetal hemoglobin of this fish differs from the adult in its affinity for oxygen, physico-chemical properties, and amino acid composition. Manwell studied the comparative biochemistry and physiology of fetal and adult hemoglobins of the dogfish *Squalus suckleyi*, whose embryos develop for almost two years within the maternal organism, and throughout this period contain a special hemoglobin with high affinity for oxygen (Manwell, 1963b). The fetal hemoglobin of this fish consists of a special protein differing from ordinary hemoglobin in its amino acid composition and sequence (peptide analysis), although the extent of the differences between the HbF and HbA in this case was less than in the case of human HbF and HbA.

Comparative studies of fetal and adult hemoglobins have also been undertaken on several other aquatic animals, including the salamander *Amblystoma tigrinum* (Hahn, 1962), the trout *Salmo irideus* (Ostroumova, 1962), etc. Metamorphosis of the hemoglobins has been studied in particular detail in the frog (various species genus Rana) (Herner and Frieden, 1961; Chieffi et al., 1960; Trader et al., 1963; Trader and Frieden, 1966; Kurata and Arakawa, 1963; Namada et al., 1964). The system of hemoglobins in the tadpole and frog is more complex than in mammals. The tadpole contains four electrophoretically different fractions of hemoglobins. The frog also contains four, but these hemoglobins differ from each other biochemically. Baglioni and Sparks (1963) compared hemoglobin fractions of tadpoles and frogs by electrophoresis and peptide analysis ("fingerprints"). They found that even those fractions of tadpole hemoglobin which possess identical electrophoretic mobility with fractions of frog hemoglobin differ sharply in their chromatographic peptide spectrum.

In amphibians the switch from synthesis of fetal hemoglobins to synthesis of adult hemoglobins is unquestionably under the control of thyroxin, like other processes of metamorphosis (Moss and Ingram, 1965), although this takes place at the level of erythropoiesis and not at the level of young circulating erythrocytes, which continue to synthesize in accordance with their predetermined program.

A number of comparative studies have been made of fetal and adult hemoglobins of chickens and other domestic birds (Fraser, 1961; Wilt, 1962; Manwell et al., 1963).

4) **Genetics of the control system of morphogenetic switchover from HbF synthesis to HbA synthesis and its congenital anomalies.** At a certain moment of development, culminating in birth, the γ^F gene which controls synthesis of the γ chain of HbF in erythropoietic cells is switched off and the β^A gene, controlling synthesis of β chains, which interact with α chains to give adult HbA, is switched on. Neither the switching on of the β^A gene nor the switching off of the γ^F gene takes place suddenly, but they are spread over a period of time (several months in man) and consist essentially of a reduction in the activity of the γ^F gene and an increase in the activity of the β^A gene.

This process is a very convenient model for the study of the action of a genetic control system working at the level of individual genes. It is evident that ultimately most acts of morphogenesis and differentiation are brought about by the switching on of some genes and switching off of other genes in the required place at the required time, and in the required relationships, so that discovery of the molecular-genetic mechanism of switching the genes during synthesis of the different hemoglobins would be an important link in the investigation of general mechanisms of action of genetic systems programming development. All the facts, explanations, and hypotheses concerned with the mechanism of this switch must therefore be examined.

We must first consider a series of observations showing that, besides the α, β, γ and δ genes controlling synthesis of the corresponding polypeptide chains of human A, A_2 and F hemoglobins, there are also certain regulator genes (or a gene) controlling (switching on) the system of changing from synthesis of the γ chain of HbF to synthesis of the β chain of HbA. This conclusion, in accordance with classical genetic schemes of localization of mutations, is deduced from the fact that, as well as numerous mutations of the structural genes of the hemoglobins, leading to the appearance of abnormal hemoglobins, with substitutions and rearrangements of amino acids, there are other mutations disturbing the actual system responsible for switching synthesis from HbF to HbA, but having no effect on the structure of their polypeptide chains.

Two main types of genetic abnormalities are found in which the content of HbF in adult human blood is sharply increased: thalassemia, associated with anemia, and a condition known as "high HbF," in which no anemia is observed, oxygen exchange is undisturbed, and individuals possessing this mutation appear outwardly healthy.

In thalassemia (thalassemia major) erythrocytes do not contain normal concentrations of hemoglobin, the decrease being attributable to HbA. In this case HbF is synthesized in increased quantities and sometimes accounts for 80-90% of the total hemoglobin, yet it is insufficient to compensate for the inhibition of HbA synthesis and this leads to anemia, sometimes very severe.

The external manifestations of thalassemia differ sharply from the high HbF syndrome, and in genetic combinations these two characters behave extremely differently. It is therefore generally agreed that a "thalassemia gene" exists. In thalassemia, synthesis of the β chain is mainly affected (inhibited). There are many different variations of thalassemia, which I shall not examine. These variants of thalassemia, as well as many complex cases (combinations of thalassemia with other anomalies) are adequately described in a number of surveys (Chernoff, 1961; Baglioni, 1963; Jonxis, 1962).

The precise character of the genetic defect in thalassemia is not yet known, but the possibility is not ruled out that it may be complex in character, particularly if the multiplicity of forms of thalassemia is borne in mind. One hypothesis (Chernoff, 1961) postulates that the thalassemia factor is not directly connected with hemoglobin (β chain) synthesis but acts on the manifestation of β gene activity as a genetic repressor or disturbs the normal type of interaction between α and β chains. In heterozygosis at the thalassemia gene only half of the required number of β chains are formed, and the excess α chains in this case interact with γ chains, increasing the HbF concentration. In homozygosis at the thalassemia gene, HbA formation is almost completely disturbed and only a compensatory increase in HbF formation takes place, but this is insufficient to prevent anemia, and the animal is usually incapable of surviving for long and dies before the age of puberty. The HbF concentration also rises in compensation in many hemoglobinopathies (Beaven et al., 1961), so that the fundamental lesion in thalassemia is inhibition of synthesis of β chains and not the increase in HbF.

Ingram and Stretton (1959) have put forward another hypothesis of the genesis of thalassemia. They claim that the decrease in synthesis of the β chain of HbA is associated with a change in the structural gene which alters the amino acid sequence, without altering the electrophoretic mobility of the abnormal hemoglobin, thus creating difficulties in its separation from the normal α chain (in heterozygosis). However, no structural anomalies of the α chain have yet been discovered in any form of this disease.

The existence of selective inhibition of HbA synthesis in thalassemia was demonstrated by experiments to study comparative HbA and HbF synthesis by ribosomes of reticulocytes from patients with various forms of hemoglobinopathies (Burka and Marks, 1963). In fact, although the total concentration of ribosomes and polyribosomes in the reticulocytes is unchanged in thalassemia, the ability of these ribosomes to incorporate C^{14}-leucine was sharply reduced below the control level. Meanwhile, incorporation of isoleucine (present in HbF but absent from HbA) was increased in this case. Burka and Marks postulate that these anomalies are evidently associated with structural changes in messenger RNA for the α chain or with a decrease in the rate of synthesis of this mRNA.

A number of interesting schemes have been suggested to explain the genetic basis of anomalies such as thalassemia and high HbF, mainly using the concepts of regulator genes and operons as developed previously for bacteria and now applied for the first time to genetic analysis of biochemical mutations in mammals, namely in systems of hemoglobin synthesis.

The first attempts to apply the operon model to systems regulating hemoglobin synthesis were made by Neel (1961) and Wheeler and Krevans (1961a). These ideas were subsequently extended by Motulsky (1962) and Baglioni (1963). The hypothesis of the existence of genes regulating the switch from HbF to HbA was also formulated by Manwell and co-workers (1963) on the basis of studies of regulation of HbF and HbA synthesis in birds.

However, the most detailed theoretical analysis of genetic aspects of thalassemia and of the high HbF mutation was undertaken by Zuckerkandl (1964), who developed genetic models of these syndromes by using the concepts of regulator genes, the operon, operators, and so on suggested by Jacob and Monod, and hitherto used only for the analysis of genetic phenomena in bacteria and viruses.

Two possible hypothetical schemes (a and b) of hemoglobin operons are illustrated in Fig. 75 (Zuckerkandl, 1964).

In these schemes Zuckerkandl proposes the following variants of regulation:

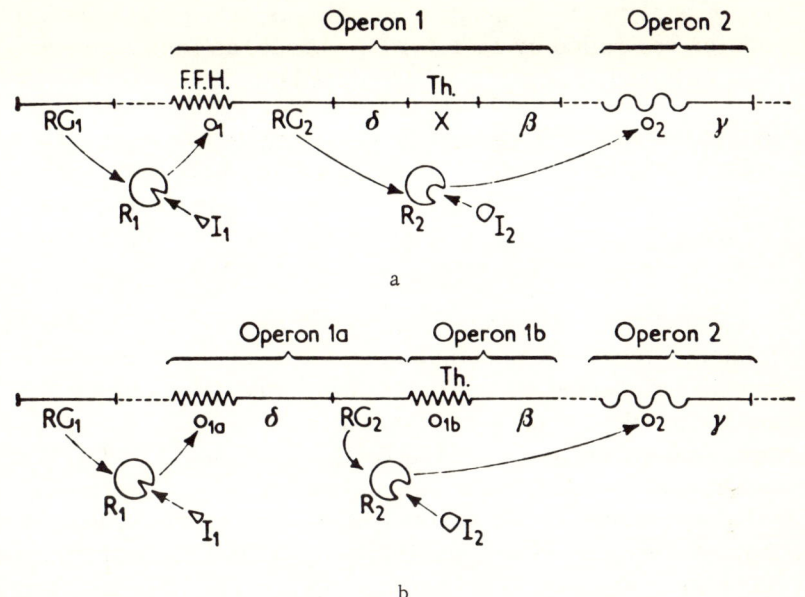

Fig. 75. Model of genetic regulation of synthesis of different hemoglobin chains (Zuckerkandl, 1964). Two variants: a and b. Explanation in text. Legend: O_1 and O_2 represent operators; RG_1 and RG_2 regulator genes; R repressor molecule synthesized under the control of RG; I_1 and I_2 inducers; β, γ, δ structural genes of hemoglobin chains; X) unknown structural gene; FFH zone responsible for mutation of familial fetal hemoglobinemia; Th zone possibly responsible for β-thalassemia mutation.

1. Genes of both β and δ chains are dependent on the operator O_1. The latter is controlled by regulator RG_1, the product of which, joining the operator O_1, stops synthetic activity of the entire operon 1. At a certain stage of development the inducer I_1 is formed in increased quantity and, joining the repressor R_1, synthesized by regulator gene RG_1, prevents the action of this repressor on operator O_1 and thus induces synthetic activity of the operon.

2. The gene of the γ chain is possibly associated with other unknown genes dependent on another operon O_2, located on the same chromosome in this scheme. Operator O_2 can join the product of regulator gene RG_2, i.e., with repressor R_2. When this union takes place, synthesis of messenger RNA for the γ

chain is stopped. Regulator gene RG_2 is incorporated into the operon controlled by operator O_1. Hence, as soon as operon 1 is switched on, operon 2 is switched off.

3. R_2, the product of regulator gene RG_2, joins inducer I_2, if this diffusive factor is present, and its action on operator O_2 is thus prevented. Operon 2 can thus be activated to a greater or lesser degree depending on the concentration of inducer I_2, even when operon 1 is also in a state of activity.

Zuckerkandl bases these theoretical models on a number of experimental facts. For example, genetic analysis of inheritance of a number of anomalies has in fact shown that genes of the β chains and of the γ and δ chains are closely linked on one chromosome (research in this direction is summarized in Zuckerkandl's paper). For these three genes (β, γ, δ), he postulates two operators, on the basis of facts showing that synthesis of the β and δ chains is usually correlated quantitatively and that synthesis of both chains is switched on at the same time in the period before birth.

The gene of the γ chain is also shown very close to the genes of the β and δ chains in Fig. 75, but even if this does not in fact take place, this does not seriously detract from the theory, because an independent regulator gene is assumed to be involved in the control of synthesis of the γ chain. This regulator gene RG_2 is also in close proximity to the genes of the β and δ chains. The spatial proximity of the regulator gene to the target gene is important for determination of the necessary number of "hits" on the target and, correspondingly, for the definite rate of synthesis of repressor molecules. If it is subsequently shown that structural genes for formation of γ chains and β chains (or δ chains) in fact segregate independently, in Zuckerkandl's opinion this will prove the existence of regulator gene RG_2, connected with the genes of the β and δ chains.

So far as the sequence of gene δ and regulator gene RG_2 and the order of action of operators O_{1a} and O_{1b} (Fig. 75b), and the localization of regulator gene RG_1 are concerned, the scheme presented must be regarded as speculative.

Analysis of the possible nature of regulator gene RG_2 (an attempt to associate its function with a known hemoglobin gene) led Zuckerkandl to conclude that regulator gene RG_2 is an independent locus of the genome, different from the structural hemoglobin genes hitherto known.

On the question of regulation of synthesis of the γ chains of fetal hemoglobin, Zuckerkandl postulates that this synthesis takes place under dual control: by the regulator gene and by a hypothetical inducer I_2, a diffusible factor. This follows from the ability of the organism to maintain increased production of fetal hemoglobin in certain blood diseases, from familial inconstancy of residual HbF synthesis, from correlation between the level of HbF synthesis and the severity of thalassemia (compensatory synthesis), and so on. Zuckerkandl accepts that physiological influences of this type on the HbF level may be effected through interaction between the product of the regulator gene and the inducer.

In Zuckerkandl's scheme, the system of coordinated switching on of the genes of the β and δ chains takes place as follows: operon 1, containing genes of the β and δ chains, is progressively activated during embryonic development in connection with the appearance of increasing quantities of inducer I_1. This inducer joins the repressor molecule synthesized under the control of regulator gene RG_1. Since regulator gene RG_2, which has for its target operon 2, is incorporated into operon 1, as a result of this the gradual and progressive activation of synthesis of β and δ chains correlates with inhibition of synthesis of the γ chains.

On the basis of this model Zuckerkandl drew up a table of possible mutations of structural and regulator genes, operators, inducers, and so on, included in the model, and gave a hypothetical description of the phenotypic manifestations of these mutations (mutations of I_1, RG_1, O_1, I_2, RG_2, and so on). Many of these mutations still await description, and could be quite unperceived in heterozygotes. However, some of the known mutations agree well with the theoretical predictions.

Two types of mutations of regulation of hemoglobin synthesis have received most study at the present time: hereditary thalassemia and the high HbF mutation.

From the genetic point of view, β-thalassemia is a complex hemoglobinopathy and cannot be an operator-negative mutation of operator O_1 (Fig. 75a). In the first place, activation of the gene of the γ chain only begins when formation of the β chains, when expressed per individual cell, falls to a small fraction of its earlier value. In addition, the gene of the delta chain is not inhibited and production of δ chains may also be increased. In Zuckerkandl's opinion, thalassemia also cannot be a mutation of regulator gene RG_1 or of the gene controlling inducer I_1, because inhibition of the gene of the β chain is found in the genetic cis-position, as has been shown in the case of heterozygotes and β-thalassemia and a structural anomaly of the β chain (the cis-position is meant when two independent mutations of one character are located in one of a pair of homologous chromosomes; the trans-position is meant when they are located in different chromosomes:

$\frac{MM_1 +}{+++}$ = cis; $\frac{M++}{+M_1+}$ = trans). This mutation in typical cases cannot therefore be identified as a mutation of the regulator or structural genes. To overcome this difficulty, but without altering the model of regulation, Zuckerkandl used the principle of polarity of the operon (Jacob and Monod, 1961), developed in connection with properties of the lactose operon of *Escherichia coli* (Jacob and Monod, 1961). These workers showed that the operon possesses polarity of a type in which mutations found in a zone lying proximally to the operator gene can reduce synthetic activity of structural genes located distally in the same operon. In some cases the reduction in synthetic activity reached 95%. When applying this principle to hemoglobin synthesis in β-thalassemia, Zuckerkandl postulated that the gene of the β chain, unlike the gene of the δ chain, lies distally to the operator O_1 and to the unknown gene X.

This arrangement is shown in Fig. 75a. The inactivating mutation of β-thalassemia in this case is located in the X gene. Depending on the nature of the mutation in the X gene, transcription of the β gene must be inhibited to a greater or lesser degree.

An explanation of a different type is shown in Fig. 75b. In this case it is assumed that the genes of the β chain and δ chain are controlled by two separate, but structurally identical, operators. In this case β-thalassemia may be explained by operator-negative mutation in the operator controlling the β gene.

Phenotypically, in accordance with the model we are discussing, the high HbF mutation may be the result of mutations of different loci, for example, mutations of genes controlling the quantity and quality of inducer I_1, constitutive-regulatory mutation of regulator gene RG_2, super-repressive regulator mutation of regulator gene RG_1, constitutive-operator mutation in operator O_2, and operator-negative mutation of operator O_1.

Besides what are described as molecular diseases (anomalies of structural genes) it is evident that diseases (mutations) of regulator genes (Zuckerkandl and Pauling, 1962), on which the rate of synthesis of other proteins lying at the basis of chemical organization of the cell depend, may evidently play an equally important role in evolution and variation.

Baglioni (1963, 1966) also has studied possible genetic mechanisms of mutant persistance of a high HbF concentration in adults (20-30% of the total blood hemoglobin in heterozygotes). He states that this mutation is linked with one gene and that its inheritance in families is Mendelian in type. Homozygotes with respect to this gene (a very rare phenomenon) contain only HbF (Wheeler and Krevans, 1961; Baglioni, 1963). Baglioni also considers that the high HbF mutation affects the system of genetic control of hemoglobins working under the operon principle rather than the regulator gene working as an operator gene. Besides differences in the composition of their hemoglobins, fetal and adult blood also show differences in the composition of their enzymes, the antigenic properties of the membranes, and so on. However, Baglioni (1966) has described a mutation in man in which 100% of the hemoglobin was in the fetal form, yet in all other properties (biochemical) the erythrocytes were of adult type.

Some very interesting information on localization of hemoglobin loci in human somatic chromosomes has recently been obtained in two studies of the hemoglobin pattern in certain chromosomal anomalies in man (trisomy) (Huehns et al., 1964; Powars et al., 1964).

Regarding the mechanisms of morphogenetic switching from HbF synthesis to HbA synthesis, it must not be forgotten that at different stages of embryonic development erythropoiesis is concentrated in different parts of the embryo. Cells containing hemoglobin appear first in the mesenchyme of the embryo; later the liver and spleen serve as foci of most active erythropoiesis (until birth). At birth the main center of erythrocyte formation is the bone marrow; after birth active erythropoiesis takes place only in the red marrow of the sternum and vertebrae (Maximov and Bloom, 1957).

However, no correlation has been established between morphological areas of erythropoiesis and the switch from HbF synthesis to HbA synthesis, and both HbF and HbA can be produced in the bone marrow and spleen (Betke, 1958; Thomas et al., 1960).

Differentiation of cells of the erythroid series is controlled by the hormone erythropoietin. Anemia stimulates erythropoietin formation, and this stimulates erythropoiesis. Erythropoietin also increases hemoglobin synthesis by stimulating synthesis of the corresponding messenger RNA (Krantz and Goldwasser, 1965). However, the possible link between erythropoietin and activation of adult hemoglobin (HbA and not HbF) has not yet been studied.

An interesting attempt has been made to determine the direct cause of the switch from HbF synthesis to HbA synthesis on the basis of morphogenetic principles and the level of differentiation in the course of erythropoiesis. These hypotheses are briefly summarized by Baglioni (1963, 1966). This group of hypotheses is based on the assumption that differentiation of cambial cells into

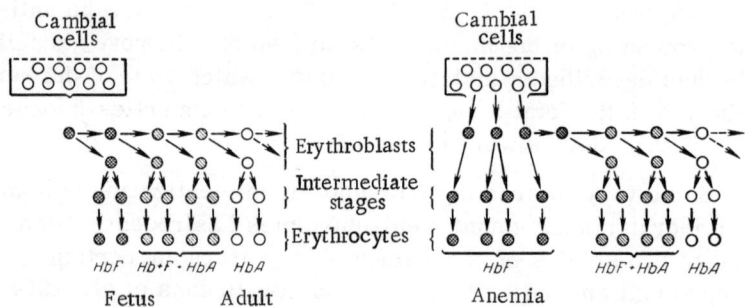

Fig. 76. Model of differentiation of erythrocytes. Switch from HbF synthesis to HbA synthesis at the end of gestation is shown on the left; mechanisms of HbF formation in severe anemia on the right. Density of shading reflects relative concentration of HbF synthesized by precursors of erythrocytes at different stages of differentiation or contained in erythrocytes (Baglioni, 1963).

mature erythrocytes may follow a different course in adults and fetuses.

By analogy with observations on exsanguinated animals, it can be assumed that in patients with anemia differentiation of cambial cells into mature erythrocytes is associated with a limited number of cell divisions, whereas in normal erythropoiesis the number of divisions required for the erythrocytes to reach maturity is considerable. According to one report, differentiation of cambial cells into erythrocytes sometimes takes place without division (Suit, 1957). Similarly, in the fetus, erythrocytes enter the blood stream after several divisions of their precursors, the number of these divisions being limited by the age of the fetus itself and, perhaps, by increased activity of its erythropoiesis. Hence, according to Baglioni (1963) a definite correlation exists between the number of divisions unaccompanied by morphological differentiation taking place in the erythrocyte precursors or, in other words, between the length of time during which these cells remain in the bone marrow as erythroblasts, and the type of hemoglobin found in the daughter erythrocytes. According to this hypothesis, erythrocytes formed from cambial cells as a result of several divisions may probably contain more HbF than erythrocytes formed from erythroblasts as a result of several divisions. Information concerned with synthesis of hemoglobin chains changes with ageing and (or) with multiplication of cells of the erythroid series. A possible scheme of this relationship is given in Fig. 76.

In another variation of this hypothesis (Burka and Marks, 1964) it was suggested that during erythropoiesis, with a gradual decrease in ability of the cells to synthesize hemoglobin and a decrease in the number of polyribosomes, the decrease in number of templates for synthesis of the various globins takes place at different speeds. In this case the increased HbF content in some cases may be associated, it is considered, with delay in destruction of the templates for synthesis of this hemoglobin and also, more important still, with the fact that differentiation of erythrocytes takes place from earlier erythroid cell forms. Burka and Marks support their views with data indicating that the ratio between rates of synthesis of hemoglobins A and F is lower than the ratio between the HbA and HbF concentrations in the circulating blood.

e) **Morphogenetic Changes in the Patterns of Enzymes and Isozymes in Tissues and Organs.** Two distinct trends can be detected in recent research into morphogenetic changes in enzymes in different animal species: the study of changes in the pattern and time of appearance of different types of enzymes and the study of changes in the pattern of subfractions of individual enzymes, i.e., of isozymes. The second trend in morphogenetic enzymology is very closely interwoven with the genetics of development, and we shall therefore examine the facts which have been obtained as a result of these two lines of study separately.

1) **Morphogenetic changes in spectra of different enzymes.** The study of changes in enzyme activity in connection with embryonic and postembryonic development of mammals, with metamorphosis in insects, with morphogenetic processes in amphibians, and so on have intensified in recent years. Although these findings have recently been discussed in a special survey (Moog, 1965) and in other publications (Medvedev, 1965; Kretchmer et al., 1963), it is necessary to draw a general picture of these phenomena in the present book because enzymes are the principal intermediate link between the genetic program, in the form of the DNA-nucleotide text, and its translation into concrete morphogenetic processes. Enzymes, with their varied patterns, constitute the "string" reproducing the musical nucleotide note.

As mentioned above, almost every cell of the multicellular organism contains two types of enzyme systems: a universal, including enzymes essential for all forms of specialization and responsible for universal life processes (glycolysis, oxidation, deamination, phosphorylation, and so on), and a special type, including enzymes responsible for strictly specialized functions of a particular organ (kidneys, liver, nerve tissue, and so on). This subdivision, however, is arbitrary because some enzymes of the universal group, such as ribonuclease, can be special for certain cells of the pancreas producing this enzyme in large quantities for secretion. In other cases a special enzyme may be adaptive, if the function with which its action is connected is periodic in character, and so on. Nevertheless, this general subdivision into groups is useful for classification and for providing an order of examination.

Enzymes of the universal group change their activity during morphogenesis, mainly by increasing or decreasing it depending on whether the general metabolism is increased or decreased in the particular differentiating tissue. The number of enzymes in this group is very large, the number of organisms for study is practically limitless, and it is therefore obvious that despite many investigations, the results so far obtained are still fragmentary in character.

In many investigations the intensity of respiration and glycolysis and the activity of individual oxidative and glycolytic enzymes have been studied in embryogenesis either in individual organs — in the developing rat's brain (Rappoport et al., 1963; Murthy and Rappoport, 1963; Pigareva, 1962), in the rat liver (Johnson et al., 1964), in the muscles of the cricket (Brosemer, 1965), and so on — and in total homogenates of embryos of different animals: fishes (Tatarskaya et al., 1958), amphibians (Wilde and Crawford, 1963; Kurata, 1962), silkworms (Nikitin and Morozova, 1956), and so on.

Examples of work in this direction could be multiplied many times over. It is very difficult to summarize the variegated pattern of this research. All that we can say is that some enzymes have unexpectedly high maxima of activity in early stages of development, while the activity of others, on the contrary, starts to increase gradually at the moment of birth. As an example we can show a graph of changes in activity of three enzymes in the early stages of development of the frog, which is taken from the interesting paper by Kurata (1962) (Fig. 77). All these enzymes were typical of early embryogenesis, evidence possibly of a type of biochemical recapitulation, for they characterize the early forms of development of life.

Another typical example of individual behavior of enzymes during embryonic development is shown in Fig. 78, demonstrating the dynamics of activity of four enzymes in the first week of development of the chick embryo (Moog, 1965).

The embryonic morphogenesis of alkaline and acid phosphatases in fish (Tatarskaya et al., 1958), in the rat kidney (Pinkstaff et al., 1962), the rat lung (Zawistowska et al., 1963), the mouse brain (Lierse, 1963), the chick brain (Lee et al., 1961), in insects (Chaudhary and Lemonde, 1964), etc; and of several other enzymes of phosphorus and nucleotide metabolism (Donath, 1962; Baker and

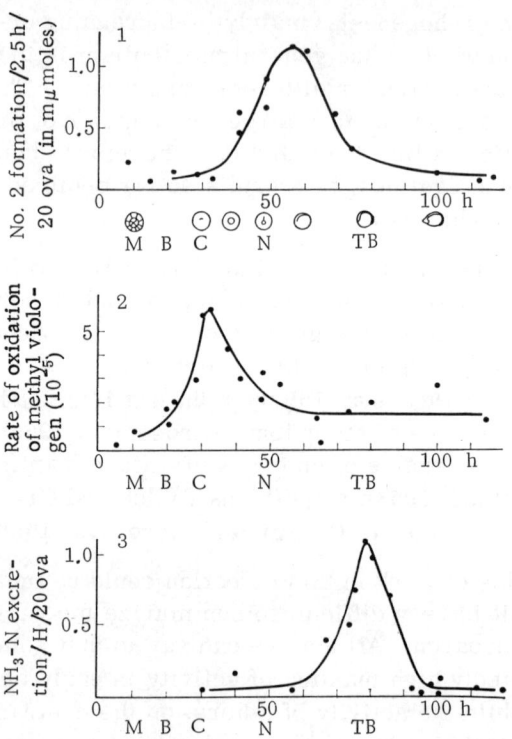

Fig. 77. Graph showing changes in activity of three enzymes at embryonic stages of development of the frog (Rhacophorus schlegeli). 1) Nitrate reductase activity during embryonic development of the frog: M, morula; B, blastula; G, gastrula; N, neurula, TB, tailbud; 2) hydrogenase activity during embryonic development of the frog; 3) aspartase activity during embryonic development of the frog (Kurata, 1962).

Newburgh, 1963) has been the subject of many investigations. In this case, too, the dynamics of enzyme activity varied widely depending on the organism, and on the phase of development but the character of cell differentiation also had a strong influence on enzyme activity.

The dynamics of activity of the enzymes concerned in amino acid activation (formation of aminoacyladenylates) has also been investigated (Maggio and Catalano, 1963; Hayashi et al., 1964; Norton et al., 1965). In this case enzyme activity was clearly correlated with intensity of protein synthesis and with growth of the embryonic organs. Conversely, desoxyribonuclease and ribonuclease activity was usually low in the earliest stages of development, increasing with differentiation almost until the stage of the adult organism (Finamore, 1955; Solomon, 1964; Waravdekar and Griffin, 1964).

Enzymes of the specialized group. Enzymes of this group usually appear at a certain stage of morphogenesis in connection with the formation of the organ or tissue (liver, kidney, brain, muscle, etc.) with whose function their activity is associated. In early stages of embryonic development, activity of these enzymes is either absent or very small. Later, in a definite place and at a definite time the corresponding gene or group of genes is evidently "switched on" and activity of the enzyme begins to increase, usually parallel to the degree of specialization and differentiation of the corresponding tissue.

Enzyme systems of the embryonic and postembryonic liver have been studied in particular detail from this aspect. Lea and

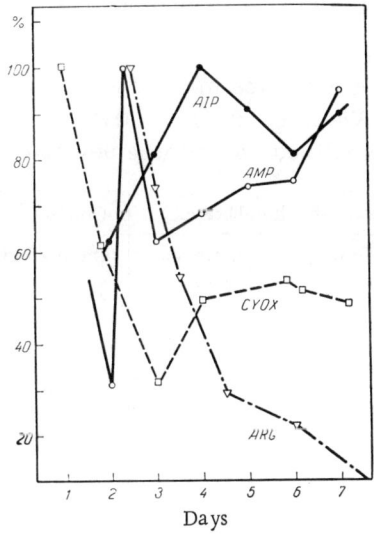

Fig. 78. Dynamics of activity of four enzymes in chick embryos during the first days of development. Activities shown as changes relative to maximal activity in this period. AIP alkaline phosphatase; AMP aminopeptidase; ARG arginase; CYOX cytochrome oxidase (Moog, 1965).

Walker (1964) determined the activity of six enzymes of carbohydrate metabolism (glucose-6-phosphatase, phosphoglucomutase, phosphoglucose-isomerase, glucose-6-phosphate dehydrogenase, etc.) in the liver and kidneys of the guinea pig at various phases of embryonic and postembryonic development. Each of these organs had a specific activity for each enzyme. Glucose-6-phosphatase activity is connected with glycogen metabolism, mainly typical of the postembryonic period. Intensive activity of this enzyme in the liver characteristically did not appear until shortly before birth, reaching a maximum rapidly during the first days after birth and then gradually decreasing (Table 1). Activity of glucose-6-phosphate dehydrogenase, as Table 1 shows, changed in an almost opposite manner.

The qualitative indices for glucose-6-phosphatase, it will be noted, did not vary with the phases of development or in different organs. Absence of glucose-6-phosphatase in the embryonic liver and its appearance shortly before birth were first demonstrated in 1954 (Nemeth, 1954). The activity of hepatic tyrosine aminotransferase showed changes of the same type in ontogenesis in

TABLE 1. Activity of Enzymes of the Developing and Adult Guinea Pig Liver (Lea and Walker, 1964). Activity Shown in Micromoles of Substrate Metabolized per Minute per 100 g Fresh Weight

Age of animals	Glucose-6-phosphatase		Glucose-6-PO_4-dehydrogenase	
	µmoles/min/g fresh weight	% of activity of adult tissue	µmoles/min/g fresh weight	% of activity of adult tissue
Embryos				
34-47 days	-	-	9.1	505
48-54 days	-	-	11.0	611
52-61 days	8.0	15	6.0	333
62-68 days	13.1	25	6.2	344
Newborn				
0-3 days	79.8	154	3.1	172
4-10 days	56.5	109	1.7	94
Adult				
4 months	51.9	100	1.8	100

CHANGES IN THE PROTEIN SPECTRUM OF CELLS AND TISSUES

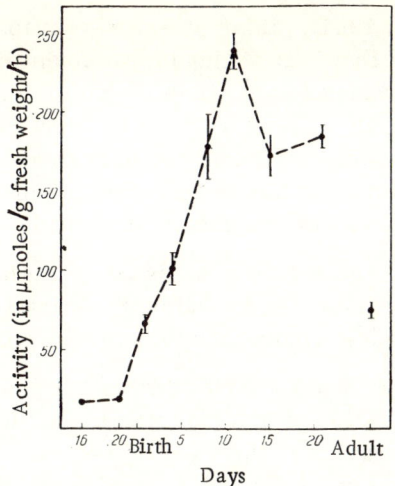

Fig. 79. Activity of fructose-1,6-diphosphatase in the rat liver after birth (Ballard and Oliver, 1962).

several animals (Litwack and Nemeth, 1965). This enzyme was absent in the liver in the period of embryonic development and appeared immediately after birth, reaching the adult level in the first 24 h of postembryonic life (in rabbits and guinea pigs). In this case the hormone hydrocortisone was responsible for initiating activity. Before birth, neither hydrocortisone nor the substrate, tyrosine, could induce formation of the enzyme, and it is thus evident that ability to undergo induction may be seen as a new characteristic of certain enzymes. Similar results were obtained in other investigations of the ontogenetic dynamics of this enzyme (Lin and Knox, 1958; Sereni et al., 1959).

Hence, starting at birth or shortly before birth, activity of fructose-1,6-diphosphatase (Ballard and Oliver, 1962) (Fig. 79), UDP-glucose-glycogen transglucosylase, phosphorylase, and UDP-glucose-pyrophosphorylase, and other enzymes of glycogen metabolism (Kornfield and Brown, 1963; Ballard and Oliver, 1963; Zheludkova, 1966), arginase (Eliasson, 1962, 1963), p-hydroxyphenylpyruvic acid oxydase (Goswami and Knox, 1961), phosphoenolpyruvate carboxylase (Il'in et al., 1966), detoxication enzymes (Brandt, 1964), glutamate-alanine transaminase (Ponomareva and Drel', 1964), and glucokinase (Walker and Holland, 1965) appears for the first time in the mammalian liver and increases rapidly. In every case the corresponding system of enzyme synthesis was "switched on" in accordance with the morphogenetic program, and not by the principle of substrate induction. Substrate could influence only an already active system.

The increase in activity consisted essentially of the synthesis of a new protein and it was inhibited by inhibitors of protein synthesis (puromycin, fluorophenylalanine, actinomycin).

Changes in enzyme activity during morphogenesis in connection with specialization of other organs takes place in approximately the same way as in the liver. The brain is the system which has received most study from this point of view. As an example we can point to the dynamics of acetylcholinesterase (Fig. 80), Burdick and Strittmatter, 1965). Activity of this enzyme clearly increases independently of the substrate concentration.

The various organs of an animal contain a large group of enzymes of inducible type, whose demonstrable activity is very

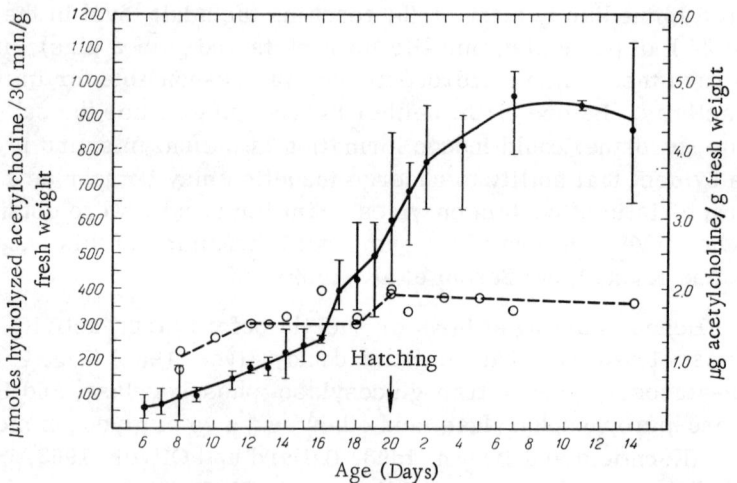

Fig. 80. Normal dynamics of acetylcholinesterase and acetylcholine in the brain of the chick embryo (Burdick and Strittmatter, 1965): ●—● acetylcholinesterase activity; O--O acetylcholine concentration.

strongly dependent on the presence and concentration of substrate. However, as we have seen above, ability to undergo adaptive changes appears along with genetically determined ability of the cells to synthesize this particular enzyme. The system of synthesis of tryptophan pyrrolase in the liver has been most thoroughly studied from this point of view. This enzyme in the liver of the adult animal is clearly activated and induced by substrate (Knox, 1951; Greengard and Feigelson, 1961; Feigelson and Greengard, 1962a, b). If, however, this substrate is introduced into the body at different periods of development (Nemeth, 1959), the property of substrate induction itself undergoes "morphogenesis." This investigation showed that injections of L-tryptophan had no action on the tryptophan pyrrolase activity of the liver in embryos and newborn animals, and induction began to appear. Moreover, it did so very suddenly, only one day after birth. The action of adrenocorticotropic hormones on activity of this enzyme also appeared only after birth.

This investigation thus showed not only that certain enzymes appear for the first time and are synthesized at certain phases of development, but also that inducibility of the enzyme also arises at certain phases.

Later work (Nemeth, 1962; Nemeth and Haba, 1962) showed that both the embryonic and the inductive formation of tryptophan pyrrolase is dependent on systems of synthesis of proteins and RNA (i.e., is not simple activation of the proenzyme → enzyme type) and is inhibited by inhibitors of synthesis of RNA and proteins. Neither the qualitative characters nor the homogeneity of the enzyme changed during ontogenesis and induction (Nemeth, 1961).

It is interesting to note that the ability of certain enzymes to undergo induced changes is not increased, but diminished, with the change from embryonic to postembryonic development. The enzyme apparently changes from the adaptive to the constitutive type. This was demonstrated particularly clearly for p- hydroxy phenylpyruvate oxidase (Goswami and Kox, 1961). In the embryonic liver about 70-80% of the enzyme is in a form capable of activation. In sexually mature animals only 10% of the enzyme is activated. In guinea pigs in the same cycle of development the percentage of activatable enzyme fell from 88 to 1.

Moog (1965), in his survey, gives a very thorough examination of recent evidence concerning the dynamics of different enzymes in connection with the morphogenesis of certain organs: liver, kidneys, intestinal mucous membrane, retina, heart, and so on. He showed not only that different organs possess an absolutely individual enzymologic profile at different periods of development, not only that each enzyme changes independently of the rest, but also that within the same group of functionally interconnected enzymes, each one has its own individual dynamics. Furthermore, even specialized organelles such as mitochondria have several different patterns of enzymes and their activities in different organs at different stages of development.

2) **Morphogenetic changes in isozyme groups and their significance in differentiation.** During recent years considerable attention has been paid in enzymology to complex forms of certain enzymes consisting of groups of "isozymes," each of which, while possessing identical specificity, differs from its partners in electrophoretic mobility and in certain other characteristics. In most cases these complex forms are enzymes whose molecules consist of subunits, each of which is determined by its own genetic locus.

Heterogeneity of this type in enzyme protein has been discovered during recent years in very many enzymes. Besides the case of lactate dehydrogenase, which has received the most study, similar heterogeneity has been established for alkaline phosphatase, glutamate dehydrogenase, succinate dehydrogenase, malate dehydrogenase, aspartate transaminase, amylase, tyrosinase, collagenase, leucylaminopeptidase, cholinesterase, acetylcholinesterase, transaminase, and many other enzymes. The problem of the distribution properties, structure, and functional role of isozyme groups has recently received a detailed examination in a special monograph (Wilkinson, 1965) and several surveys (Stadtman, 1963; Markert, 1963), so that we can concentrate our attention here mainly on the morphogenetic aspects of isozyme biochemistry. The isozyme group which has been most thoroughly studied from the biochemical and genetic points of view and also in relation to changes during development, is lactate dehydrogenase, an enzyme catalyzing the interconversion of pyruvic and lactic acids and occupying a key position in carbohydrate metabolism.

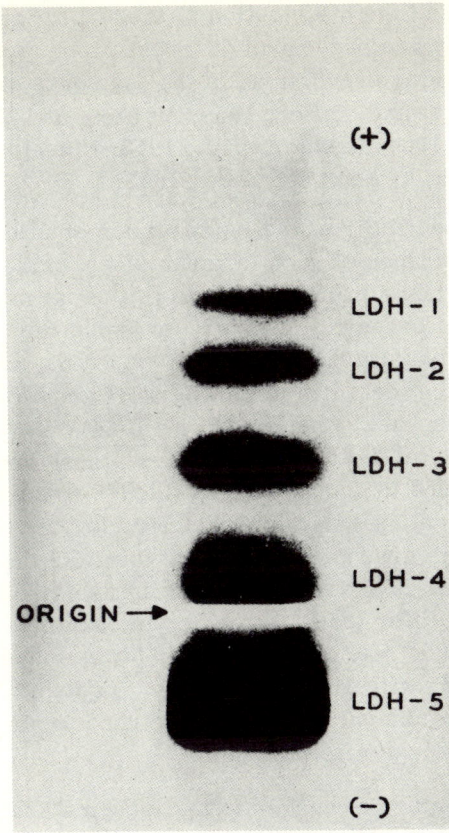

Fig. 81. Photograph of starch gel after electrophoresis of crystalline preparations of lactate dehydrogenase from bovine muscles (special reaction for enzyme LDH-1, LDH-2, etc., isozyme fractions) (Markert and Appella, 1963).

A typical spectrum of individual lactate dehydrogenases (LDH) of muscles after electrophoresis on starch gel is shown in Fig. 81 (Markert and Appella, 1963). These five fractions, as mentioned previously, are determined by the existence of two genetic loci for synthesis of two types of protein subunits, specific as regards amino acid composition and antigenicity, designated A and B, and by the presence of a tetramer structure of the principal protein (AAAA-LDH-5; AAAB-LDH-4; AABB-LDH-3; ABBB-LDH-2; BBBB-LDH-1).

The study of the distribution of these LDH subfractions showed that they are present in different proportions in different organs, and that the proportions of the isozymes are tissue-specific (Markert and Møller, 1959; Markert and Ursprung, 1962; Markert, 1963, 1964; Ressler et al., 1963; Nisselbaum and Bodansky, 1963a, b; Koen and Shaw, 1965).

Tissue specificity of the isozyme combinations naturally assumes the appearance of such differentiation in the period of embryogenesis. This served as the basis for a large series of investigations of changes in the isozyme spectrum in each organ at the different phases of morphogenesis. These changes were studied in greatest detail in mice (Markert and Møller, 1959; Markert and Ursprung, 1962; Markert, 1964). Despite the definite tissue specificity of the isozyme pattern, most embryonic mouse tissues were found to contain mainly LDH-5 (lying closest to the anode). As the animals developed, other isozymes appeared; the enzyme spectrum gradually broadened toward LDH-1 (nearest the cathode). Studies of the properties of the isolated isozymes have shown that they differ from each other in certain kinetic properties (different substrate optima). LDH-5 was found in the greatest amount in tissues with a poor O_2 supply. Lactate dehydrogenases located nearer to the cathode after electrophoresis were predominant in intensively aerated tissues.

The character of changes in the isozyme spectrum in a number of mouse organs is shown schematically (zymograms) in Fig. 82 (Markert and Ursprung, 1962), while the analogous data for the heart muscle, in the form of photographs of the starch gel after electrophoresis of enzyme isolated at different phases of development, are shown in Fig. 83.

These workers consider that ontogenetic changes of this type are due to successive changes in activity of genetic loci A and B, as a result of which different numbers of enzyme subunits are formed, thus spontaneously creating different stochastic relationships for the formation of tetramers, so that in their opinion the final distribution of the tetramer types is established epigenetically. *In vitro*, if LDH-1 and LDH-5 are mixed together in high concentrations of salts, spontaneous hybridization of subunits of each isozyme is observed (Vesell, 1965).

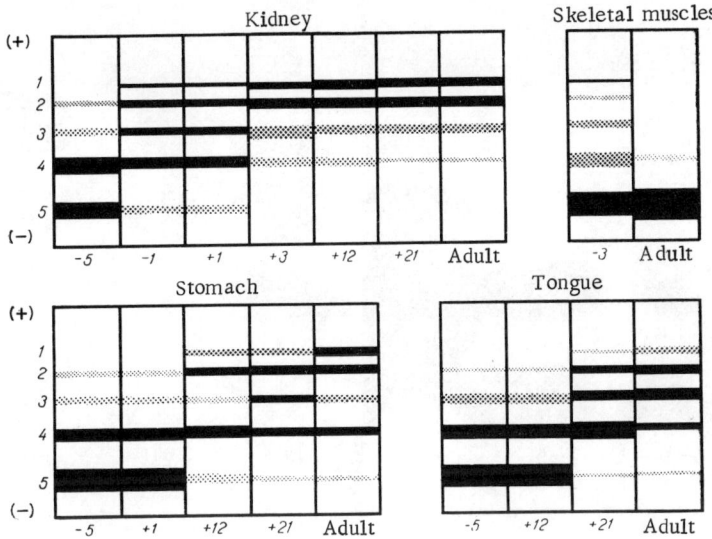

Fig. 82. Scheme (zymograms) of changes in lactate dehydrogenase isozymes during development of mouse organs. Numbers along abscissa indicate days before (-) and after (+) birth (Markert and Ursprung, 1962).

These morphogenetic changes are of definite functional significance, because the different types of LDH differ in their catalytic properties, especially as regards maintenance of a particular ratio between NAD and $NAD.H_2$, and they differ in their behavior toward optimal concentrations of substrates.

Basically similar changes in the isozyme pattern of lactate dehydrogenases during embryonic and postembryonic development have been obtained in many other studies of organs from rabbits (Dawson et al., 1964), rats (Fine et al., 1963), chickens (Cahn, 1964; Lindsay, 1963), sheep and cattle (Hinks and Masters, 1964), and many other animals and also man.

It is interesting to note that in some animals of evident hybrid origin, many more isozyme subfractions of lactate dehydrogenase are found. In the salamander (*Amblystoma gracile*) 11 or 12 isozymes were clearly identified on starch gel (Adams and Finnegan, 1965), and the same number of fractions was discovered in the brain of the crucian carp (*Carassius carassius*), the number

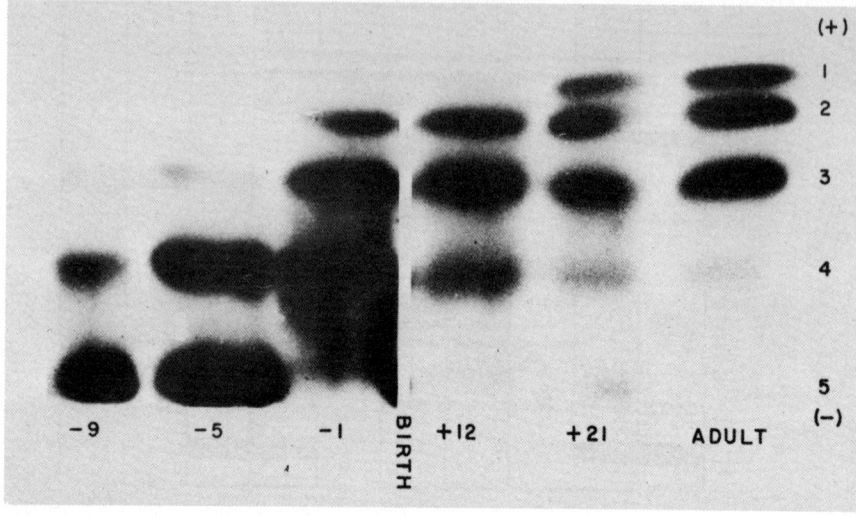

Fig. 83. Ontogenesis of lactate dehydrogenase isozymes of the mouse heart (starch gel electrophoresis). Numbers denote days before (-) and after (+) birth (Markert and Ursprung, 1962).

of subfractions being greater in animals kept in the cold than in those kept under warm conditions (Hochachka, 1965).

Following the investigation of lactate dehydrogenase, morphogenetic changes in the isozyme spectrum were studied in other isozyme groups: malate dehydrogenase (two subfractions in mammals and three in invertebrates) (Wiggert and Villee, 1964; Moore and Villee, 1963), dehydrogenases (Nakano and Whiteley, 1965), creatine kinase (three forms: BB, BM, MM) (Eppenberger et al., 1964), alkaline phosphatase of mice (Moog et al., 1966) and Drosophila (Schneiderman et al., 1966) and several other enzymes. Like lactate dehydrogenase, alkaline phosphatase shows distinct tissue specificity in its isozyme spectrum, and in *Drosophila* its activity in different organs is controlled by genes located in different chromosomes.

3) Regulation of enzyme activity in ontogenesis. Many mechanisms exist in biological systems for the activation and inhibition of activity of enzymes and enzyme cycles, for enzyme function is the most labile and dynamic component of the living system enabling constant adaptation to constantly changing

external and internal environmental conditions. It is this continuous fluctuation and balancing of enzyme activity which creates the level of constancy and stability of all kinds of biological structures and the level of equilibrium between opposing processes that are essential for the prolonged survival of individual systems.

Enzyme activity is controlled by changes in the conformation of the molecules, masking and unmasking of the active centers, the relative proportions of isozymes, substrate concentration, concentration of the end product, activators and inhibitors at the genetic level, activators and inhibitors at the functional level, hormones, coenzymes, fluctuations in pH and ionic strength of the medium, and many other factors. Of these other factors the most important are the velocity of synthesis and breakdown of enzymes, determining the enzyme concentration in the cytoplasm and other media of the organism.

The level of activity of enzyme synthesis and, consequently, the level of activity of the corresponding genes are mainly responsible for determining the changes in enzyme systems during embryonic development which we have just examined.

Many factors which influence enzyme activity at this period, such as embryonic hormones, inducers, and so on, do so in fact through their action on the DNA → RNA → enzyme chain, i.e., through activation of a particular gene. In normal morphogenesis this type of activation and repression can thus be regarded as elements of the strict program of development.

On the other hand, changes such as an increase in activity of some, perhaps many, enzymes results in a series of morphogenetic processes normally taking place when certain phases of development give way to others. For example, at a certain stage of morphogenesis, tadpoles excreting ammonia into the surrounding water switch over to the formation and excretion of urea, i.e., they form a new "terrestrial" method of detoxication of ammonia formed in the body. It has been shown (Brown et al., 1959) that the activity of enzymes of the urea cycle increases several times over at a certain phase of metamorphosis. The hormone thyroxin, which accelerates metamorphosis, increases the activity of these enzymes by 10-12 times, but only before metamorphosis. Immunochemical experiments have shown that the increase in enzyme activity is associated with stimulation of the synthesis of their proteins. If this synthesis is inhibited by thiouracil, even in the presence of thyroxin, metamorphosis is delayed (Paik and Cohen, 1961).

Another interesting phenomenon must be mentioned. Adrenocortisone stimulates alkaline phosphatase activity of the duodenum in mice during the first days after birth, before this enzyme shows a similar increase during normal development (on the 18th day) (Moog, 1962). However, although this degree of premature stimulation depends on the hormone concentration, the maximum which it reaches is never as high as that observed during normal development. Delay in certain phases of development caused by the action of inhibitors of protein and RNA synthesis together with inhibition of the synthesis of certain enzymes has also been demonstrated by other investigations (Greengard et al., 1963; Scarano et al., 1964).

These facts, and all the material we have previously examined on ontogenetic changes in the protein and enzyme spectrum, make it clear that the key to the understanding of the programming of biochemical development lies at the level of DNA-dependent (genetic) RNA synthesis.

f) The Immunology of Development. Changes in the protein and enzyme spectrum naturally modify the antigenic properties of the tissues at different periods of morphogenesis. The number of antigenic components of differentiating cells changes, and it changes differently in different forms of differentiation. This is quite obvious, and in this respect studies of the immunology of development have provided no new evidence to assist our understanding of mechanisms of morphogenesis. We shall not therefore examine the immunomorphogenetic material in detail. Some of the relevant facts have been described by Vyazov (1962).

All that need be said is that immunochemical methods as applied to the study of changes in the morphogenetic protein pattern have proved particularly useful in the case of determination of the patterns of protein components of the cell organelles, notably the ribosomes (D'Amelio et al., 1965; Okada and Yamamura, 1964; Okada and Sato, 1963).

§2. Morphogenetic Changes in the Pattern of Ribonucleic Acids Synthesized in the Nucleus

A new trend in biochemical embryology, the study of changes in the pattern of ribonucleic acids and, in particular, of messenger RNA, throughout the phases of embryogenesis, starting from the earliest, is of considerable interest. As is now evident, changes in the pattern of proteins synthesized by the cells in association with differentiation and morphogenesis are connected with changes in the pattern of templates of protein synthesis (molecules of mRNA) synthesized in the nucleus, i.e., they are a reflection of changes in the functional activity of chromosomes which we examined in the preceding chapter.

Messenger RNA molecules are products of the work of various genes, and it is therefore clear that any alteration of the program, any differentiation must be reflected primarily in the composition of the messenger RNA. Moreover, a variable length of time may elapse between changes in the RNA pattern determining the biochemical fate of the cell and the actual changes in its protein composition and functions. This can be observed particularly clearly during maturation of oocytes. During oogenesis, a special phase of development, an inactive store of mRNA molecules programmed for protein synthesis and essential for cleavage processes as far as the blastula stage accumulates. This phenomenon is now being studied intensively and it is worth while examining it separately from the processes of biochemical differentiation in the later period of embryogenesis.

a) Changes in the Pattern of RNA during Oogenesis and Early Embryogenesis. Although the individual development of every organism can be divided very precisely into the period before fertilization (maturation of the gametes) and the period after fertilization (true embryogenesis), changes which largely determine the character of primary differentiation of the embryo take place during maturation of the gametes which, in some cases, occupies as long a period as embryogenesis itself. This became clear many years ago, after it had been shown that during parthenogenesis the oocyte from which the nucleus has been removed can undergo cleavage and form a blastula. The primary program of its development is thus contained in the cytoplasm.

However, interest in changes in RNA during oogenesis and early embryogenesis became intensified with the interesting discoveries of Neifakh (1959), who formulated his results initially into the hypothesis of periodicity of morphogenetic activity of the nuclei in embryogenesis. This hypothesis and the results of Neifakh's experiments in this direction provide as a basis for many important biochemical investigations undertaken with the object of discovering at least the external biochemical manifestations of the morphogenetic programming function of the nucleus.

Neifakh carried out his initial investigations on the oocytes of the loach (*Misgurnus fossilis*) in the first phases of embryogenesis. He irradiated the oocytes and early embryos with large lethal doses of x-rays (20,000-60,000 R) at different periods of development and determined the time between irradiation and subsequent arrest of development, followed by death of the embryo. He found that this time varied with the period of irradiation, whereas arrest of development after irradiation at different times usually occurred at a particular phase of development, which depended on the time of preliminary irradiation. For example, irradiation of the embryo during cleavage or at various periods of the early blastula stage arrested development only before the beginning of gastrulation (6th hour of development), regardless of whether irradiation was given 1, 2, 3, 4, or 5 h previously. On the other hand, if irradiation was given in the blastula stage (6-8.5 h of development), the time when development stopped depended precisely on the moment of irradiation: the later during this period irradiation was given, the more retarded was the arrest of development in the period of gastrulation, and the later the manifestation of the lethal action of irradiation. After 8.5 h of embryogenesis this effect disappeared, and if irradiation was given in the late blastula or early or middle gastrula stages the embryo completed a normal gastrulation and the lethal action of irradiation did not appear until 18 h after the beginning of organogenesis. Irradiation in the late gastrula period (after 14 h of development) produced the same results as in the period between 6 and 8.5 h: the later the irradiation the longer organogenesis continued.

In subsequent experiments (Neifakh and Rass, 1960; Neifakh, 1961a, b; 1962) the same response to irradiation was found in the early embryogenesis of other animals. At the time when these experiments were carried out, the concepts of messenger RNA and

DNA-dependent RNA synthesis were still incompletely developed, although the localization of a considerable part of RNA synthesis in the nucleus and the morphogenetic role of the nucleus has been established, or particularly in experiments with transplantation of the nucleus under various experimental conditions.

Without specifying the precise biochemical basis of the phenomena observed, Neifakh suggested for their explanation the hypothesis of periodicity of the morphogenetic function of the nuclei, implying by the morphogenetic time function of the nucleus the transmission of biochemical information to the cytoplasm determining the direction of cell differentiation. According to Neifakh, in early embryogenesis there are certain periods during which the nuclei actively process morphogenetic information and supply it to the cytoplasm (for example, between 6 and 8.5 h of development and between 14 and 18 h). In between these periods the nucleus apparently makes preparation for a new cycle of activity and does not influence cytoplasmic processes taking place on the basis of the program it has already supplied. During these periods, therefore, lethal injury to the nucleus is not reflected, Neifakh postulates, in the character of differentiation until the exhaustion of certain morphogenetic products of the nucleus. In particular, until the late blastula stage development of the embryo proceeds on the basis of information already contained in the cytoplasm and accumulated there during oogenesis, i.e., before fertilization. Nuclear function comes into operation again only after the beginning of gastrulation.

It is well known to embryologists that the cytoplasm of the oocyte undergoes complex differentiation, in the form of a series of gradients and morphological structures, during oogenesis, and that this is precisely reflected in the early differentiation of the embryo. Extensive material on this subject has been summarized by Raven (1961) in his monograph. Neifakh's experiments, and his subsequent hypothesis, directed attention in this matter to the importance of certain products periodically liberated into the cytoplasm and changing its program.

As soon as methods of identification and study of the forms of messenger RNA synthesized by the nucleus and controlling cytoplasmic syntheses in accordance with the nuclear genetic program were available, attempts were naturally made to study Neifakh's model of nucleo-cytoplasmic relationships in development from the point of view of nuclear synthesis of messenger RNA. The first aim of these experiments was to determine whether synthesis of messenger RNA by the nucleus is also periodic and varies qualitatively at different periods of morphogenesis.

A series of important investigations in this direction was undertaken in Spirin's laboratory, in some cases with Neifakh's assistance (Belitsina et al., 1963a, b; Aitkhozhin et al., 1964; Spirin et al., 1964; Belitsina et al., 1964; Spirin and Belitsina, 1965; Spirin, Belitsina, and Aitkhozhin, 1965). These workers studied RNA synthesis by nuclei of the loach at various stages of early embryogenesis, under normal conditions and after irradiation, in accordance with the scheme used in Neifakh's experiments described above, using different methods for fractionating the RNA (ultracentrifugation in a density gradient with determination of molecular weight and of rate of incorporation of labeled precursors). In this way the RNA could be separated into sRNA, rRNA, and mRNA. These experiments clearly showed that synthesis of all forms of RNA takes place in the nucleus without fluctuation of activity both in the period of "morphogenetic activity" of the nucleus and in the period of "interruption of morphogenetic function." In other words, the periodicity of RNA synthesis in the nucleus demanded by Neifakh's hypothesis could not be found. Studies of fractions of total RNA and cytoplasmic RNA at different stages of development of the loach embryo showed that during Neifakh's period of "interruption of morphogenetic function of the nuclei" the pattern of distribution of labeled RNA fractions was qualitatively the same as in the "active" periods. The same RNA fractions were synthesized at all these stages, and the intensity of their synthesis simply grew quantitatively in the course of development, irrespective of any "periodicity in morphogenetic function of the nuclei."

It was further shown that lethal irradiation (50,000 R) during the investigated intervals of early embryogenesis did not stop the synthesis of different forms of RNA and had no effect on the molecular characteristics of the isolated nucleic acids. However, DNA synthesis in the nuclei stopped almost completely soon after irradiation. Continuation of synthesis of mRNA after these high doses of irradiation did not mean, of course, that this synthesis was normal from the biochemical and genetic points of view. As an explanation of the possible character of radiation damage to the nuclei in these experiments, Spirin, Belitsina, and Aitkhozhin (1965) assume that although such injury does not stop mRNA synthesis, it brings about certain disturbances of the genetic apparatus of the cells (chromosomes) and thus interferes with the regulation of its activity, causing disorganized depression of certain genes. Synthesis of mRNA begins soon after irradiation in the cells of the embryo on genes which, under normal conditions, would be inactive and this leads to synthesis of the wrong proteins for that stage of development and to disorganization of biochemical processes, resulting in arrest of development. These workers also, on the basis of their experimental results, suggested a new and apparently more objective explanation of the principles governing development which were revealed by Neifakh's experiments. In their opinion, during early development of the embryo periodicity is observed, not in the RNA-synthesizing activity of the nuclei, but in the "reprogramming" of the ribosomes in the cytoplasm. This means that at certain periods the bulk of the mRNA associated with ribosomes is stable and is not broken down, and newly synthesized mRNA, liberated into the cytoplasm, is not bound to the ribosomes. At other periods, however, "reprogramming" of the ribosomes takes place and the new mRNA binds itself to the ribosomes, while the old mRNA, attached to the ribosomes, starts to break down at a varying speed. From this point of view the period of cleavage and of the early blastula is a period of relative mRNA stability; the ribosomes are programmed throughout this period by maternal mRNA stored in the oocyte during oogenesis. In the period preceding gastrulation, the ribosomes are reprogrammed with new mRNA, synthesized by the nuclei of the embryo itself. Development naturally continues under normal conditions only when normal mRNA for the next stage of development is bound to the ribosome. If irradiation at any stage preceding the reprogramming led to the synthesis of the "wrong" mRNA, during reprogramming of the ribosomes these wrong mRNA molecules would bind themselves to the ribosomes, and soon after completion of the reprogramming period this would lead to the arrest of development. Conversely, so long as no reprogramming of the ribosomes took place, the presence of the "wrong" mRNA in the cytoplasm would have no effect on development.

The facts obtained by Neifakh concerning the action of radiation on development of the loach embryo clearly fit in much better with this explanation than with the original hypothesis of periodicity in morphogenetic activity of the nuclei. Neifakh himself has put forward a similar explanation of his early results (Neyfakh, 1964; Korolev and Neifakh, 1965).

The periodicity in this case is associated essentially with the length of functional service of the messenger RNA already attached to the ribosomes of the cytoplasm before fertilization. As soon as the store of templates for primary synthesis is exhausted, the ribosomes join with new fractions of mRNA entering the cytoplasm.

The process can take place in this way naturally only if no free ribosomes are formed in the period of early embryogenesis and no ribosomal RNA is synthesized, but the whole development of the early embryo can take place on the basis of ribosomes and ribosomal RNA accumulated during oogenesis.

That this is in fact true was demonstrated by one of the investigations already cited (Aitkhozhin et al., 1964) and also by the experiments of Timofeeva and Kafiani (1964) using the same model of early development of the loach. These workers showed that before the end of gastrulation synthesis of ribosomes and ribosomal RNA is absent in the oocytes of the loach, and accumulation of RNA directly in the embryo itself takes place as a result of transfer of ready-made ribosomes from the yolk formed before fertilization. Nuclei of the embryonic cells synthesize mainly only sRNA and mRNA in the early phases. However, the new (labeled) mRNA did not replace the mRNA of the cytoplasmic polysomes before the late blastula stage, and incorporation of activity into this fraction began after 8 h of development. New RNA synthesized by the nucleus, before transfer to the ribosomes, was linked to certain components which Spirin called "informosomes." He postulated that informosomes differ from ribosomes in the type of their ribonucleoprotein structures intended for transferring information from the nucleus to the cytoplasm. Under certain conditions they may exist for a time in a free state, unattached to ribosomes.

In later investigations Neifakh (1965) showed that the action of inhibitors of RNA and DNA synthesis on early embryogenesis, with some qualifications, obeys the same rule of a d i s p l a c e d effect as was discovered for the effect of radiation.

Of course, not all the conclusions deduced from investigations of cleavage of oocytes of the sea urchin, loach, and other aquatic animals can be extrapolated to mammals. The first cleavage division, which in the sea urchin and loach is insensitive to actinomycin and other inhibitors, was found to be extremely sensitive to them in the mouse (Thomson and Biggers, 1966).

Considerable accumulation of ribosomes during maturation of oocytes as a special type of reserve for early embryogenesis has been established by a number of investigations (Hultin, 1961; Nemer, 1962; Brown and Littna, 1964). However, these ribosomes possessed only weak protein-synthesizing activity. Activation of these ribosomes took place almost immediately after fertilization, and parthenogenetic stimulators, inducing cleavage, also caused a rapid increase in the protein-synthesizing power of the ribosomes (Hultin, 1961).

Later experiments provided the initial evidence to suggest that the synthetic passivity of the ribosomes of unfertilized oocytes could be due to absence of interaction between templates of protein synthesis and mRNA molecules, or to the presence of too few templates, as a result of which the ribosomes were "uncharged" and did not form polysomes (Nemer, 1962; Wilt and Hultin, 1962; Nemer and Bard, 1963; Hultin, 1964). When synthetic systems were created *in vitro* and mRNA molecules were replaced by synthetic polynucleotides, no significant differences were found between the ribosomes of fertilized and unfertilized oocytes. It was soon shown, however, that the messenger RNA synthesized during oogenesis can combine with ribosomes, but cannot exercise its functions because of special protein membranes, which evidently break up during activation of the oocytes as a result of increased protease activity (Monroy et al., 1965).

These results are in good agreement with those of other investigations showing that large reserves of templates of protein synthesis (messenger RNA molecules) are in fact formed before fertilization, but that a factor is present which prevents them from forming active complexes with ribosomes. This hypothesis was suggested by several workers (Gross, Malkin, and Moyer, 1964; Brachet et al., 1963; Maggio et al., 1964; Malkin et al., 1964; Gross and Cousineau, 1964; Slater and Spiegelman, 1966; Detlaf, 1966), showing that in unfertilized oocytes there is a reserve of messenger RNA adequate for the first phases of development and that this reserve provides for cleavage and blastula formation and for active protein synthesis during the first few hours after fertilization, even if the synthesis of new messenger RNA was blocked by actinomycin before fertilization. Mirsky and co-workers (Crippa, Davidson, and Mirsky, 1967) showed that the messenger RNA which accumulates during oogenesis in amphibians is not completely used up and can be identified as late as the middle blastula stage. After this there is a sudden change in the population of RNA molecules.

The nature of this inhibition of protein synthesis in unfertilized oocytes of the sea urchin has been studied in some interesting investigations (Glišin and Glišin, 1964). These workers studied the character of incorporation of P^{32} and labeled uridine into various forms of RNA (tRNA, mRNA, rRNA) in earlier and more finely subdivided periods after fertilization (the stage of 2, 4, 32, etc., blastomeres until the end of the blastula stage). These experiments showed that until the stage of four blastomeres, P^{32} incorporation was observed only into the tRNA of the amino acid activating system (sRNA), and only in the form of regeneration of the terminal acceptor pCpCpA group of this RNA. Incorporation of P^{32} into mRNA began at the phase of 32 cells, and no incorporation into ribosomal RNA occurred before the end of the blastula stage. These workers suggest that the inactivity of the terminal acceptor group of sRNA molecules transferring amino acids to ribosomes is responsible for the inhibition of synthesis in the oocyte, and the restoration (or completion) of these groups with a special enzyme is the trigger mechanism for synthetic processes after fertilization.

The nature of the inactive state of mRNA in the unfertilized oocyte was examined by Spirin (1966) in a special investigation. On the basis of a theoretical analysis of existing data he considers that messenger RNA is present in the oocyte in a masked or reserve form. In his opinion, the informosomes which he found in his investigations, being concerned with the transfer of mRNA from nucleus to cytoplasm, may contain an mRNA molecule in a form which is protected against nucleases by means of a special protein membrane. According to Spirin's hypothesis, the accumulation of messenger RNA during oogenesis to provide the programs for early embryogenesis takes place in the form of inactive informosomes of this type. Fertilization operates a certain mechanism of activation of the chromosomes and their liberation from the membranes (activation of proteases) in the same way as RNA molecules of the RNA viruses are freed from protein membranes inside the cells.

In a comprehensive survey of the results of analysis of an extensive series of investigations into the nature of the inactive state of RNA in the oocyte, Brachet (1966) also concludes that the messenger RNA of the oocyte is bound up in special capsules which are broken soon after fertilization. Inactive capsules of this type, with reserves of RNA, were recently discovered in unfertilized sea urchin oocytes (Stavy and Gross, 1967). Nemer (1963) showed that RNA is present in unfertilized sea urchin oocytes mainly as 17S and 22S fractions. Before the beginning of gastrulation the character of the RNA changed and a peak with sedimentation constant of about 10S began to predominate. Nemer suggests that this change is connected with intensive synthesis of messenger RNA. During further development the polydispersity of the RNA increased.

In later investigations (Spirin and Nemer, 1965; Nemer and Infante, 1965) forms of messenger RNA with different sedimentation constants were isolated from the sea urchin blastula and their link with the ribosomes was studied. In the period of blastula formation in the sea urchin, essentially only messenger RNA is formed (Comb, Katz, et al., 1965), and synthesis of sRNA and rRNA starts only with gastrulation. In the gastrulation period in Zenopus, synthesis of mRNA is strongly activated (Bachvarova ct al., 1966). Subsequent investigations (Glišin et al., 1966) showed that messenger RNA in the blastula stage of the sea urchin is indistinguishable in its profile from the messenger RNA of the unfertilized oocyte. In the period between the blastula and gastrula, about 40% of the mRNA disappears and is replaced by another mRNA typical of the more completely formed state of the embryo. Similar results were obtained by Whiteley and co-workers (1966). It is interesting to note that the appearance of enzymes of RNA methylation and the process of RNA methylation itself in sea urchin embryos does do not begin until late gastrulation (Comb, 1965).

b) Pattern of Ribonucleic Acids during Differentiation Processes. Appearance of Tissue Differences in RNA Composition. Variability of the

RNA fractions in later periods of embryonic development have not yet been adequately studied, and there is still very little accurate information regarding morphogenetic changes in RNA and tissue specificity of RNA. This is because of a number of technical and theoretical reasons. First, it is now obvious that soluble forms of sRNA with low molecular weight, responsible for the transport of activated amino acids to ribosomes, and high-molecular weight ribosomal ribonucleic acids do not and cannot possess tissue specificity because of their universal role, and even the species specificity of these forms of RNA is ill defined. However, these fractions of RNA in mammalian cells account for not less than 90-95% of the total cell RNA, and this naturally masks the morphogenetic changes in RNA affecting only the messenger RNA, a special and rapidly metabolizing fraction carrying genetic information associated with protein synthesis.

Second, messenger RNA was discovered in the cells of organs and tissues only in 1961, and its intensive study in connection with morphogenesis has therefore only recently begun, at least many years later than in relation to protein differentiation. Third, protein differentiation is much sharper than differentiation of the corresponding forms of RNA. Proteins have a complex, highly specific tertiary and quaternary structure which is not found in polynucleotides, they possess different functions, and consist of 20 amino acids. These properties facilitate their differentiation and fractionation. On the other hand, fractionation of RNA from this point of view is much more difficult. Fourth, and finally, the life span of messenger RNA molecules is usually much shorter than that of protein molecules synthesized on its basis. For all these reasons and many others, we can understand why the study of the morphogenetic pattern of the ribonucleic acids is much less advanced than that of the general picture described above for protein morphogenesis.

For a long time much of the work on morphogenetic changes in nucleic acids during the period of active morphogenesis, and especially in the embryonic period, was associated with the study of quantitative changes in RNA in cells of different specialization and at different periods of development. Much of this work carried out before 1959 was reviewed by Brachet (1960) in his monograph. General conclusions drawn from these investigations and also from others concerned with the effect of inhibitors of RNA synthesis (analogs of nucleotides, etc.) on morphogenetic processes were that RNA synthesis is connected with morphogenetic processes, that RNA is important for differentiation, and that a concentration gradient of RNA is present in the embryo.

There is no need to review these investigations in detail here, because research into the role of RNA in morphogenetic processes is now faced with other and more complex problems. All that need be said is that during recent years the study of the purely quantitative dynamics of RNA content and synthesis in relation to morphogenesis of different animals have of course continued, and in many cases, especially when purely chemical methods and general fractionation (into ribosomal, nuclear, and sRNA) were used, has extended our knowledge and enabled us to draw a better biochemical picture of differentiation (Zybina, 1963; Gluck and Kulovich, 1962; Wilt, 1964; Linzen and Wyatt, 1964; Finamore, 1964; Tocco et al., 1963; Moureaux, 1963; Duck-Chong et al., 1964).

In other investigations an attempt was made to detect changes in the nucleotide composition of RNA in connection with the phases of embryogenesis and tissue differentiation (Makhin'ko and Blok, 1961; Mazurov and Orekhovich, 1963; Nemeth and Dickerman, 1960; Burch and Von Dippe, 1964; Comb and Brown, 1964; Deuchar and Bristow, 1965). However, no precise principles were established, because determinations of the total nucleotide composition of RNA may fail to detect differences in the actual information carried on account of degeneracy of the code and because of the triplet (trinucleotide) alphabet used to record the information. Obviously, therefore, only a much finer and more functional fractionation of the ribonucleic acids can provide an accurate indication of the regulatory functions of the cell nucleus in morphogenesis.

The first work in this direction was published by Scholtissek, (1962) who studied the rapidly renewed nuclear RNA fraction (corresponding to the messenger RNA fraction) isolated from various organs (nuclei of the spleen, kidneys, and liver) and found that this RNA differs in type in different organs. The heterogeneity of this RNA fraction was determined from the distribution of labeled phosphorus between oligonucleotides and nucleotide products of incomplete RNA breakdown after injection of P^{32} into animals. Scholtissek accordingly suggests that the nuclei of different cells produce different types of messenger RNA in the cytoplasm.

Another investigation of these lines (Dingman and Sporn, 1962) showed that the microsomal RNA of the brain and liver of rats is uniform in type both during development and in the adult animals, whereas the nuclear RNA differs in these organs and these differences are increased in ontogenesis.

Further advances in the indirect study of morphogenetic differentiation of messenger RNA may be made possible by the study of morphogenetic reactions to inhibition of messenger RNA synthesis. Exposure of the early gastrula of *Rana pipiens* to actinomycin, followed by transfer into normal conditions, for example, delayed development of certain systems, while most organs and tissues of later or, conversely, of earlier specialization were normally formed (Flickinger, 1963). This demonstrates the existence of a time sequence for synthesis of different types of messenger RNA. More recently an attempt has been made to use tissue-specific forms of RNA to induce the onset of formation of the corresponding organs in embryos (Hillman and Niu, 1963).

Interesting work on the study of morphogenetic changes in the RNA pattern in organs of triton embryos (specificity of RNA composition during comparison of different organs and at different periods of embryogenesis) has been carried out by Waddington and Perkowska (1965). They determined the sedimentation profile of RNA and incorporation of tritium-labeled uridine into forms of RNA with different sedimentation constants during development of embryos. In both cases (tissue specificity and changes in time) the changes in RNA profile were very distinct and affected all sedimentation groups (from 2 to 50S). The changes in the region of messenger RNA are natural, but Waddington and Perkowska consider that the specificity which they found in the zones of sRNA and rRNA is unexpected.

In the late phases of embryogenesis there is also evidence of cytoplasmic determination of a number of differentiations by a similar mechanism of accumulation of long-living "masked" messenger RNA in the cytoplasm and not functioning immediately after its formation. This has been shown, in particular, by the work of Kirk (1965), who studied the connection between RNA synthesis and synthesis of glutamine synthetase in the embryonic retina of chick embryos *in vivo* and *in vitro*. Humphreys and co-workers (1964) discovered nonfunctional ribosomal aggregates in the skin of the chick embryo. These aggregates contained reserves of messenger RNA which did not function and were incorporated into protein synthesis later, in the period of keratinization. Intermediate, temporary differentiations are characterized by the formation of rapidly renewed, short-living forms of messenger RNA. However, as soon as the final specialization process, for example, synthesis of lens crystals or of hemoglobin by erythroblasts begins, long-living, stable forms of RNA appear. This gives stability to the final differentiation and quickly insures quantitative predominance of stable forms of RNA over all the rest. Ultimately only two types of stable RNA are formed in the initial cells of the lens of the eye (Reeder and Bell, 1967).

§ 3. Biochemical Composition of DNA during Morphogenesis and Embryonic Differentiation

The oocyte usually contains very large reserves of superfluous nongenetic DNA which accumulates in oogenesis mainly in the cytoplasm and is utilized as the nucleotide stock during cleavage. In addition, the immediate reserves of desoxyribonucleotides and ribonucleotides in the oocyte are very considerable. The content of reserve cytoplasmic DNA is about 99% of the total DNA of the oocyte, and the content of nuclear DNA of the oocyte is also greater than the stock of DNA in a diploid set of chromosomes. Determinations on different species of animals using different methods of analysis give different values, but in the sea urchin and in amphibians — classical organisms for investigations of this type — the DNA content in the mature oocyte varies within limits from 100 to 25,000 times higher than the DNA content in diploid cells of the same species (Hoff-Jørgensen and Zeuthen, 1952; Gregg and Løvtrup, 1960; Sze, 1953; Raven, 1961; Zbarskii and Mil'man, 1963; Haggis, 1964; and others).

However, this cytoplasmic DNA possesses a number of properties distinguishing it sharply from the genetic DNA of the chromosomes. The work of Zbarskii and Mil'man (1963) showed that this DNA from frog's eggs is polydispersed in type, and mainly of low molecular weight (about 50,000). The molecular weight of cytoplasmic DNA in the oocytes of the sturgeon is 800,000; (Shmerling, 1965). Dawid (Dawid, 1965) has recently studied the properties of this DNA isolated from oocytes of the frogs *R. pipiens* and *Xenopus laevis*. In its physico-chemical properties, this cytoplasmic DNA of the oocytes showed little difference from the DNA of the somatic cells of the same species (from liver cells and erythrocytes). However, the nucleotide sequence of this DNA, when determined by the method of hybridization on agar columns charged with somatic DNA, was found to be highly specific. Only about 5% of the sequences of the somatic and oocyte DNA were complementary, and this was about at the limit of accuracy of experiments of this type. RNA synthesized under the control of cytoplasmic DNA in experimental systems *in vitro* likewise was not complementary to somatic DNA. Clearly, therefore, changes in the DNA composition in the early period of embryogenesis are mainly associated with utilization of this reserve or metabolic DNA in metabolic processes and with changes in the ratio chromosomal DNA/cytoplasmic DNA in the oocyte. The function of the cytoplasmic DNA of the oocyte has still not been precisely established. Besides the widely held view of its role as reserve of ready-made desoxyribonucleotides for the rapid synthesis of chromosomal DNA during cleavage, information has been obtained (Dawid, 1966; Mickenthaler and Mahowald, 1966) to show that much of this DNA is of mitochondrial origin, and performs certain synthetic functions during oogenesis when the large stock of reserve materials is built up in the oocyte. However, the second hypothesis does not contradict the first. The work of Shmerling and Platova (1966) showed that cytoplasmic DNA in the oocytes of the sturgeon family is largely of nuclear origin. Similar results were obtained when they studied DNA in the ooplasm of Drosophila. However, this does not rule out the possibility of its mitochondrial localization.

Shmerling (1967) showed that cytoplasmic DNA in enucleated frog oocytes is used as templates for cytoplasmic RNA synthesis. She suggests that some of this DNA may consist of additional replicas of genes required for the early stages of embryogenesis.

So far as changes in DNA in connection with differentiation are concerned, it is clear that no substantial changes can be expected in the qualitative composition of DNA because the somatic cells retain their diploid set of chromosomes and the DNA content per diploid nucleus usually is unchanged during differentiation. The extensive data in the literature on this subject has been summarized by Mirsky and Osawa (1961).

The vast quantity of information concerning the quantitative changes in DNA in embryonic development (concentration in different organs, concentration gradients, DNA/RNA ratio, and so on) which has been published, although important for the general description of embryogenesis, is not essential for our purpose of understanding the mechanisms of differentiation, and I shall not therefore examine this material here.

Only one paper has been published during recent years in which the possibility of tissue differences in the composition of DNA fractions has been noted (Bendich et al., 1956). These workers studied the elution profile of DNA from chromatographic columns (this elution fractionates DNA as a series of peaks on diagrams of ultraviolet absorption of the eluates). However, Kondo and Osawa (1959), when they repeated this work, did not confirm the results showing tissue specificity of the composition of the DNA fractions. They found that even organs with such widely different functions as the kidneys and brain gave a virtually identical picture of the fraction separated by chromatographic fractionation. Kit (1960 a, b, c) confirmed the absence of tissue differences in DNA composition as a result of the study of six organs of mice and rats and their tumors. The character of distribution of the subfractions and the kinetics of denaturation of the DNA specimens were practically identical for all studied organs. Kok (1960) studied the nucleotide composition of DNA from various organs of the carp, ox, rat, and mouse. No appreciable differences were found in the nucleotide composition of the total DNA from different organs and tissues of the same animal.

However, it is not impossible that in certain forms of intensive specialization, definite differences may develop in the DNA also, particularly as regards its physico-chemical properties. For example, reports have been published indicating differences between DNA of lymphocytes and neutrophils in their ability to be hydrolyzed by desoxyribonuclease I and II (Gökcen, 1962).

A very interesting approach to the study of morphogenetic variation in DNA has recently been used by Flickinger and coworkers (1965). They studied the ability of DNA and chromatin isolated from frog embryos at various stages of development and from adult frogs (*R. pipiens*) to act as templates of RNA synthesis under the influence of RNA-polymerase (in the presence of four ribonucleoside-triphosphates and $MnCl_2$). The velocity of RNA synthesis in these systems was studied by the incorporation of labeled ribonucleoside-triphosphates into the synthesized RNA. If chromatin was used instead of pure DNA, weak RNA synthesis was found even without the addition of RNA-polymerase, but only on account of proteins contained in the chromatin.

In this case (chromatin without addition of RNA-polymerase), the template properties of the DNA were reduced during passage through the phases of embryogenesis (from gastrula to the stage of the independently feeding larva). Conversely, if RNA-polymerase was added to the system, the template properties of the chromatin increased during the phases of development. Chromatin preparations from late stages of development contained a high proportion of protein in the DNA-protein complex.

Chromatin from different parts of the embryo also differed in its activity. DNA preparations (containing protein), like chromatin, increased its template activity during the stages of development. Incorporation of labeled nucleoside into RNA in a system DNA + RNA-polymerase was as follows (mean values): in the case of DNA from the gastrula 13,934 cpm; neurula 21,692 cpm; tail bud stage 33,615 cpm; stage 25 (larva) 41,988; DNA from adult liver 65,850 cpm.

However, the discoveries resulting from this work are concerned mainly, not with morphogenetic changes in information carried by DNA, but changes in the opportunities for detecting this information as a result of changing forms of interaction between DNA and protein or of certain physical states of the DNA (for example, the number of zones of intertwining of the double helix).

In some types of specialization more profound morphogenetic changes in the nuclear system of the cells are possible. Three types of specialization are known, one of which, associated with a multinuclear state (in muscles, for example), does not in fact change the quality of the total genetic information in the cells. The best known types of differentiations with a change in total information are sex differentiation and erythrocyte formation in mammals. Sex differentiation is connected with the existence of the sex X- and Y-chromosomes. During reduction division in the period of spermatogenesis, the formation of haploid gametes is associated with separation of the X-and Y-chromosomes into different gametes.

Besides these chromosomes, differing in their size and shape, other sex-linked genes also enter these cells. Sex differentiation of the cells connected with genetic differences in character thus arises. Spermatozoa with the Y-chromosome determine male specialization of the organism in the progeny (XY-diploid set), while spermatozoa with the X-chromosome determine female specialization (XX-diploid set).

However, the appearance of genetic heterogeneity can also take place at later stages of ontogenesis. A typical case of this type is the formation of anuclear erythrocytes in mammals. When reticulocytes are formed, the cells lose their nucleus. Only the messengers carrying genetic information of DNA, i.e., mRNA molecules, remain in the reticulocytes and these act as templates for synthesis of hemoglobin and certain other special proteins. The total potential of the cells is lost, they lose their ability to divide, but genetic determination of synthesis of particular proteins remains as a result of the presence of gene replicas in the form of DNA-like (messenger) RNA.

The possibility is not ruled out that similar phenomena of partial or complete loss of total genetic potential may also be found in certain other processes of intensive specialization accompanied by complete loss of the power of mitotic division.

I am not concerned here with phenomena relating to somatic mutation of cells, the formation of chromosomal mosaics and chimeras, of tumors, and so on, or with other cases going beyond the bounds of normal development.

Literature Cited

Adams, E., and Finnegan, C. V., 1965, J. Exp. Zool., 158:241.
Aitkhozhin, M. A., Belitsina, N. V., and Spirin, A. S., 1964, Biokhimiya, 29:169.
Aizawa, K., Kaboyashi, M., and Abe, F., 1960, J. Sericult. Sci., Japan, 29:197.
Alekseenko, L. P., 1966, In: Molecular Basis of Pathology, ed. V. N. Orekhovich, Izd. Meditsina, Moscow, p. 42.
Ames, B. N., and Hartman, Ph. E., 1963, Cold Spring Harbor Symp. Quant. Biol., 28:349.
Arfors, K. E., Beckman, L., and Lundin, L. G., 1964, Acta Genet. Statist. Med., 13:89.
Ashton, G. C., 1965, Genetics, 51:431.
Askoy, M., and Lehmann, H., 1957, Nature, 179:1248.
Atwater, J., Schwartz, I. R., and Tocantins, L. M., 1960, Blood, 15:901.

Bachvarova, R., Davidson, E. H., Allfrey, V. G., and Mirsky, A. E., 1966, Proc. Nat. Acad. Sci., USA, 55:358.
Baglioni, C., 1963a. In: Molecular Genetics, ed. J. H. Taylor, Academic Press, New York.
Baglioni, C., 1963b, Nature, 198:1177.
Baglioni, C., 1966, J. Cell Comp. Physiol., 67, Suppl. I:169.
Baglioni, C., Ingram, V. M., and Sullivan, E., 1961, Nature, 189:467.
Baglioni, C., and Sparks, Ch. E., 1963, Develop. Biol., 6:272.
Baker, W. W., and Newburgh, R. W., 1963, Biochem. J., 89:510.
Ballard, F. J., and Oliver, I. T., 1962, Nature, 195:498.
Ballard, F. J., and Oliver, I. T., 1963, Biochim. Biophys. Acta, 71:578.
Beaven, G. H., Ellis, M. J., and White, J. C., 1961, Brit. J. Haematol., 7:169.
Beckman, L., and Holmgren, G., 1963, Acta Genet. (Basel), 13:361.
Belitsina, N. V., Gavrilova, L. P., Aitkhozhin, M. A., Neifakh, A. A., and Spirin, A. S., 1963a, Dokl. Akad. Nauk SSSR, 153:464.
Belitsina, N. V., Gavrilova, L. P., Neifakh, A. A., and Spirin, A. S., 1963b, Dokl. Akad. Nauk SSSR, 153:1204.
Belitsina, N. V., Aitkhozhin, M. A., Gavrilova, L. P., and Spirin, A. S., 1964, Biokhimiya, 29:363.
Bendich, A., Pahl, H. B., and Beiser, S. M., 1956, Cold Spring Harbor Symp. Quant. Biol., 21:31.
Betke, K., 1958, Schweiz. Med. Wochschr., 88:1005.
Blanco, A., Zinkham, W. H., and Kupchyk, L., 1964, J. Exp. Zool., 156:137.
Boyer, S. H., Porter, I. H., and Weilbacher, R. G., 1962, Proc. Nat. Acad. Sci., USA, 48:1868.
Brachet, J., 1960, The Biochemistry of Development, Pergamon Press, London—New York.
Brachet, J., 1966, Zh. Obshch. Biol., 27:523.
Brachet, J., Decroly, M., Ficq, A., and Quertier, J., 1963, Biochim. Biophys. Acta, 72:660.
Brandt, I. K., 1964, Develop. Biol., 10:202.
Braunitzer, G., 1966, J. Cell. Comp. Physiol., 67, Suppl. 1, 1.
Brosemer, R. W., 1965, Biochim. Biophys. Acta, 96:61.
Brown, G. W., Brown, W. R., and Cohen, P. P., 1959, J. Biol. Chem., 234:1775.
Brown, D. D., and Littna, E., 1964, J. Mol. Biol., 6:688.
Buettner-Janusch, J., 1962, Ann. N.Y. Acad. Sci., 102:235.
Burdick, C. J., and Strittmatter, M. E., 1965, Arch. Biochem. Biophys., 109:293.
Burch, H. B., and Von Dippe, P., 1964, J. Biol. Chem., 239:1898.
Burka, E. R., and Marks, P. A., 1963, Nature, 199:706.
Burka, E. R., and Marks, P. A., 1964, Nature, 204:659.
Cahn, R. D., 1964, Develop. Biol., 9:327.
Chaudhary, K. D., and Lemonde, A., 1964, Experientia, 20:86.
Chernoff, A. I., 1961, Am. J. Human Genet., 13:151.
Chieffi, G., Siniscalo, M., and Adinoifi, M., 1960, Acad. Naz. Lincei, 28:233.
Coleman, J. R., Herrmann, H., and Bess, B., 1965, J. Cell Biol., 25:69.
Comb, D. G., 1965, J. Mol. Biol., 11:851.

LITERATURE CITED

Comb, D. G., and Brown, R., 1964, Exp. Cell Res., 34:360.
Comb, D. G., Katz, S., Branda, R., and Pinzino, C. J., 1965, J. Mol. Biol., 14:195.
Craig, M. L., and Russell, E. S., 1964, Develop. Biol., 10:191.
Crippa, M., Davidson, E. H., and Mirsky, A. E., 1967, Proc. Nat. Acad. Sci., USA, 57:885.
Crockett, J. R., Koger, M., and Chapman, H. L., 1963, 22:173.
D'Amelio, V., Mutolo, V., and Piazza, E., 1963, Exp. Cell Res., 31:499.
Davidson, R. G., Nitowsky, H. M., and Childs, B., 1963, Proc. Nat. Acad. Sci., USA, 50:481.
Dawid, J. B., 1965, J. Mol. Biol., 12:581.
Dawid, J. B., 1966, Proc. Nat. Acad. Sci., USA, 56:269.
Dawson, D. M., Goodfriend, T. L., and Kaplan, N. O., 1964, Science, 143:929.
Detlaf, T. A., 1966, Zh. Obshch. Biol., 27:401.
Deuchar, E. M., and Bristow, D. A., 1965, Nature, 205:1321.
Dingman, W., and Sporn, M., 1962, Biochim. Biophys. Acta, 61:164.
Donath, R., 1962, Naturwissenschaften, 49:609.
Dorfman, V. A., 1958, Uspekhi Sovr. Biol., 45:313.
Duck-Chong, C., Pollak, J. K., and North, R. J., 1964, J. Cell Biol., 20:25.
Dustin, J. P., Shapira, G., Dreyfus, J. C., and Hestermans-Medard, O., 1954, Compt. Rend. Soc. Biol., 148:1207.
Edds, M. V., 1958, Proc. Nat. Acad. Sci., USA, 44:296.
Efroimson, V. P., 1961, In: Problems in Cybernetics, ed. A. A. Lyapunov, No. 6, Izd. Fiz.-Mat. Lit., p. 161.
Eliasson, E., 1962, Exp. Cell Res., 26:175.
Eliasson, E., 1963, Exp. Cell Res. 30:74.
Eppenberger, H. M., Eppenberger, M., Richterich, R., and Aebi, H., 1964, Develop. Biol., 10:1.
Escobar, M. A., Heller, P., and Trobaugh, F. E., 1964, Arch. Intern. Med., 113:428.
Feigelson, P., and Greengard, O., 1962a, J. Biol. Chem., 237:1908.
Feigelson, P., and Greengard, O., 1962b, J. Biol. Chem., 237:3714.
Finamore, F. J., 1965, Exp. Cell Res., 8:533.
Finamore, F. J., 1964, J. Biol. Chem., 239:1882.
Fine, I. H., Kaplan, N. O., and Kuftinec, D., 1963, Biochemistry, 2:116.
Flickinger, R., 1963, Science, 141:608.
Flickinger, R. A., Coward, S. J., Mivagi, J., Moser, C., and Rollins, E., 1965, Proc. Nat. Acad. Sci., USA, 53:783.
Fraser, R. C., 1961, Exp. Cell Res., 25:418.
Glišin, V. R., and Glišin, M. V., 1964, Proc. Nat. Acad. Sci., USA, 52:1548.
Glišin, V. R., Glišin, M. V., and Doty, P., 1966, Proc. Nat. Acad. Sci., USA, 56:285.
Gluck, L., and Kulovich, M. V., 1962, Science, 138:530.
Gökcen, M., 1962, Proc. Nat. Acad. Sci., USA, 48:978.
Goswami, M. N. D., and Knox, W. E., 1961, Biochim. Biophys. Acta, 50:35.
Gratzer, W. D., and Allison, A. C., 1960, Biol. Rev., 35:459.
Greengard, O., and Feigelson, P., 1961, J. Biol. Chem., 236:158.
Greengard, O., Smith, M. A., and Asc, G., 1963, J. Biol. Chem., 238:1548.

Gregg, J. R., and Lovtrup, S., 1960, Exp. Cell Res., 19:621.
Gross, P. R., and Cousineau, G. H., 1964, Exp. Cell Res., 33:368.
Gross, P. R., Malkin, L. I., and Moyer, W. A., 1964, Proc. Nat. Acad. Sci., USA, 51:407.
Haggis, A. J., 1964, Develop. Biol., 10:358.
Hahn, W. E., 1962, Comp. Biochem. Physiol., 7:55.
Hamada, K., Sakai, Y., Shukuya, R., and Kaziro, K., 1964, J. Biochem., 55:636.
Hartshorne, D. J., and Perry, S. V., 1962, Biochem. J., 85:171.
Hayashi, Y., Ishihama, A., and Kondo-Mizuno, N., Exp. Cell Res., 35:142.
Herner, A. E., and Frieden, E., 1960, J. Biol. Chem., 235:2845.
Herner, A. E., and Frieden, E., 1961, Arch. Biochem. Biophys., 95:25.
Herrmann, H., 1963, In: Cytodifferentiation and Macromolecular Synthesis, ed. M. Locke, Academic Press, New York, p. 85.
Hillman, N. W., and Niu, M. C., 1963, Proc. Nat. Acad. Sci., USA, 50:486.
Hinks, M., and Masters, C. J., 1964, Biochemistry, 3:1789.
Hochachka, P. W., 1965, Arch. Biochem. Biophys., 111:96.
Hoff-Jørgensen, E., and Zeuthen, E., 1952, Nature, 169:245.
Hopkinson, D. A., Spencer, N., and Harris, H., 1963, Nature, 199:969.
Hosoi, T., 1965, Exp. Cell Res., 37:680.
Huehns, E. R., Hecht, F., Keil, J. V., and Motulsky, A. G., 1964, Proc. Nat. Acad. Sci., USA, 51:89.
Huisman, T. H. J., 1954, Arch. Intern. Physiol., 62:564.
Hultin, T., 1961, Exp. Cell Res., 25:405.
Hultin, T., 1964, Exp. Cell Res., 10:305.
Humphreys, T., Penman, S., and Bell, E., 1964, Biochem. Biophys. Res. Comm., 17:618.
Hunt, J. A., 1959, Nature, 183:1373.
Hunt, J. A., and Ingram, V. M., 1958, Nature, 181:1062.
Hutton, J. J., Bishop, J., Schweet, R., and Russell, E. S., 1962, Proc. Nat. Acad. Sci., USA, 48:1718.
In'il, V. S., Usatenko, M. S., and Evstratova, L. A., 1966, Zh. Evol. Biokhim. Fiziol., 2:185.
Ingram, V. M., 1959, Nature, 183:1795.
Ingram, V. M., 1961, Hemoglobin and Its Abnormaties, C. C Thomas, Springfield.
Ingram, V. M., 1963, The Hemoglobins in Genetics and Evolution, Columbia Univ. Press, New York.
Ingram, V. M., and Stretton, A. O. W., 1959, Nature, 184:1903.
Itano, H. A., 1957, Advances in Protein Chem., 12:215.
Itano, H. A., 1963, Proc. Symp. "Abnormal Haemoglobins" (Ibadan, Nigeria), Black-. well, Oxford.
Itano, H. A., and Robinson, E. A., 1960, Proc. Nat. Acad. Sci., USA, 46:1492.
Ivanov, I. I., Yur'ev, V. A., Kadykov, V. V., Krymskaya, B. M., Moiseev, V. P., and Tukhachinskii, S. E., 1956a, Biokhimiya, 21:612.
Ivanov, I. I., Yur'ev, V. A., Kadykov, V. V., Krymskaya, B. M., Moiseev, V. P., and Tukhachinskii, S. E., 1956b, Dokl. Akad. Nauk SSSR, 111:649.
Jacob, F., and Monod, J., 1961, Cold Spring Harbor Symp. Quant. Biol., 26:193.

Javid, J., 1964, Proc. Nat. Acad. Sci., USA, 52:663.
Johnson, B. E., Walsh, D. A., and Sallach, H. J., 1964, Biochim. Biophys. Acta, 85:202.
Jonxis, J. H. P., 1962, In: Erbliche Stoffwechselkrankheiten (Herausgegeben von F. Linnewch), Urban und Schwarzenberg, Munich-Berlin, p. 49.
Jonxis, H. P., 1963, Ann. Rev. Med., 14:297.
Kadykov, V. V., 1963, Vopr. Med. Khim., 9:311.
Kasavina, B. S., 1950, In: Problems in Medical Chemistry, Vol. 2, Izd. Akad. Med. Nauk. SSSR, p. 165.
Kasavina, B. S., 1954, Trudy Vses. Obshch. Fiziolog. Biokhimik. Farmakolog., 2:151.
Kasavina, B. S., and Umanskaya, M. V., 1958, Dokl. Akad. Nauk SSSR, 118:340.
Keech, M. K., and Reed, R., 1957, Ann. Rheumatic Diseases, 16:35.
Kirk, D. L., 1965, Proc. Nat. Acad. Sci., USA, 54:1345.
Kit, S., 1960a, Biochem. Biophys. Res. Comm., 3:361.
Kit, S., 1960b, J. Biol. Chem., 235:1756.
Kit, S., 1960c, Arch. Biochem. Biophys., 88:1.
Knox, W. E., 1951, Brit. J. Exp. Pathol., 32:462.
Koen, A. L., and Shaw, C. R., 1965, Biochim. Biophys. Acta, 96:231.
Kok, I. P., 1960, Dokl. Akad. Nauk SSSR, 133:1216.
Kondo, M., and Osawa, S., 1959, Nature, 183:1602.
Kornfeld, R., and Brown, D. H., 1963, J. Biol. Chem., 238:1604.
Korolev, M. B., and Neifakh, A. A., 1965, Zh. Obshch. Biol., 26:352.
Korzhuev, P. A., 1964, Hemoglobin: Comparative Physiology and Biochemistry, Izd. Nauka, Moscow.
Korzhuev, P. A., 1966, In: Current Problems in Age Physiology and Biochemistry, ed. V. N. Nikitin, Izd. Meditsina, Moscow, p 205.
Krantz, S. B., and Goldwasser, E., 1965, Biochim. Biophys. Acta, 103:325.
Kraus, A. P., and Neely, C. L., 1964, Science, 145:595.
Kretchmer, N., Greenberg, R. E., and Sereni, F., 1963, Ann. Rev. Med., 14:407.
Kurata, Y., 1962, Exp. Cell Res., 28:424.
Kurata, Y., and Arakawa, W., 1963, Blut., 9:42.
Lai, L., Nevo, S., and Steinberg, A. G., 1964, Science, 145:1187.
Laufer, H., 1960, Ann. N. Y. Acad. Sci., 89:490.
Laufer, H., 1964, In: Taxonomic Biochemistry and Serology, ed. C. A. Leone, Ronald Press, New York, p. 171.
Lea, M. A., and Walker, D. G., 1964, Biochem. J., 91:417.
Lee, R. H., Angeletti, P. U., and Caramia, F. G., 1961, Growth, 25:393.
Lehmann, H., 1960, In: The Metabolic Basis of Inherited Disease, ed. J. B. Stanbury, et al., McGraw-Hill, New York, p. 1086.
Lewis, H. W., and Lewis, H. S., 1963, Ann. N. Y. Acad. Sci., 100:(11) 827.
Lierse, W., 1963, Z. Mikrosk.-Anat. Forsch., 70:48.
Lin, E. C. C., and Knox, W. E., 1958, J. Biol. Chem., 233:1186.
Lindsay, D., 1963, J. Exp. Zool., 152:75.
Linzen, B., and Wyatt, G. R., 1964, Biochim. Biophys. Acta, 87:188.
Litwack, G., and Nemeth, A. M., 1965, Arch. Biochem. Biophys., 109:316.
Maggio, R., and Catalano, C., 1963, Arch. Biochem. Biophys., 103:164.

Maggio, R., Vittorelli, M. L., Rinaldi, A. M., and Monroy, A., 1964, Biochem. Biophys. Res. Comm., 15:436.
Makhin'ko, V. I., and Blok, L. N., 1961, Biokhimiya, 26:993.
Malkin, L. I., Gross, P. R., and Romanoff, P., 1964, Develop. Biol., 10:378.
Malt, R. A., and Bell, E., 1965, Nature, 205:1081.
Manwell, C., 1963a, In: The Biology of Myxine, ed A. Brodal and R. Fange, Universitets—Porlaget, Oslo, Norway, p. 372.
Manwell, C., 1963b, Arch. Biochem. Biophys., 101:504.
Manwell, C., 1964, In: Oxygen in the Animal Organism, Symp., Pergamon Press, Oxford, p. 49.
Manwell, C., Baker, A., and Childers, W., 1963, Comp. Biochem. Physiol., 10:103.
Manwell, C., Baker, C. M., Roslansky, J. D., and Foght, V., 1963b, Proc. Nat. Acad. Sci., USA, 49:496.
Markert, C. L., 1963, In: Cytodifferentiation and Macromolecular Synthesis, ed. M. Locke, Academic Press, New York, p. 65.
Markert, C. L., 1964, In: Second Internat. Conference on Congenital Malformation, Int. Med. Congress Ltd., New York, p. 163.
Markert, C. L., and Appela, E., 1963, Ann. N. Y. Acad. Sci., 103:915.
Markert, C. L., and Møller, F., 1959, Proc. Nat. Acad. Sci., USA, 45:753.
Markert, C. L., and Ursprung, H., 1962, Develop. Biol., 5:363.
Marks, P. A., 1966, in: Current Topics in Developmental Biology, Vol. 1, ed. A. A. Moscona and A. Monroy, Academic Press, New York.
Matioli, G., Brody, S., and Thorell, B., 1962, Acta Haematol., 28:73.
Maximov, A. A., and Bloom, W., 1957, Text Book of Histology, Seventh Edition, Saunders, Philadelphia.
Mazurov, V. I., and Orekhovich, V. N., 1963, Vopr. Med. Khim., 9:434.
Medvedev, Zh. A., 1963, Protein Biosynthesis and Problems in Ontogenesis, Moscow.
Medvedev, Zh. A., 1965a, Uspekhi Sovr. Biol., 59:333.
Medvedev, Zh. A., 1965b, In: Problems in Medical Genetics, ed. D. A. Biryukov, Izd. Meditsina, Leningrad, p. 45.
Minnich, V., Cordonnier, J. K., Williams, W. J., and Moore, C. V., 1962, Blood, 19:137.
Mirsky, A., and Osawa, S., 1961, In: The Cell, ed. J. Brachet and A. Mirsky, Academic Press, New York—London, p. 677.
Mohler, D. N., and Crockett, C., 1964, Blood, 23:427.
Monroy, A., Maggio, R., and Rinaldi, A. M., 1965, Proc. Nat. Acad. Sci., USA, 54:107.
Moog, F., 1962, Federation Proc., 21:51.
Moog, F., 1965, In: The Biochemistry of Animal Development, Vol. I, ed. B. Weder, Academic Press, New York, p. 307.
Moog, F., Vire, H. R., and Grey, 1966, Biochim. Biophys. Acta, 113:336.
Moore, R. O., and Villee, C. A., 1963, Science, 142:389.
Moureaux, T., 1963, Ann. Biol. Anim. Biochim. Biophys., 3:33.
Moss, B., and Ingram, V. M., 1965, Proc. Nat. Acad. Sci., USA, 54:967.
Motulsky, A. G., 1962, Nature, 194:607.
Muckenthaler, F. A., and Mahowald, A. P., 1966, J. Cell Biol., 28:199.

LITERATURE CITED

Murthy, M. R. V., and Rappoport, D. A., 1963, Biochim. Biophys. Acta, 74:51, 328.
Mutolo, V., D Amelio, V., and Piazza, E., 1965, Exp. Cell Res., 37:597.
Nakano, E., and Whiteley, A. H., 1965, J. Exp. Zool., 159:167.
Nass, M. M. K., 1962, Develop. Biol., 4:289.
Neel, J. V., 1956, Ann. Human Genet., 21:1.
Neel, J. V., 1961, Blood, 18:769.
Neifakh, A. A., 1959, Zh. Obshch. Biol., 20:202.
Neifakh, A. A., 1961a, Zh. Obshch. Biol., 22:42.
Neifakh, A. A., 1961b, Dokl. Akad. Nauk SSSR, 136:1248.
Neifakh, A. A., 1962, Problems in Relationships between Nucleus and Cytoplasm in Development, Inst. Morphol. Akad. Nauk SSSR, Moscow.
Neifakh, A. A., 1965, In: Cell Differentiation and Mechanisms of Induction, ed. G. V. Lopashov et al., Izd. Nauka, Moscow, p. 38.
Neifakh, A. A., and Rass, I. G., 1960, Dokl. Akad. Nauk SSSR, 135:1557.
Nemer, M., 1962, Biochem. Biophys. Res. Comm., 8:511.
Nemer, M., and Bard, S. G., 1963, Science, 140:664.
Nemer, M., and Infante, A. A., 1965, Science, 150:217.
Nemeth, A. M., 1954, J. Biol. Chem., 208:773.
Nemeth, A. M., 1959, J. Biol. Chem., 234:2921.
Nemeth, A. M., 1961, Biochim. Biophys. Acta, 48:189.
Nemeth, A. M., 1962, J. Biol. Chem., 237:3703.
Nemeth, A. M., and Haba, G. de la, 1962, J. Biol. Chem., 237:1190.
Nemeth, A. M., and Dickerman, H., 1960, J. Biol. Chem., 235:1761.
Neyfakh (Neifakh), A. A., 1964, Nature, 201:880.
Nikitin, V. N., and Morozova, V. F., 1956, Transactions of the Research Institute of Biology, Khar'kov University, Vol. 24, p. 153.
Nisselbaum, J. S., and Bodansky, O., 1963a, Ann. N.Y. Acad. Sci., 103:930.
Nisselbaum, J. S., and Bodansky, O., 1963b, J. Biol. Chem., 238:969.
Norton, S. J., Key, M. D., and Scholes, S. W., 1965, Arch. Biochem. Biophys., 109:7.
Ohno, S., Poole, J., and Gustavsson, I., 1965, Science, 150:1737.
Okada, T. S., and Sato, A. G., 1963, Exp. Cell Res., 31:251.
Okada, T. S., and Yamamura, H., 1964, Embryologia, 8:115.
Orlovskaya, G. V., Zaides, A. L., and Tustanovskii, A. A., 1956, Dokl. Akad. Nauk SSSR, 111:1396.
Osawa, Y., 1958, Nature, 182:1312.
Ostroumova, I. N., 1962, Dokl. Akad. Nauk SSSR, 147:263.
Paigen, K., 1959, J. Histochem. Cytochem., 7:248.
Paigen, K., 1961, Proc. Nat. Acad. Sci., USA, 47:1641.
Paik, W. K., and Cohen, P. P., 1961, J. Biol. Chem., 236:531.
Pantelouris, A. M., and Hale, P. A., 1962, Nature, 195:79.
Pauling, L., Itano, H. A., Singer, S. J., and Wells, I. C., 1949, Science 110:543.
Pels, S. F., 1959, Exp. Cell Res. Suppl., 6:97.
Pigareva, Z. D., 1962, Acta Univ. Carolin. Med., 8:523.
Pinkstaff, C. A., Sandler, M., and Bourne, G., 1962, J. Gerontol., 17:267.
Pinaev, G. P., 1965, Biokhimiya, 30:20.
Poglazov, B. F., 1965, Structure and Functions of Contractile Proteins, Izd. Nauka, Moscow.

Ponomareva, T. F., and Drel', K. A., 1964, Biokhimiya, 29:185.
Popp, R. A., 1962, In: Symp. on Methodology in Mammalian Genetics, ed. W. Burdette, Holden-Day Inc., San Francisco.
Popp, R. A., 1962a, J. Heredity, 53:73.
Popp, R. A., 1962b, J. Heredity, 53:142.
Popp, R. A., 1963, Science, 140:893.
Popp, R. A., 1965, Federated Proc., 24:1252.
Powars, D., Rohde, R., and Graves, D., 1964, Lancet, p. 1363.
Prockop, D. J., Peterkofsky, B., and Udenfriend, S., 1962. J. Biol. Chem., 237:1581.
Rall, D. P., Schwab, P., and Zubrod, Ch. G., 1961, Science, 133:279.
Ranney, H. M., 1954, J. Clin. Invest., 33:1634.
Rappoport, D. A., Fritz, R. R., and Moraczewski, A., 1963, Biochim. Biophys. Acta, 74:42.
Raven, C. P., 1961, Oogenesis, The Storage of Developmental Information, Pergamon Press, New York and London.
Reeder, R., and Bell, E., 1967, J. Mol. Biol., 23:577.
Robson, E. B., and Harris, H., 1965, Nature, 207:1257.
Ressler, N., Schulz, J. L., and Joseph, R., 1963, Nature, 197:872.
Roche, J., Derrien, Y., and Rogues, M., 1953, Bull. Soc. Chim. Biol., 35:933.
Robinson, D. S., 1952, Biochem. J., 52:633.
Rucknagel, D. L., and Neel, J. V., 1961, In: Progress in Medical Genetics, ed. A. G. Steinberg, Vol. I, Grune and Stretton, New York–London, p. 158.
Schneiderman, H., Young, W., and Childs, B., 1966, Science, 151:461.
Scholtissek, C., 1962, Nature, 194:353.
Schroeder, W. A., Jones, R. T., Shelton, J. R., Shelton, J. B., Cormick, J., and McCalla, K., 1961, Proc. Nat. Acad. Sci., USA, 47:811.
Schroeder, W. A., Shelton, J. R., Shelton, J. B., and Cormick, J., 1963a, Biochemistry, 2:1353.
Schroeder, W. A., Shelton, J. R., Shelton, J. B., Cormick, J., and Jones, R. T., 1963b, Biochemistry, 2:992.
Scarano, E., Petrocellis, B., and Augusti-Tocco, G., 1964, Biochim. Biophys. Acta. 87:174.
Sereni, F., Kenney, F. T., and Kretchmer, N., 1959, J. Biol. Chem., 234:609.
Shaw, C., 1965, Science, 149:936.
Shaw, C. R., and Barto, E., 1963, Proc. Nat. Acad. Sci., USA, 50:211.
Shaw, C. R., and Barto, E., 1965, Science, 148:1099.
Shepard, M. K., Weatherall, D. L., and Lockard, C. C., 1962, Bull. Johns Hopkins Hospital, 110:293.
Shmerling, Z. G., 1965, Biokhimiya, 30:113.
Shmerling, Z. G., 1967, Molekul. Biol., 1:83.
Shmerling, Z. G., and Platova, T. P., 1966, In: Structure, Properties, and Genetic Functions of DNA, I. V. Kurchatov Institute of Atomic Energy, Moscow, p. 215.
Shmerling, Z. G., and Uspenskaya, V. D., 1955, Biokhimiya, 20:31.
Siakotos, A. N., 1960, J. Gen. Physiol., 43:1015.
Sick, K., and Nielson, T., 1964, Hereditas, 51:292.
Slater, D. W., and Spiegelman, S., 1966, Proc. Nat. Acad. Sci., USA, 56:164.
Solomon, J. B., 1964, Nature, 201:618.

LITERATURE CITED

Solomon, J. B., 1965, In: The Biochemistry of Animal Development, Vol, 1, ed. R. Weber, Academic Press, New York.
Spirin, A. S., 1966, Zh. Evol. Biokhim. Fiziol., 2:285.
Spirin, A. S., and Belitsina, N. V., 1965, Usp. Sovr. Biol., 59:187.
Spirin, A. S., Belitsina, N. V., and Aitkhozhin, M. A., 1964, Zh. Obshch. Biol., 25:321.
Spirin, A. S., Belitsina, N. V., and Aitkhozhin, M. A., 1965, In: Cell Differentiation and Mechanisms of Induction, ed. G. V. Lopashov et al., Izd. Nauka, Moscow, p. 18.
Spirin, A. S., and Nemer, M., 1965, Science, 150:214.
Stadtman, E. R., 1963, Bacteriol. Rev., 27:170.
Stavy, L., and Gross, P. R., 1967, Proc. Nat. Acad. Sci., USA, 57:735.
Strekalov, A. A., 1967, Genetika, No. 7.
Suit, H. C., 1957, J. Clin. Pathol., 10:267.
Sze, L. C., 1953, J. Exp. Zool., 122:577.
Tashian, R. E., Plato, C. C., and Shows, T. B., 1963, Science, 140:53.
Tatarinov, Yu. S., Afanas'ev, A. V., and Parfenova, L. F., 1963, Vopr. Med. Khim., 9:403.
Tatarskaya, R. I., Kafiani, K. K., and Kanopkaite, S. I., 1958, Biokhimiya, 23:527.
Thomas, E. D., Lochte, H. L., Greenough, W. B., and Wales, M., 1960, Nature, 185:396.
Thomson, J. L., and Biggers, J. D., 1966, Exp. Cell Res., 41:411.
Timofeeva, M. Y., and Kafiani, K. A., 1964, Biokhimiya, 29:110.
Tocco, G., Orengo, A., and Scarano, E., 1963, Exp. Cell Res., 31:52.
Trader, C. D., and Frieden, E., 1966, J. Biol. Chem., 241:357.
Trader, C. D., Wortham, J. S., and Frieden, E., 1963, Science, 139:918.
Vesell, E. S., 1965, Proc. Nat. Acad. Sci., USA, 54:111.
Villafranca, G. W., 1954, J. Exp. Zool., 127:367.
Vyazov, O. E., 1962, The Immunology of Development, Medgiz, Moscow, p. 327.
Waddington, C. H., and Perkowska, E., 1965, Nature, 207:1244.
Walker, D. G., and Holland, G., 1965, Biochem. J., 97:845.
Walter, M., 1963, Science, 141:123.
Waravdekar, V. S., and Griffin, C. C., 1964, Exp. Cell Res., 33:450.
Weber, R., ed., 1965, The Biochemistry of Animal Development, Vol. I, Academic Press, New York.
Weller, E. M., and Schechtman, A. M., 1962, Develop. Biol., 4:517.
Wheeler, J. T., and Krevans, J. R., 1961a, Clin. Res., 9:168.
Wheeler, J. T., and Krevans, J. R., 1961b, Bull. Johns Hopkins Hospital, 109:217
Whiteley, A. H., McCarthy, B. J., and Whiteley, H. R., 1966, Proc. Nat. Acad. Sci., USA, 55:519.
Wiggert, B. O., and Villee, C. A., 1964, J. Biol. Chem., 239:444.
Wilde, C., and Crawford, R., 1963, Develop. Biol., 7:578.
Wilkinson, J. H., 1965, Isoenzymes, 168 pp., Spon.
Wilt, F., 1962, Proc. Nat. Acad. Sci., USA, 48:1582.
Wilt, F. H., 1964, Develop. Biol., 9:299.
Wilt, F. H., and Hultin, T., 1962, Biochem. Biophys. Res. Comm., 9:313.

Winnick, T., and Goldwasser, R., 1961, Exp. Cell Res., 25:428.
Woods, K. R., Yudowitz, B. S., Engle, R. L., 1961, Proc. Soc. Exp. Biol. Med., 107:616.
Zawistowska, H., Zawistowski, S., and Kedzia, H., 1963, Folia Morphol., 22:49.
Zbarskii, I. B., and Mil'man, L. S., Zh. Obshch. Biol., 24:380.
Zheludkova, Z. P., 1966, Zh. Evol. Biokhim. Fiziol., 2:193.
Zinkham, W. H., Blanco, A., and Kupchyk, L., 1964, Science, 144:1353.
Zipursky, E., Neelands, P. J., Janet, P., Bruce, Ch. and Israels, Z., 1962, Pediatrics, 30:252.
Zuckerkandl, E., 1964, J. Mol. Biol., 8:128.
Zuckerkandl, E., and Pauling, L., 1962, in: Horizons in Biochemistry, ed. M. Kasha and B. Pullman, Academic Press, New York, p. 189.
Zybina, E. V., 1963, in: Collected Transactions of the Institute of Cytology, Academy of Sciences of the USSR, No. 5, Izd. Akad. Nauk SSSR, p. 34.

Chapter 6

Chromosomal Proteins and Their Role in Regulation of Selective Gene Activity during Differentiation and in Specialized Cells

Introduction

 The material dealing with functional activity of chromosomes and changes in the spectrum of proteins and nucleic acids during morphogenesis, described in the previous chapters, have revealed the pattern of purposive differentiated activity of the genes in ontogenesis and has indicated the external manifestation of this activity in the form of synthesis of essential products at a definite time and in a definite place.

 However, we have not yet examined the logically inescapable problem of which biochemical system is responsible for this differentiated activity of particular genes selectively picked out from the total specific complement of genes which is the same for all or nearly all cells, i.e., what concrete conditions determine the active or inactive state of genes in the chromosomes. In bacteria, as we saw in our examination of the material in Chapter 3, the change from an active to a passive state of the genes is associated theoretically with the appearance of repression and with the action of repressors, substances which are still largely hypothetical. The need for the change from an active to a repressed state is associated in this case with environmental conditions, with substrate induction. The biochemical composition of the medium is itself the genetic inducing agent, maintaining the purposive character of work of the genes which are grouped hereditarily in the operons in such a way that the character of their reaction to the inducing agent is complex and purposive.

However, in bacteria a few genes are usually found in a repressed state, namely those responsible for adaptive syntheses. In specialized cells, on the other hand, a few genes remain in the active state; most are permanently repressed in cells with this type of specialization, although in cells with another type of specialization the patterns of active and repressed genes will be different. Since some of the genes in the tissues of multicellular organisms are active, while others are passive, and the pattern of these gene groups varies with differentiation, it is natural that attempts were being made many years ago (much earlier than in bacterial genetics) to explain the nature of the active and passive states of genes in chromosomes by means of the simplest schemes of interaction between genetic loci of the chromosomes and other substances. The complex of gene with substance X implied a passive state while absence of substance X implied the active state, or vice versa. As soon as the highly intimate connection between genetic material and certain of the principal chromosomal proteins (protamines and histones) became understood, it was to these proteins that attention was directed as possible regulators of gene activity.

The hypothesis concerning the importance of histones and protamines in the regulation of gene activity was put forward several years ago (Stedman and Stedman, 1950; Danielli, 1953), and in the succeeding years it has often been discussed from different angles and used in experimental research, mainly with confirmatory results. In a comparatively short space of time the study of the biochemical and genetic functions of histones have grown into an extensive branch of science, and we cannot hope to consider its aspects. These problems concerning histones have been the subject of numerous surveys (Peacock, 1960; Leslie, 1961; Phillips, 1962; Moore, 1961; Sidorova, 1967; Bonner, 1965; Khesin, 1965; Rappoport, 1965) and monographs (Busch, 1965; Bonner and Ts'o, 1964), so that there is no need to examine all the literature concerned with the genetic functions of histones in this book. I shall mention only some of the fundamental problems which have a bearing on the possible molecular mechanism of the regulating influence of histones on manifestation of the activity of the genetic loci of DNA.

As we shall see, examination of these problems still does not give the answer to the fundamental question of the genetics and biochemistry of development: the mechanism of purposive regulation of gene activity in the period of development in space and time and mechanisms of preservation of specialization. It is only a milestone along the road to analysis of this problem. If the histones are the product (the instrument) of a special programming genetic system through which it manifests its action, the study of genetic aspects of the functional biochemistry of the histones will indeed be a method of studying definite manifestations of this programming system which somehow or other will enable us to probe into the precise nature of its automation.

The problem of selective manifestation of genetic information on the basis of biochemical factors, it must be emphasized, differs significantly from the problem of embryonic inducing agents, which are also very important in the phenomena of differentiation. The main difference is that embryonic inducing agents are substances which are external to the cell, while genetic repressors are factors actually built into the genetic system for maintaining a particular type of specialization.

§ 1. Chemical Characteristics of Histones
and Protamines and the Nature of Their
Complexes with DNA in the Chromosomes

The protamines are strongly basic proteins, usually electrophoretically heterogeneous, of comparatively low molecular weight (about 6000), and composed to the extent of nearly 60% of arginine residues. They are typical proteins of the nuclei of the sperm cells of fish and certain other animals, but are usually not found in somatic cells. Protamines contain no tyrosine, tryptophan, cysteine, or methionine.

The histones are nuclear proteins of somatic cells and are intermediate in their properties between the protamines and tissue proteins. They are less strongly basic than the protamines and their arginine content is smaller. Histones can be divided into two fractions, which also are heterogeneous: lysine-rich (up to 40% lysine) and arginine-rich (about 20% arginine). Histones likewise do not contain tryptophan or sulfur-containing amino acids. The mean molecular weight of the histones is higher than that of the protamines, ranging between 5000 and 37,000.

Protamines and histones are usually linked in the nuclei with DNA to form special nucleoprotein compounds, although they differ somewhat in their ability to combine with DNA. Modern views of the chemical character of complexes of DNA with protamines and histones are based on the classical experiments of Wilkins and coworkers (Felix et al., 1956; Wilkins, 1956; Wilkins and Zubay, 1959).

The polypeptide chain in nucleoprotamine has the apparent function of rewinding the DNA molecule along the base of the shallow groove (Fig. 84). The side groups branching from the polypeptide chain lie almost at right angles, so that the basic groups of arginine and other basic proteins can combine with phosphate groups of DNA. One-third of the amino acid residues of protamine consists of other amino acids, and these workers suggest that these residues lie in the folds of the polypeptide chain, so that all the basic groups can combine with phosphate. A single residue cannot form folds, but two are sufficient to do so. This suggests that these residues cannot lie singly in the chain, but in pairs; evidence has been obtained to confirm the possibility of this hypothesis (Felix et al., 1956).

However, nucleoprotamines are found only in the sperm cells of most species of animals, and not in the cell nuclei of tissues and organs. Wilkins (1956) initially postulated that this type of structure may correspond to the inert state of genetic material, lasting until the beginning of its genetic function in the fertilized oocyte. The structure of the DNA nucleoproteins in cells which are growing, differentiating, and synthesizing proteins may, in Wilkin's opinion, be correlated more directly with the dynamic interactions between DNA, RNA, and proteins. In these cells DNA is linked to histones, but other, acid proteins are also present in the nuclei, and in some cases they are also components of nuclear nucleoproteins. Wilkins obtained definite x-ray structural data to show that histones are bound in a different manner from protamines

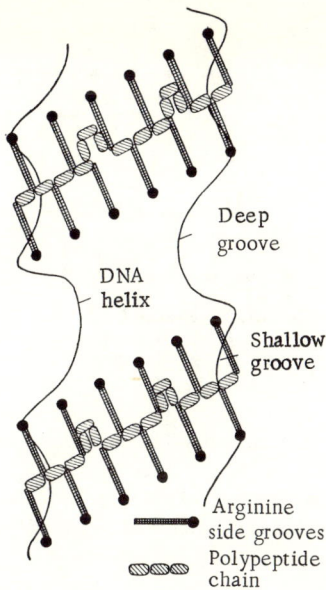

Fig. 84. Diagram showing how protamine rewinds the DNA molecule. Phosphate groups are shown as black circles (1) and coincide with basic arginine side groups (2) of polypeptides. Nonbasic residues are shown in pairs in the folds of the polypeptide chain (3) (Wilkins, 1956; Felix et al., 1956).

to DNA. According to his findings, histone covers both grooves of the DNA helix, both shallow and deep (Fig. 85).

An important investigation of the character of the change from protamine to histone and from histone to protamine in nuclear structures of the snail (*Helix aspersa*) was carried out by Bloch and Hew (1960a, b). In the first of these investigations they examined, by various methods, the transition from somatic histone to protamine during spermatogenesis in this animal. This transition was accompanied by an essential change in the amino acid composition, because whereas lysine is predominant in the composition of histone, arginine is predominant in protamine. They noted that the change from histone to protamine is in fact accompanied by the synthesis of a new protein, and that this synthesis does not correlate with replication of DNA, i.e., it is an independent process. In a second investigation, Bloch and Hew examined the

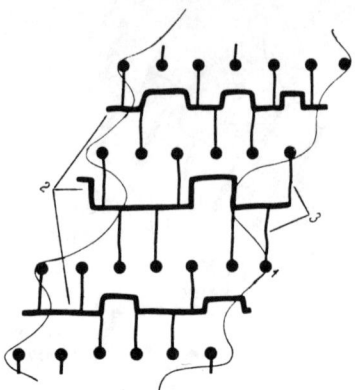

Fig. 85. Diagram showing a possible method of rewinding of the DNA molecule by the histone chain. Black circles (1) show the position of phosphate groups on the DNA helix. Lines of the polypeptide chain (2) with its folds and the side groups of lysine and arginine (3) are indicated faintly.

reverse process; the change from protamines to histones in the early phases of embryonic development. A series of subsequent x-ray crystallographic and spectral investigations (in polarized infrared rays) largely confirmed these fundamental findings concerning the structure of nucleoprotamines and nucleohistones (Bradbury et al., 1962; Busch, 1965; Zubay and Wilkins, 1962).

The space between the DNA molecules in chromosomal structures is filled by hydrated histone gel, and, although it is linked with phosphate groups of DNA, this bond is not very strong.

The mass of histone in the chromosomes is larger than that necessary purely for structures of this type, as shown in Fig. 85. Considering the total mass of histone in the chromosomes, these workers conclude from the absence of orientation of the diffraction spectrum and the absence of dichroism of infrared absorption of histone that the polypeptide chain of histones is not all arranged in the same way, along or perpendicular to the length of the DNA molecules.

After dissociation of nucleohistone into DNA and histone in concentrated NaCl solutions, the nucleohistone complex can be restored by interaction between DNA and histone in dilute salt solution (Zubay and Wilkins, 1964). The x-ray diffraction spectrum of the reconstituted nucleohistone shows that it is identical with native nucleohistone. The sedimentation coefficients of the reconstituted complex are also identical with those of native preparation. This demonstrates that the structures of the original native and reconstituted complexes are identical. This property of the histones is evidently of great importance for the performance of their function.

In their discussion of these findings, Wilkins and his collaborators point out that it might be expected that when DNA is sedimented during reconstitution of histone, a large number of cross linkages should be formed between the DNA molecules, and this would lead to a change in the sedimentation coefficient of the nucleohistone newly formed *in vitro*. However, contrary to these expectations, these coefficients remain the same as for the native preparations. Their explanation is that each histone molecule can attach itself preferentially to one particular specific DNA molecule, so that no effective linkages are formed between the DNA molecules. Cross-linkages are perhaps formed only by interaction between concentrated solutions of histone and DNA, and after dilution of the solutions for ultracentrifugation, a specific regrouping

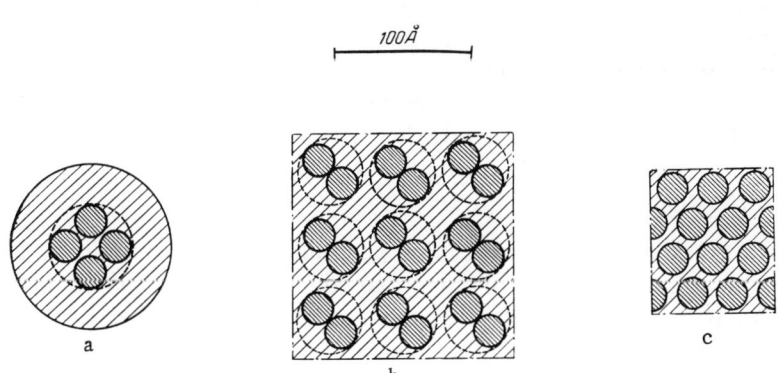

Fig. 86. Structure of nucleohistone shown as a section cut perpendicularly to the long axis of the particles. Darkly shaded circles denote DNA, lightly shaded circles regions containing histone. a) Isotropic solutions, structure of one multistranded particle; 4 DNA molecules are located in the center of the micelle; b, c) other types of possible structures (Luzzati and Nicolaieff, 1963).

of the molecules can be observed in accordance with their mutual selectivity (Huang et al., 1964).

A rather different model of the linkage between DNA and protein in nucleohistone has been suggested by Luzzati and Nicolaieff (1963) on the basis of their investigations. They used a new technique of x-ray diffraction analysis, which they developed themselves, suitable for studying biological molecules under conditions close to those found *in vivo* . As a rule, no aggregation of DNA molecules took place in solutions of different concentrations, but in the presence of histones, bundles of DNA, cross-linked by histone gel, were formed. These workers interpret their results concerning the character of the nucleohistone gel in accordance with the following scheme (Fig. 86).

Phillips and Simpson (1962) carried out an interesting investigation to determine the degree of periodicity of base (arginine and lysine) residues in arginine-rich histone containing on the average three nonbasic amino acid residues between two basics (lysine or arginine). They found that the distribution of arginine and lysine base residues (neutralizing phosphates of DNA) is irregular, and in some cases groups of basic amino acids are found (these zones must be linked with DNA as in nucleoprotamine), while in other zones of histone long peptide links are found (containing up to 7 amino acid residues, with neither lysine nor arginine residues. According to the model of Wilkins, Zubay, and Wilson (1959), these links must form large loops.

§ 2. Heterogeneity and Specificity of Histones in Connection with Their Possible Genetic Function

The hypotheses concerning possible function of histones as genetic regulators (repressors) are based on several assumptions. First, it is evident that before there can be any serious discussion that histones fulfill such functions, in addition to our knowledge of the structure of nucleohistones and the nature of the links between DNA and histones, the following must also be proved.

a) That histones are a highly heterogeneous group of protein subfractions, for as repressors they must recognize the very many different specific polynucleotide information-carrying zones in DNA.

b) That the spectrum of these fractions possesses tissue specificity and undergoes definite changes.

c) That different fractions of histones actually inhibit the synthesis of different forms of messenger RNA and different proteins.

In this and the next section of this chapter we shall examine the modern evidence on these problems.

a) Heterogeneity of Histones in Relation to Amino Acid Composition, Molecular Weight, and Other Biochemical Properties. It was shown several years ago that the total histone isolated from cell nuclei (thymus histone has received most study from this aspect) can be split up by simple biochemical methods into a series of fractions differing in their amino acid composition. The subfractions thus obtained can be divided into two principal groups: arginine-rich histone and lysine-rich histone, depending on which of these amino acids is predominant in the fraction. Separation of the histones into these groups was the chief method of their fractionation for very many years, until new methods in extensive use since the end of the 1950's made further protein fractionation possible.

Most of the work in this field has been carried out in Butler's laboratory (Johns et al., 1960, 1961; Hnilica et al., 1962; Johns and Butler, 1962). As a result of their investigations, histone from thymus desoxyribonucleoprotein (DNP) was first separated by electrophoresis on starch into three fractions: arginine (f_3), lysine-rich (f_1), and a fraction with a lower lysine content (f_2). In fraction f_3, alanine usually occupied the terminal position. Histones of the f_2 group were separated by electrophoresis on starch into subfractions f_{2a} and f_{2b}, differing in their N-terminal groups. The lysine-rich fraction also was separated into f_{1a} and f_{1b}, likewise differing in their N-terminal amino acids. Distinct differences between these fractions were also found for other amino acids.

This classification of the histones has been accepted and confirmed by many other workers.

The use of polyacrylamide gel instead of starch for fractionation of histones enabled an even finer fractionation of total histone and its principal fractions to be carried out (Driedger

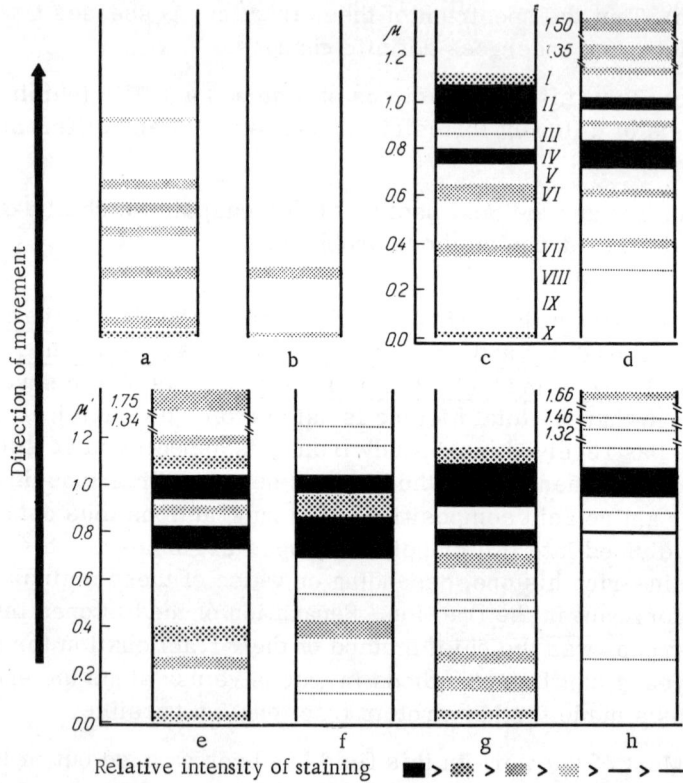

Fig. 87. Scheme of electrophoretic separation of thymus fractions. a) Electrophoresis of unfractionated histone on starch gel; b) electrophoresis of histone very rich in lysine on starch gel; c-h electrophoresis in polyacrylamide gel; c) unfractionated histone; d) histone very rich in lysine; e) unfractionated histone showing several faintly stained bands; f) histone II; g) histone I; h) ribonuclease (McAllister et al., 1963).

et al., 1963; McAllister et al., 1963). In the latter of these two investigations a particularly fine degree of separation of total histone and its fractions was achieved. The results of comparative fractionation of thymus histones obtained in this investigation are shown schematically in Fig. 87.

The electrophoretic heterogeneity of the histones is mainly associated with differences in the amino acid composition of their fractions. However, fractionation on Sephadex, based on separation

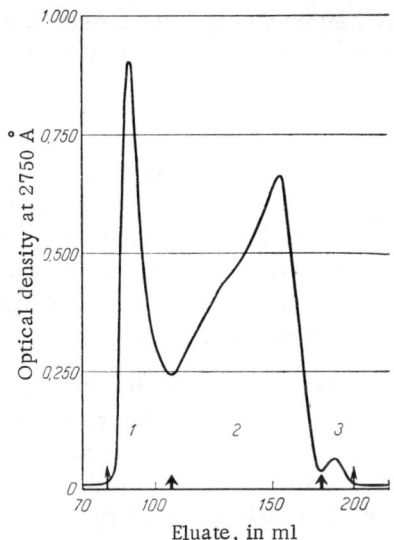

Fig. 88. Curve of elution and separation into three fractions of 200-mg arginine-rich histone f_{2a} and f_3 by chromatography on Sephadex G-75 (Hnilica and Bess, 1964).

of fractions differing in molecular weight, also reveals definite heterogeneity of the histones. Arginine-rich histone, which is practically homogeneous during electrophoresis, separates on Sephadex G-75 into three distinct fractions (Hnilica and Bess, 1964) (Fig. 88). Sephadex G-75 separates proteins within a molecular-weight range from 10,000 to 40,000. Slight differences in amino acid composition were found between the three fractions. Total thymus histone was separated on Sephadex G-75 into five fractions (Johnson et al., 1964). This method showed that histone includes a fraction of peptides with molecular weight of about 2000. Whether these are degradation products of histones or normal components of these proteins is not yet clear. By the use of a specially improved technique of electrophoresis on polyacrylamide gel, Shepherd and Gurley (1966) separated total thymus histone into 34 fractions.

The use of other methods of fractionation of histones (carboxymethylcellulose, Amberlite IRC-50, DEAE-cellulose) has also demonstrated their heterogeneity, but these methods do not possess any greater resolving power than electrophoresis on starch and polyacrylamide gel.

b) Tissue and Morphogenetic Differences between Patterns of Heterogeneity of Histones (Tissue Specificity of Histones). The existence of tissue and morphogenetic differences in the histone spectrum has been studied and discussed very intensively during recent years. However, the information available in this field is highly contradictory and largely dependent on the methods used.

Investigations have shown (Hnilica et al., 1962; Laurence et al., 1963; Hnilica, 1966; etc.) that histones from some tissues (tumors, liver, spleen, thymus) gave a practically identical pattern of their fractions and identical amino acid composition of the fractions. However, this still does not provide an adequate basis for the conclusion that all forms of tissue specificity in general are absent between all tissues. Considerable evidence has now been obtained to show that specificity of this type does in fact exist, although it is not very well defined.

Mauritzen and Stedman (1959, 1960) found slight but definite tissue differences in the amino acid composition of nuclear histones isolated from different organs. These differences concerned only some amino acids (aspartic and glutamic acids, alanine, leucine, isoleucine, and valine). They postulate that histones may be tissue-specific proteins. The relative quantities of the same histone fractions also differ appreciably from one tissue to another.

In another investigation (Neelin and Butler, 1961) specificity and a high degree of heterogeneity of histones were also discovered in the organs of the chicken. Histones of the spleen and liver, when tested by zonal electrophoresis on starch gel, formed 18 zones, the arrangement of which varied. Histones of the erythrocytes, however, possessed a much smaller assortment of components.

Tissue differences in the distribution of histone fractions have also been discovered by chromatography on carboxymethyl-cellulose (Davis and Busch, 1959; Starbuck and Busch, 1960). Of later work in this direction, we need mention only the very detailed investigation made by Driedger and co-workers (1963). Using electrophoresis on polyacrylamide gels, these workers showed that histones from different organs give very different numbers of clearly defined fractions after electrophoresis: thymus histones gave 12-15 clear bands, spleen histones 12-14, testis histones 2-4, liver histones 2-4, and kidney histones 4-6 bands. Differences of the same type were found by electrophoresis of histones in starch gel.

In another investigation (Neidle and Wallsch, 1964) using the method of electrophoresis on polyacrylamide discs, species differences were found in the composition of the fractions and in their relative percentages, when histones of the brain and liver were compared in three animals: rats, guinea pigs, and rabbits. Within the limits of the species, histones of the brain, liver, and kidneys showed no difference in electrophoretic spectra, but thymus histone differed in the presence of two weak additional fractions. Characteristically, however, precise differences were found between the liver and brain histones of newborn animals and the corresponding histones of adult animals: the histones of the newborn were more heterogeneous.

Lindsay (1964) studied the electrophoretic spectra (on starch) of histones from the liver of the developing chicken and differences in the histones of three tissues of the adult animal. Development of the animals correlated with precise differences in relation to certain fractions (Fig. 89). The liver of the adult animal had 14 histone-like fractions, while the liver in the initial period of development had only 10. Some fractions passed through the whole of development unchanged, while others changed quantitatively. Differences in certain fractions of histones between some organs (liver, kidneys, spleen) were very slight affecting only one of the 14 fractions. Lindsay considers that the extent of these differences does not correspond to what would be expected on the assumption that the histones are the differentiating factor during tissue specialization. This is quite true, because the general spectrum of differences between the organs compared is much wider and sharper than the slight differences between histone spectra found in all these investigations.

However, this still does not provide a basis for drawing generalizations. Too few systematic studies have yet been made of all aspects of heterogeneity and tissue specificity of the histones. In particular, when the quantitative histone/DNA ratios were

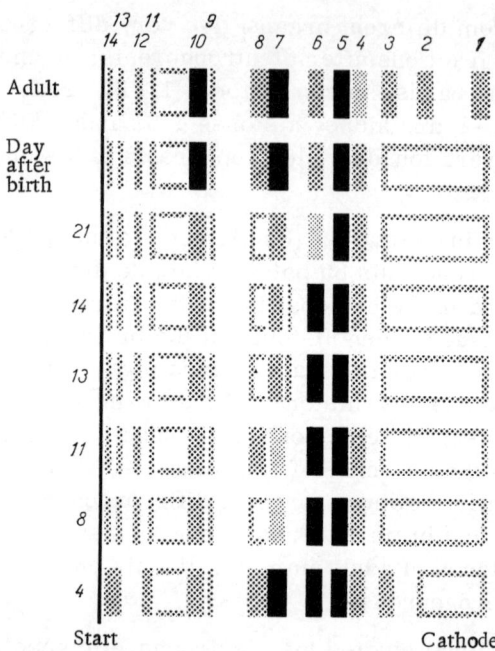

Fig. 89. Electrophoretic spectra of histones from developing chicken liver. Intensity of stippling corresponds to intensity of staining of proteins. Regions surrounded by dotted lines show very weak staining. Numbers near starting line of electrophoresis indicate days from beginning of incubation of eggs. Numbering of histones begins with fastest fraction (Lindsay, 1964).

determined in the nuclei of different organs of the chicken, marked differences were found between nuclei of the erythrocytes and of other tissues (Sporn and Dingman, 1963).

In another investigation (Agrell and Christensson, 1965), also devoted to changes in the composition of histones in organs of the chick during embryogenesis, it was clearly shown that the ratio between the histone fractions varied with the phases of development (considerably in the erythrocytes, brain, and liver, and slightly in the heart and eyes).

Everybody who has worked with electrophoresis will understand that a protein which divides during electrophoresis into 12, 14, or 18 fractions is exceptionally heterogeneous. This heterogeneity can provide for many specific combinations. Considering the heterogeneity of molecular weight of the histones, and the existence of low-molecular weight fractions, it can be assumed that repression of a particular gene does not take place by a one gene – one histone molecule mechanism, but rather by that of a one gene – two, three, and so on of histone molecules. In this case the existence of only 10-12 different types of histone molecules could provide for a tremendous number of different species- or tissue-specific combinations.

It must also be remembered that, according to the current concepts, the histones are not activators but repressors of genes, and of the total balance of genes for a given species and connected with tissue specificity, only one small fraction is active in each organ (along with tissue-universal genes). There is, therefore, no reason to expect any sharply defined tissue specificity of the histone spectrum. In view of the gross nature of the methods used, it must be very slight. The results obtained in this field cannot therefore be regarded as in conflict with the role hypothetically ascribed to the histones.

c) Effect of Histones on the Transcription of Genetic Information of DNA (on Synthesis of RNA and Proteins by Chromosomal Structures). The Histones as Repressors. The view that the histones were perhaps connected with differential transcription of genetic information in the chromosomes originated on the basis of purely structural analogies to do with the type of interaction between histones and DNA in the composition of the chromosomes and were subsequently cross-checked as our knowledge of biochemical functions of the chromosome and the synthetic processes taking place in them increased. Naturally, as soon as it was discovered that the fundamental synthetic process in the chromosomes concerned with transcription of their genetic information is DNA-dependent synthesis of messenger RNA, it became clear that the influence of histones on this process and, consequently, their influence on protein synthesis in a wider context should be studied.

There were other important reasons why systematic experiments should be carried out in this direction. It has been known for a comparatively long time that both protamines and histones, if acting in isolation on the cell, inhibit protein synthesis and that this is due to the formation of nucleoprotein complexes (Stedman et al., 1944; Zbarskii and Perevoshchikova, 1954; Mirsky and Allfrey, 1958; Becker and Green, 1960).

However, the classical experiments of Bonner and coworkers (Huang and Bonner, 1962; Bonner and Huang, 1962; Bonner, Huang, and Gilden, 1963; Huang and Bonner, 1964; Bonner, 1965) showed that histones cannot simply suppress all DNA-dependent RNA synthesis, but that this suppression may be selective, differential in character, in relation to different genetic loci forming functionally different proteins. They carried out most of their experiments to study synthesis of mRNA and proteins by chromatin isolated from pea embryos or, more correctly, young seedlings, a system which has been found very convenient for research of this type. These seedlings (about 1 cm long) were collected in large numbers by a special device, and as a first step their chromosomal material (chromatin) was isolated after homogenization (about 0.5 g/kg) of the embryos. This chromatin was purified from accompanying proteins by gradient centrifugation in 2 M sucrose. Pure chromatin isolated by this method contained 31% DNA, 17.5% RNA, 33% histones, and 18% of nonhistone protein.

Chromatin isolated by this method, as the initial experiment showed, was capable of synthesizing RNA from riboside triphosphates (Huang el at., 1960; Bonner et al., 1961), and the presence of all four riboside triphosphates (adenyl, guanyl, cytosyl, and uridyl) was necessary for this synthesis. This synthesis was DNA-dependent.

The enzyme for this synthesis, DNA-dependent RNA-polymerase, is present also in chromatin, being very firmly bound with DNA. A very remarkable property of this chromatin, as Bonner and co-workers showed, is its ability to synthesize specific proteins under suitable conditions (in the presence of amino acids, an activating system, and ribosomes), and the chromatin isolated from the cotyledons, in which the main protein is globulin, was much more capable of synthesizing globulin than chromatin of the embryos, in which no globulin of reserve type is formed *in vitro*.

Clearly, therefore, protein synthesis in the chromatin material isolated by this method preserves, although perhaps only partially, the specialization of synthesis typical for differentiated cells in which the mRNA of part of the genome participates in protein synthesis.

The chromatin in the experiments of Huang and Bonner (1962) was subjected to further fractionation, the principal stage of which was separation of DNA and histones. For this purpose the chromatin was dispersed in 4 M cesium chloride solution, which ruptured all the ionic bonds in the chromatin. In this case the proteins came to the surface while the nucleic acids were sedimented during centrifugation. Separation of histones from acid proteins in the protein layer was achieved by treating them with a weak alkaline buffer solution, pH 8, in which all the nonhistone proteins including RNA-polymerase, dissolved. The DNA obtained in the residue was of the high-polymer type (molecular weight 6×10^6). However, about half the total content of RNA-polymerase remained bound to DNA even after treatment with cesium chloride.

Characteristically, after the addition of chromosomal RNA-polymerase which had separated together with the histones, the histone-free DNA was much more active as regards RNA synthesis than the original chromatin (Bonner, 1965) (Table 2).

Even without addition of RNA-polymerase, DNA purified from histone possessed considerably (five times) more activity as regards RNA synthesis on account of the polymerase still remaining in it (Huang and Bonner, 1962). These workers concluded from their findings that histones repress mRNA synthesis on DNA. They calculated that, judging from the chemical structure of nucleohistone, if DNA were to be completely bound by histone the

TABLE 2. Effect of Removal of Histone on Ability of Chromosomal DNA to Maintain RNA Synthesis (Bonner, 1965)

Preparation	RNA synthesized (in 10 min) μμmoles incorporated labeled nucleotide/mg DNA
Unpurified chromatin	1,175
Purified chromatin	1,175
Deproteinized DNA + chromosomal polymerase	90,000

ratio between histone and DNA must be about 1.35/1. In chromatin this ratio was close to 1/1, indicating that some DNA is free from histone blocking. By the use of a special preparative method, enabling DNA-histone complexes to be separated from free DNA (Zubay and Doty, 1959), Huang and Bonner (1962) found that in chromatin from pea embryos about 70-80% of all the DNA is bound with histone. This DNA was only just able to maintain RNA synthesis even after the addition of RNA-polymerase, whereas the histone-free DNA was very active (Bonner and Huang, 1963). These results show that histones somehow repress DNA-dependent synthesis of messenger RNA and, consequently, synthesis of proteins transcribing genetic information. The ability of DNA to maintain RNA synthesis showed a very precise linear correlation with its percentage content of histone, i.e., with the nucleohistone/DNA ratio (Fig. 90) (Huang and Bonner, 1962).

Despite their obvious clarity, these experiments, however, gave no information on the specificity of the repressive action of histones. For example, they did not show that it is in fact histones which, in the case of preferential synthesis of a particular protein, close the DNA loci containing information for synthesis of the other proteins but do not close loci synthesizing the required specialized proteins, and that if histones are separated from the DNA the spectrum of protein synthesis is in fact disturbed,

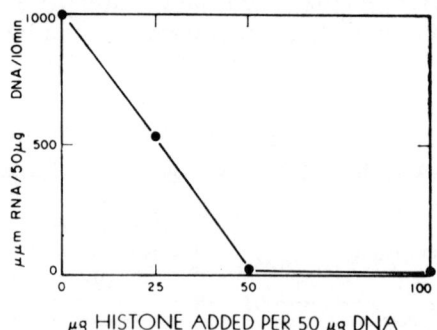

Fig. 90. Effect of chromatin histone from pea shoots on DNA-dependent RNA synthesis; 50 µg DNA to 0.5 ml of standard reaction mixture containing enzyme (1.04 mg) + histone, as shown on the graph (Huang and Bonner, 1962).

while attachment of histones to DNA not only inhibits the synthesis of RNA and proteins, but also modifies their pattern. To answer this question, Bonner and co-workers devised and performed another series of extremely demonstrative experiments (Bonner, Huang, and Gilden, 1963). They chose for these experiments chromatin from cotyledons of pea seedlings, producing a special cotyledon globulin not formed by the chromatin of embryonic seedlings. In cotyledon chromatin, it is assumed, the gene (cistron) for synthesis of the globulin messenger RNA is "open" (not repressed).

If, however, the chromatin of the cotyledons and apical buds is deproteinized and only the DNA templates are used for synthesis of RNA and protein, the differences in specificity between the synthesized products disappear, but the relative globulin content in the proteins synthesized by cotyledon DNA falls from 7 to 0.4%, whereas in the products synthesized on DNA of the apical buds, it increases from 0.1 to 0.4-0.5%. This is natural, because in deproteinized DNA all the genes are active and derepressed. Whereas in the case of cotyledon DNA this leads to a decrease in the relative volume of the DNA templates for synthesis of RNA templates for the cotyledon proteins, in the case of DNA of the apical buds, however, it depresses the gene for synthesis of cotyledon protein.

In a series of experiments (Huang, Bonner, and Murray, 1964) endogenous RNA-polymerase from the seedlings was inactivated by gentle heating (to 60°) and replaced by highly active RNA-polymerase from *Escherichia coli*, which was added in different quantities. In this way the synthesis of messenger RNA could be sharply intensified without changing the properties of the nucleohistones observed in the first experiments, i.e., maintaining synthesis of RNA purely on derepressed sites. In particular, with RNA-polymerase from *E. coli* also, DNA contained in the nucleohistone was approximately only one-tenth as active as the same mass of DNA after deproteinization.

In this way, by providing all the enzymes necessary for synthesis of RNA and by comparing the synthetic activity of DNA before and after deproteinization, it is possible to determine how much of the DNA in the chromatin is in a free state and how much is bound with histone (repressed). However, it was found that these determinations could be carried out very clearly and demonstratively in another way, by a physico-chemical method. Double helical DNA, as we know, is denatured and "melts" on heating, i.e., it is unwound into single polynucleotides. This melting of DNA takes place in the form of a precise temperature reaction somewhere about 70°C. This process can be detected by the sharp increase in viscosity in DNA solutions within these temperature limits (the appearance of two single polynucleotide chains instead of one double chain).

Histone stabilizes DNA also when it is present in the form of nucleohistone. The process of DNA denaturation starts at a higher temperature, at 80-84°C (Fig. 91), (Huang and Bonner, 1962). If, however, chromatin from growing shoots is examined, its DNA has two melting (denaturation) points: the first at 70°C, the second at about 84°C, corresponding to the presence of free DNA and of DNA in nucleohistone form (Fig. 92). Considering the relative contribution of each of these denaturation levels to the total change in viscosity of the solution, we can calculate the relative proportions of free and bound DNA in the chromatin. There is also another physico-chemical method of determining these indices; intensive mechanical mixing of the chromatin causes the free DNA to aggregate into clumps while the nucleohistones remain in solution.

Bonner and his collaborators thus used three methods of determining the relative proportions of repressed and derepressed

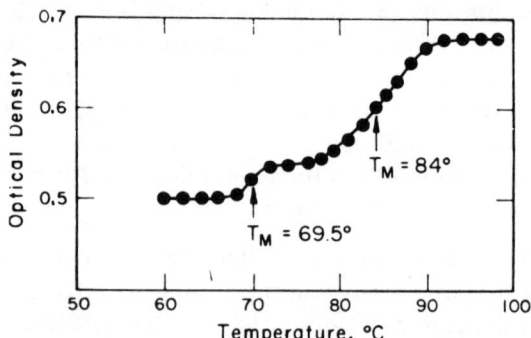

Fig. 91. Characteristics of the process of melting of purified chromatin from pea embryos. Heating carried out in 0.016 M citrate buffer (Huang and Bonner, 1962).

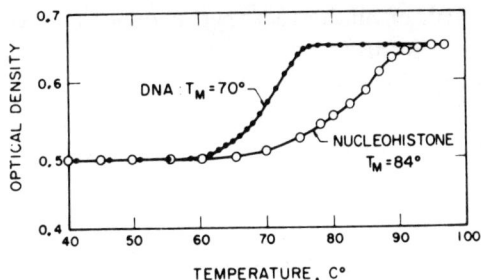

Fig. 92. Characteristics of melting of chromatin DNA from pea embryos in the form of deproteinized DNA (dark circles) or (light circles) native chromosomal nucleoprotein (Huang and Bonner, 1962).

TABLE 3. Properties of Different Chromatins

Fraction	Chromatin from pea embryos	Chromatin from developing pea cotyledons	Chromatin from duck erythrocytes
Fraction of total DNA melting at 70°C	20%	About 5%	About 0%
Fraction of total DNA found in nucleohistone form	78%	≫ 93%	98%
Fraction of total DNA found not as nucleohistone (free DNA)	22%	≫ 7%	0-2%
Effectiveness in maintaining RNA synthesis compared with synthesizing activity of pure DNA	About 20%	≫ 10%	0.1%

genome sites in the chromosomes. Comparison of these methods showed a high level of agreement and revealed characteristic differences between chromatins of different origin. The relevant data are summarized in Table 3 (Bonner, 1965).

The results in Table 3 show that the percentage of free (active) DNA in nucleohistone may vary widely. In chromatin from pea shoots the melting point method revealed about 20% of this type of DNA and about 5% in the chromatin from the cotyledons, while the chromatin of nucleated erythrocytes contains hardly any free DNA. Closely similar results are given by the method of mechanical mixing (the second and third columns) and they all correspond to the volumes of synthesis of messenger RNA. Protein synthesis is virtually absent in mature nucleated erythrocytes, and this correlates with the complete histone repression of nuclear DNA.

The discovery that histones possess repressor functions naturally raised several questions concerning the functional organization of this repression in connection with differentiation. Logically the first of these questions to arise was how different fractions of histones repress different loci in the DNA structure. This question was studied in another series of experiments carried out in Bonner's laboratory (Huang et al., 1964). In this case they fractionated calf thymus histone on Amberlite IRC-50 into a series of fractions differing in their relative contents of arginine and lysine. The following fractions were isolated: fraction Ib — lysine-rich (lysine 26%, arginine 2.2%); fraction IIb (lysine 13.5%, arginine 7.9%); arginine-rich fraction III (lysine 9.4%, arginine 12.9%); fraction IV (lysine 9%, arginine 12.8%). These histone fractions were tested for their ability to repress synthesis of messenger RNA by their action with DNA from the thymus and from pea seedlings.

The results of these experiments are given in Tables 4 and 5. They show clearly that the lysine-rich histone possesses a much stronger repressive action than the arginine-rich histone, and the degree of repression exactly matches the lysine content.

TABLE 4. Effectiveness of Various Nucleohistones as Repressors of DNA-Dependent RNA Synthesis (120 μg Pea DNA/0.5 ml Reaction Mixture in Presence of Chromosomal RNA-Polymerase) (Huang et al., 1964)

DNA in form of	RNA synthesized ($\mu\mu$moles nucleotide/10 min)
Nucleohistone Ib	0
Nucleohistone IIb	24
Nucleohistone III	80
Nucleohistone IV	216
Pure DNA	320

TABLE 5. Ability of Different Reconstituted (Thymus) Nucleohistones to Maintain RNA Synthesis by the Action of RNA-Polymerase from *E. coli* (Huang et al., 1964).

DNA (50 µg) in the form of	RNA synthesized (µµmoles nucleotide/0.5 ml/10 min
Thymus DNA	8474
Nucleohistone Ib	56
Nucleohistone IIb	140
Nucleohistone IV	4000

This difference in inhibitory functions between the histone fractions was also confirmed by other experiments. These showed that different histone fractions differ in their ability to protect DNA from thermal denaturation, and also that the level of the inhibitory properties of histones relative to DNA-dependent RNA synthesis bore a direct relationship to the level of DNA stabilization against thermal denaturation.

Characteristically, when DNA under artificial conditions reacts with a mixture of the four fractions of histones, an electrophoretically homogeneous nucleohistone is formed in which the individual DNA molecules react at random with molecules of different histones, and one DNA molecule, because of its large size, carries many different histone molecules. In native chromatin, on the other hand, the nucleohistones are heterogeneous, and in this case the individual DNA molecules are linked selectively with one particular type of histone (Bonner, 1965; Akinrimisi et al., 1965). These results also show that DNA may be active relative to mRNA synthesis even in the form of a complex with histone, and that everything depends on the type of the histone. It was accordingly postulated (Ts'o and Bonner, 1964) that genetically active DNA may form complexes with histones which possess a mild inhibitory action on mRNA synthesis, whereas genetically repressed DNA forms complexes with histones which inhibit this synthesis. According to this hypothesis, a change from the genetically inactive to the active state of the nucleohistone complex is accompanied by the exchange of inhibiting histone for noninhibiting.

Further research on the basis of this hypothesis required a more detailed study of the process whereby histones are linked with DNA (coefficients of coupling between various types of histones and DNA at different ionic strengths, effect of the conformation of DNA and of histones on linkage with DNA, nature of the bonds, etc.). These investigations, recently begun on an extensive scale (Akinrimisi et al., 1965), have already demonstrated the wide heterogeneity of the coefficients of coupling for different fractions of histones and basic proteins. In particular, at neutral pH values, coupling of the arginine-rich histone exceeded that of the lysine-rich histone.

Another investigation (Marushide and Bonner, 1966) showed that chromatin isolated from the liver possessed about one-fifth of the activity of deproteinized DNA in relation to RNA-polymerase RNA synthesis. The factor responsible for this decreased activity was the chromatin histones. RNA synthesized on purified DNA consisted of complementary polynucleotides and was similar in its nucleotide composition to the DNA. RNA synthesized by chromatin consisted of noncomplementary molecules differing from total DNA in their nucleotide composition. On this basis, these workers claim that histones inhibit (close) specific zones in liver chromatin.

The effect of histones on RNA synthesis in the chromosomes and on DNA-dependent synthesis *in vivo* has been the subject of several investigations in other laboratories also, among which the work of Allfrey and Mirsky (1962), Allfrey et al., (1963), and Frenster et al., (1963) may be mentioned in particular. These workers studied synthesis of RNA and proteins by thymocyte nuclei, and in a series of earlier experiments, carried out in 1958-1961, they showed that histones definitely inhibit this synthesis. Polylysine has a similar action. However, exogeneous histones, when added to nuclei, possessed a complex inhibitory action, some of which was due to their toxic action on ATP synthesis and on amino acid activation. Accordingly, in a series of new experiments, they used the method of studying removal of native histones from nuclei with proteases and the effect of this removal on DNA-dependent synthesis of RNA and proteins.

However, when different histones were added to the isolated nuclei, the effect of lysine-rich and arginine-rich histones in these experiments (Allfrey et al., 1963) was opposite to their effect in the experiments of Bonner and co-workers described above. In the system studied by Allfrey and co-workers, arginine-rich histone possessed more marked inhibitory properties in relation to RNA and protein synthesis by the nuclei than lysine-rich histone.

Hydrolysis of the nucleohistones with trypsin caused a sharp increase in the rate of synthesis of RNA and proteins in thymus nuclei. In interphase nuclei of thymus lymphocytes the repressed nucleohistone zones were concentrated mainly in condensed heterochromatin masses, whereas active interphase DNA was found in euchromatin fibrils (Frenster et al., 1963). A large part of the fast-labeled nuclear RNA was found in these fibrils.

The special relationship between the thymus nuclei and lysine-rich and arginine-rich histones, in contrast to the results obtained by Bonner and co-workers, is evidently attributable to differences in their effect on nuclear metabolism or to technical details. Barr and Butler (1963) obtained results which agreed practically completely with those obtained in Bonner's laboratory, namely that the lysine-rich histone possessed maximal inhibitory properties on DNA-dependent RNA synthesis when DNA from thymocytes was used. Meanwhile, in analogous systems, opposite results were obtained (Hindley, 1963): maximal inhibition of DNA-dependent RNA synthesis by arginine-rich thymus histone. By the use of a different fractionation method (extraction of histones with different concentrations of HCl) (Ord et al., 1965) a fraction richer in lysine was obtained which also had a weaker inhibitory action on DNA-dependent RNA synthesis.

Littau and co-workers (1965) studied the chromatin structure of thymus lymphocytes by biochemical and electron-microscopic methods. They found that nucleohistone chromatin masses consist of two types of complexes: compact and diffuse. The formation of compact complexes, as they showed, is associated with the lysine-rich histone, which forms cross-linkages between the DNA strands. The diffuse histone, in which synthesis of messenger RNA mainly takes place, has no such cross-linkages. It contains arginine-rich histone which is linked to DNA along its strands. The compact nucleohistone corresponds to the heterochromatin of chromosomes and the diffuse to the euchromatin. We know that heterochromatin zones of the chromosomes are inert. In the polytene chromosomes of insects, histones from different poles of the heterochromatin and euchromatin are eluted at different rates (Black and Ansley, 1964), while the heterochromatin, compact chromosomes in cells of the mealworm (*Pseudococcus obscurus*) are inactive relative to RNA synthesis and, as cytochemical investigations show, they contain a high concentration of histones (Berlowitz, 1965).

The reason for the contradictory nature of these results described above is not yet clear. Some very interesting results have recently been obtained (Liau et al., 1965) to show the effect of histones on RNA synthesis in the nucleoli, in which, as we know, synthesis of ribosomal RNA is localized. Nucleoli isolated from Novikov's hepatoma, after removal of histones by trypsin, actively synthesized RNA similar in nucleotide composition to their own DNA. When fractions of thymus histones were added to them, RNA synthesis was inhibited by 90%, but the composition of the RNA synthesized also was modified (the proportion of cytidyl and guanyl nucleotides was increased). This composition became more like that of the ribosomal RNA. Liau and co-workers consider that these results are evidence of the influence of histones on transcription of DNA in the direction of facilitating the synthesis of ribosomal RNA.

An attempt to represent the possible structure of the DNA-histone complex in repressed and active zones of DNA was undertaken by Frenster (1965) (Fig. 93). In his model he postulates the existence of special RNA-derepressors containing anticodons for messenger RNA.

The opposite aspect of the relationship between histones and DNA during RNA synthesis, i.e., the effect of DNA composition on the action of histones, has so far been discussed only in a brief report by Skalka (1964), stating that the degree of repression of RNA synthesis by histones depends on the AT/GC ratio of DNA.

Much less information has been published on the effect of histones on DNA replication, particularly in chromosomal structures. Exogenous histones added to systems *in vitro* usually inhibit DNA synthesis (Lehnert, 1964; Hnilica and Billen, 1964) and DNA-polymerase (Gurley and Irvin, 1964), although DNA synthesis takes place actively in the chromosomes during cell division despite the presence of histones.

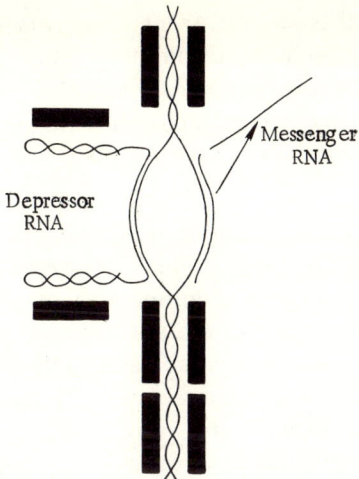

Fig. 93. Model of specific derepression of RNA-synthetase in active chromatin. Polycationic histone repressors (black rectangles) inhibit DNA function as templates for RNA synthesis. Nuclear polyanions partially remove histones from some zones of DNA, enabling special derepressor RNA to form hybrids with one of the chains of specific DNA and liberating complementary DNA for synthesis of messenger RNA (Frenster, 1965).

The histone blocking of DNA inhibits synthesis of messenger RNA to a much greater degree than synthesis (autoreplication) of DNA (Schwimmer and Bonner, 1965). However, the compact heterochromatin (lysine-rich) zones are usually among the late-replicating zones of the chromosomes (Hay and Revel, 1963; Fuyita, 1965).

d) Biosynthesis of Histones as a Genetic Problem.

1) Theoretical aspect of the problem of morphogenetic repressors. If we accept that the histones possess repressor functions, the problem of their biosynthesis, like that of the synthesis of other possible specific repressors, is of considerable morphogenetic importance. Genetic repressors at the bacterial level have in fact provided a logical explanation of genetic differentiation of the cells (activity of some and passivity

of other genes), for this was differentiation relative to environmental conditions, to the substrate. If the environment changes, so also does the character of expression of the genes, while the repressors are simply an instrument through which the environment, or changes in biochemical processes produced by the environment, act.

On the other hand, during morphogenesis of higher organisms, the concept of repressors does not provide a logical explanation of the purposive inherited automation of development, and if we assume that differences in the spectrum of active and passive genes in different tissues are associated with differences in the assortment of repressors, we are naturally faced with a question of the same order of complexity: why is it that a different assortment of repressors is formed in different tissues, and what controls the purposiveness of this phenomenon in time and in connection with its localization?

Since proteins are synthesized on RNA templates and RNA templates are synthesized on DNA sites in the chromosomes, in our answer to this question we must return to the original position: the need to explain the formation of a particular pattern of active and passive genes coding synthesis of the ordinary structural, enzyme, and other proteins with the aid of the concept of existence of a given pattern of active and passive genes determining the synthesis of different repressors. With this approach, it might appear that the concept of genetic repressors does not help to explain the genetic determination of development, does not simplify the complexity of interaction between the processes involved, and does not reveal the actual features of the controlling system, but merely returns us to our starting point with our difficulties unresolved. In fact, however, this is not so. If we assume that each individual repressor inhibits not one gene, or not even one operon, but a system of operons determining the type of cell specialization, the principle of repression and activation of genes becomes more in accordance with reality. In this case the number of types of repressors cannot exceed the number of types of specialization, and this is several orders lower than the total number of genes. In this case the region of the DNA stock controlling synthesis of repressor proteins is hundreds of times smaller than the mass of DNA controlling the synthesis of other proteins, and the illogicality of including the first and second systems in the same category of complexity thus clearly disappears, although the difficulty of explaining these two principles of purposive activity of the first controlling system still remains.

It is obvious that before we change to the concrete language of facts when discussing these relationships, we must examine the latest information concerning biosynthesis of histones.

2) Biosynthesis of Histones. Like the synthesis of many other nuclear proteins, the biosynthesis of histones has been studied from many different points of view. The existence of surveys of the literature on this problem (Mirsky and Osawa, 1961; Rapopport, 1965) frees us from the need to examine the general biochemical aspects of histone synthesis. We shall merely consider those of its other aspects which are to some degree connected with the discussion of their genetic functions.

Biochemical methods (the velocity of incorporation of label into histones, incorporation of labeled amino acids into histones in experiments with isolated nuclei) have shown that the biosynthesis of histones in different types of cells is localized in the cell nucleus (Zbarskii and Samarina, 1962; Butler and Cohn, 1963; Flamm and Birnstiel, 1964; Reid and Cole, 1964; Laurence and Butler, 1965). Although usually the processes of histone synthesis and of DNA synthesis correlate in time, selective inhibition of DNA synthesis does not bring histone synthesis to a halt (Flamm and Birnstiel, 1964; Ontko and Moorehead, 1964). The nuclear localization of histone synthesis has also been demonstrated by cytochemical methods and by microautoradiography, by determining the comparative localization of incorporation of H^3-arginine, H^3-lysine, and H^3-amino acids not present in histones (Mattingly, 1963; Prensky and Smith, 1964; Das et al., 1964). In some special systems, notably the spermatogonia of the cricket, histone synthesis has also been found in the cytoplasm (Bloch and Brack, 1964). In plant cells, active histone synthesis has also been found in the nucleolus (Birnstiel and Flamm, 1964). Hence, most of the available experimental evidence shows that histones are synthesized in the immediate vicinity of their functional locale. Different types of histones (F_1; F_{2a}; F_{2b}, F_3) in the nuclei of the functioning liver are renewed at different speeds. Meanwhile, during regeneration of the liver and in hepatomas, these differences disappear (Hnilica et al., 1965). In the cells of another tissue (the uterine endometrium) the spectrum of activity of histone renewal was different from that in the liver (Chalkley and Maurer, 1965).

e) **Experimental Approaches to Determination of the Role of Histones in Morphogenetic Processes.** Above, when we discussed the material on tissue specificity of the histones, we touched briefly on the question of changes in the histone pattern in connection with differentiation. However, this aspect of research gives only indirect evidence of their possible role in morphogenesis. Direct evidence could be obtained from the study of the action of different fractions of histones, inducing morphogenetic changes or changing the pattern of syntheses, on embryonic tissues. This line of investigation, however, still awaits exploitation and only preliminary work has yet been done. A brief report (Peck, 1963) of the differentiating action of histones on anlagen of neural and pigmented tissues in chick embryos must be mentioned. Peck treated these tissues *in vitro* with different histone preparations. In a concentration of 5 mg/ml the histones sharply inhibited incorporation of H^3-thymidine, H^3-uridine, and C^{14}-leucine (synthesis of DNA, RNA, and protein), but incorporation of S^{35}-sulfates (synthesis of mucopolysaccharides) was increased by 600 fold, and the formation of melanosomes by pigmented cells also was increased. The action of histones on the developing sea urchin embryo was also studied by Vorob'ev and Neifakh (1964). They showed that histones in solution inhibit embryonic development of the sea urchin. In a concentration of 1 mg per ml, histones produce irreversible changes and cause death of the embryos. The early stages of development were more sensitive. Other proteins did not exhibit this action.

Conclusion: Discussion of the Genetic Functions of Histones

It must not, however, be forgotten that ideas on the genetic functions of histones are still extremely contradictory. The most important statements concerning specific functions of histones as repressors and regulators have appeared in two papers (Butler, 1965; Sonnenberg and Zubay, 1965).

The principal fact to which Butler directs attention is the absence of sharply defined tissue differences in the composition of the histone fractions, for he considers that if they possess repressor functions such differences should be present. He also emphasizes the small number of electrophoretically homogeneous fractions (10-15), while the heterogeneity of DNA, chromosomes, and genes is much greater. He also considers the absence of histones from bacteria. Butler postulates that the role of genetic repressors may perhaps be played by "residual" acid proteins of the chromosome which, perhaps, are a more variable group and may be linked with DNA by bonds of many different types.

However, as already mentioned, tissues with different types of specialization may differ sharply from each other in the pattern of certain active, derepressed genes, representing a small proportion of the total stock of genes hypothetically repressed by histones. Histones which, by their composition, reflect the total volume of genetic information for all cells, can therefore have only a very weak tissue specificity, equivalent to the fraction of the specific stock of genes (not more than 3-5%) responsible for specific syntheses, typical of a given, concrete tissue, one of the scores and hundreds of different tissues in the organism. The total number of histones, as mentioned earlier, is not necessarily equivalent to the number of genes, but only to the number of types of specialization and if molecules of different histones can form aggregates with each other, their genetic variability must be very high.

Nevertheless, we cannot rule out possible genetic functions for "residual" acid proteins, but investigation of this aspect is at present in too early a stage to allow conclusions to be drawn. Only very recently Huang (1965) published details of an investigation demonstrating the marked heterogeneity of acid residual nuclear proteins.

The possibility of a more complex structure of histone molecules is very real, particularly in the light of recent work (Huang and Bonner, 1965) showing that in native nucleohistones from pea embryos a special RNA, most probably an oligonucleotide, is firmly attached to certain histone molecules to form a low-molecular weight particle consisting of about 40 nucleotides in length. One such RNA molecule, according to their findings, is included in a native complex of several histone molecules. Histone RNA contains an unusual nucleotide, dehydrouridine, in its chain.

In their discussion of these findings at a special conference on the genetic function of histones, Bonner and Huang (1966) presented additional material on the structure of this RNA—histone complex. As well as several histone molecules, it includes a special nonhistone protein, firmly bound with RNA. They postulate that this RNA plays the role of "operator," attached complementarily to zones of DNA lying next to the operons. The histone complex, in this case, inhibits the corresponding group of genes of the operon. Histone RNA thus acts as an adaptor, a carrier of repressors to different genes. This model does not require broad specificity of the histones, but it does assume the wide heterogeneity and specificity of histone forms of RNA.

However, in their attempt to find such an RNA—histone complex in a number of animal tissues, Commerford and Delihas (1966) were unable to confirm Bonner's findings exactly. In pea nucleohistone about 100 RNA—histone nucleotides were present for every thousand DNA nucleotides. In nucleohistone of the liver and intestinal mucous membrane only two RNA—histone nucleotides were present per thousand DNA nucleotides. Commerford and Delihas consider that this ratio alone is insufficient ground for assigning to this RNA fraction the role of a specific carrier of repressors.

Though, as we shall see, these objections do not overthrow the main conclusions deduced and tendencies developed in the study of genetic functions of histones, we cannot yet be completely sure that histones occupy a leading role in regulation of gene activity in multicellular organisms. In polychromosomal cells of the differentiated organism we find several different levels of regulation of genetic activity: within the operon, by groups of operons, by structural zones of chromosomes; interchromosomal, combined cytoplasmic and chromosomal, induction and hormonal levels, and so on. And it is clear that histones alone cannot make up the complex mosaic of interaction between all these genetic superstructures. Before reaching our final conclusions, we must therefore briefly examine certain other types of morphogenetic phenomena and their regulation.

Literature Cited

Agrell, I. P. S., and Christensson, E. G., 1965, Nature, 207:638.
Akinrimisi, E. O., Bonner, J., and Ts'o, P. O. P., 1965, J. Mol. Biol., 11:128.
Allfrey, V. G., Littau, V. C., and Mirsky, A. E., 1963, Proc. Nat. Acad. Sci., USA, 49:414.
Allfrey, V. G., and Mirsky, A. E., 1962, Proc. Nat. Acad. Sci., USA, 48:1590.

LITERATURE CITED

Barr, G. C., and Butler, J. A. V., 1963, Nature, 199:1170.
Becker, F. F., and Green, H., 1960, Exp. Cell Res., 19:361.
Berlowitz, L., 1965, Proc. Nat. Acad. Sci., USA, 54:476.
Birnstiel, M. L., and Flamm, W. G., 1964, Science, 145:1435.
Black, M. M., and Ansley, H. R., 1964, Science, 143:693.
Bloch, D. P., and Brack, Sh. D., 1964, J. Cell Biol., 22:327.
Bloch, D., and Hew, H., 1960a, J. Biophys. Biochem. Cytol., 7:575.
Bloch, D., and Hew, H., 1960b, J. Biophys. Biochem. Cytol., 8:69.
Bonner, J., 1965, The Molecular Biology of Development, Oxford Univ. Press.
Bonner, J., and Huang, R. C., 1963, J. Mol. Biol., 6:169.
Bonner, J., and Huang, R. C., 1966, In "Ciba Foundation Study Group," No. 24, "Histones: Their role in the transfer of genetic information," ed. A. de Reuch and J. Knight, Churchill, London, p. 18.
Bonner, J., Huang, R. C., and Maheshwari, N., 1961, Proc. Nat. Acad. Sci., USA, 47:1548.
Bonner, J., and Ts'o, P. O. P. (eds.), 1964, The Nucleohistones, Johns Hopkins Press, Baltimore.
Bradbury, E. M., Price, W. C., and Wilkinson, G. R., 1962, J. Mol. Biol., 4:39.
Busch, H., 1965, Histones and Other Nuclear Proteins, Academic Press, New York–London.
Butler, J. A. V., 1965, Nature, 207:1041.
Butler, J. A. V., and Cohn, P., 1963, Biochem. J., 87:330.
Chalkley, R., and Maurer, H. R., 1965, Proc. Nat. Acad. Sci., USA, 54:498.
Christensson, E. G., 1965, Nature, 207:638.
Commerford, S. L., and Delihas, N., 1966, Proc. Nat. Acad. Sci., USA, 56:1759.
Danielli, J. F., 1953, Cytochemistry, Wiley, New York.
Das, C. C., Kaufmann, B. P., and Gay, H., 1964, Nature, 204:1008.
Davis, J. R., and Busch, H., 1959, Cancer Res., 19:1957.
Driedger, A., Johnson, L. D., and Marco, A. M., 1963, Canad. J. Biochem. Physiol., 41:2507.
Felix, K., Fischer, H., and Kreckels, A., 1956, Progress in Biophysics, Pergamon Press, London.
Flamm, W. G., and Birnstiel, M. L., 1964, Exp. Cell Res., 33:616.
Frenster, J. H., 1965, Nature, 206:1269.
Frenster, J. H., Allfrey, V. G., and Mirsky, A. E., 1963, Proc. Nat. Acad. Sci., USA, 50:1026.
Fujita, S., 1965, Nature, 206:742.
Gurley, L. R., and Irvin, J. L., 1964, Federation Proc., 23:2, 1, 372.
Hay, E. D., and Revel, J. P., 1963, J. Cell Biol., 16:29.
Hindley, J., 1963, Biochem. Biophys. Res. Comm., 12:175.
Hnilica, L. S., 1966, Biochim. Biophys. Acta, 117:163.
Hnilica, L. S., and Bess, L. G., 1964, Analyt. Biochem., 8:521.
Hnilica, L., and Billen, D., 1964, Biochim. Biophys. Acta, 91:271.
Hnilica, L., Johns, S., and Butler, J. A. V., 1962, Biochem. J., 8:123.
Hnilica, L. S., Kappler, H. A., and Hnilica, V. S., 1965, Science, 150:1470.
Huang, R. C., and Bonner, J., 1962, Proc. Nat. Acad. Sci., USA, 48:1216.

Huang, R. C., and Bonner, J., 1965, Proc. Nat. Acad. Sci., USA, 54:960.
Huang, R. C., Maheshwari, N., and Bonner, J., 1960, Biochem. Biophys. Res. Comm., 3:689.
Huang, R. C., Bonner, J., and Murray, K., 1964, J. Mol. Biol., 8:54.
Johns, E., and Butler, J. A. V., 1962, Biochem. J., 82:15.
Johns, E. W., Phillips, D. M. P., Simpson, P., and Butler, J. A. V., 1960, Biochem. J., 77:631.
Johns, E. W., Phillips, D. M. P., Simpson, P., and Butler, J. A. V., 1961, Biochem. J., 80:189.
Johnson, L. D., Driedger, A., and Marko, A. M., 1964, Canad. J. Biochem., 42:795.
Khesin, R. B., 1965, Usp. Sovr. Biolo., 59:12.
Laurence, D. J. R., and Butler, J. A. V., 1965, Biochem. J., 96:531.
Laurence, D. J. R., Simpson, P., and Butler, J. A. V., 1963, Biochem. J., 87:200.
Lehnert, S. M., 1964, Biochim. Biophys. Acta, 80:338.
Leslie, I., 1961, Nature, 189:260.
Liau, M. C., Hnilica, L. S., and Hurlbert, R. B., 1965, Proc. Nat. Acad. Sci., USA, 53:626.
Lindsay, D. T., 1964, Science, 144:420.
Littau, V. C., Burdick, C. J., Allfrey, V. G., and Mirsky, A. E., 1965, Proc. Nat. Acad. Sci., USA, 54:1204.
Luzzati, V., and Nikolaieff, A., 1963, J. Mol. Biol., 7:142.
Marushide, K., and Bonner, J., 1966, J. Mol. Biol., 15:160.
Mattingly, S. A., 1963, Exp. Cell Res., 29:314.
Mauritzen, C. M., and Stedman, E., 1959, Proc. Roy. Soc. Biol., 150:299.
Mauritzen, C. M., and Stedman, E., 1960, Proc. Roy. Soc., Ser. B, 153:80.
McAllister, H. C., Wan, Y. C., and Irvin, H. L., 1963, Analyt. Biochem., 5:321.
Mirsky, A. E., and Allfrey, V., 1958, In: Symp. on the Chemical Basis of Development, Johns Hopkins Press, Baltimore, p. 94.
Mirsky, A., and Osawa, S., 1961, In: The Cell, Vol. II, ed. J. Brachet and A. Mirsky, Academic Press, New York—London, p. 677.
Moore, S., 1961, In: Nucleoproteins, Interscience, New York.
Neelin, J. M., and Butler, G. C., 1961, Canad. J. Biochem. Physiol., 39:485.
Neidle, A., and Waelsch, H., 1964, Science, 145:1059.
Ontko, J. A., and Moorehead, W. R., 1964, Biochim. Biophys. Acta, 91:658.
Ord, M. G., Raaf, J. H., Smit, J. A., and Stocken, L. A., 1965, Biochem. J., 95:321.
Peacock, A. R., 1960, Prog. Biophys. Biophys. Chem., 10:55.
Peck, D., 1963, J. Cell Biol., 19:2, 55A.
Phillips, D. M. P., 1962, In: Progress in Biophysics and Biophysical Chemistry, Vol. 12, Pergamon Press, Oxford—London, p. 211.
Phillips, D. M. P., and Simpson, P., 1962, Biochem. J., 82:236.
Prensky, W., and Smith, H. H., 1964, Exp. Cell Res., 34:525.
Rapoport, E. A., 1965, Usp. Sovr. Biol., 59:57.
Reid, B. R., and Cole, R. D., 1964, Proc. Nat. Acad. Sci., USA, 51:1044.
Schwimmer, S., and Bonner, J., 1965, Biochim. Biophys. Acta, 108:67.
Shepherd, G., and Gurley, L. R., 1966, Analyt. Biochem., 14:356.
Sidorova, E. V., 1967, Advances in Biological Chemistry, Vol. 8, Izd. Nauka, Moscow, p. 117.

Skalka, A., 1964, Federation Proc., 23:526.
Sonnenberg, B. P., and Zubay, G., 1965, Proc. Nat. Acad. Sci., USA, 54:415.
Sporn, M., and Dingman, C. W., 1963, Science, 140:316.
Starbuck, W. C., and Busch, H., 1960, Cancer. Res., 20:891.
Stedman, E., and Stedman, E., 1950, Nature, 166:780.
Stedman, E., Stedman, E., and Pettigrew, F., 1944, Biochem. J., 38:XXXI.
Ts'o, P. O. P., and Bonner, J., 1964, In: The Nucleohistones, ed. J. Bonner and P. O. P. Ts'o, Holden-Day, San Francisco, p. 367.
Vorob'ev, V. N., and Neifakh, A. A., 1964, Tsitologiya, 6:496.
Wang, T. Y., 1965, Proc. Nat. Acad. Sci., USA, 54:800.
Wilkins, M. H. F., 1965, Cold Spring Harbor Symp. Quant. Biol., 21:76.
Wilkins, M. H. F., and Zubay, G., 1959, J. Biophys. Biochem. Cytol., 5:55.
Wilkins, M. H. F., Zubay, G. L., and Wilson, H. R., 1959, Trans. Faraday Soc., 55:497.
Zbarskii, I. B., and Perevoshchikova, K. A., 1954, Byull. Éksper. Biol. Med., 38(1):61.
Zbarskii, I. B., and Samarina, O. P., 1962, Biokhimiya, 27:557.
Zubay, G., and Doty, P., 1959, J. Mol. Biol., 1:1.
Zubay, G., and Wilkins, M. H. F., 1962, J. Mol. Biol., 4:444.
Zubay, G., and Wilkins, M. H. F., 1964, J. Mol. Biol., 9:246.

Chapter 7

Induction of Differentiated Activity of the Genes by Cytoplasmic and Chromosomal Factors and Its Role in the General Organization of Morphogenesis

Introduction

Selective repression of genetic loci or systems of loci is only one aspect of genetic regulation. Another of its aspects is undoubtedly the existence of certain inducing factors, differentiated in time and localization and acting on DNA or on the system of repressors, or playing the role of actuator for the system of repression and derepression. Finally, there is a third part of the system of morphogenesis, as yet almost unstudied, namely its special automatic and self-generated programming apparatus.

Differentiation of chromosomal genetic systems, the main features of which have been examined in the preceding chapters, is largely controlled in bacteria, as we have seen, by inducing agents in the environment, because its purpose is to insure the most favorable system possible for interaction between the bacterial cells and environment, especially as regards metabolism of substrates. With a change to multicellular organisms, the external environment and conditions of nutrition no longer influence functional activity of the genetic systems, and substrate induction associated with periodic fluctuations in, for example, the conditions of nutrition no longer influences differentiation of the cells in the various organs. In persons adapted to a diet very rich in animal fat, such as the Eskimos, the enzyme systems of the liver and other organs differ sharply both quantitatively, and in some cases qualitatively, from the enzyme systems of persons accustomed to a carbohydrate type of diet.

During embryogenesis, however, in the period of most active morphogenesis, the external environment has no regulatory action. Nevertheless, in relation to each particular cell, to each individual nucleus and individual chromosome, many external stimuli which arise in the course of development and in connection with it can exert a specific differentiating action on its genetic controlling system.

To carry the argument further, even within the nucleus and within each individual chromosome, in relation to individual genes and groups of functionally connected genes (operons), other genes and their products in the same or other chromosomes may behave as an external factor, either repressive or inductive, and conditions external to the chromosome may influence differentiation of the genome as a whole through this system.

In this chapter we shall examine the nature of some of these interactions.

§ 1. Effect of the Cytoplasm on the Character of Functional Differentiation of Nuclear Structures

Although the programming of development is certainly a nuclear function, the property of a definite organization of the genome, it is obvious that no genetic program can be effective without some sort of feedback system. The issuing of programming instructions by the nucleus cannot be turned simply in time or in space (as in a certain type of symmetry), but it must conform to the actual morphological and physiological pattern of development, to the new conditions arising on account of the new relationships. The speed, and consequently the time of development as a whole and of its individual phases, stages, and zones, is a variable: in many species it depends on the temperature and pH of the environment, the concentration of salts, the intensity of illumination and aeration, and many other factors. Consequently, the activation of genetic loci determining a certain type of specialization, whether intermediate or final, cannot take place simply at a strictly determined time after the beginning of development, simply as a continuation of preceding changes in that chromosomal system, but it must allow for expediency and for the actual situation, so that the type of specialization fits precisely into its proper place within the system of other types of specialization, forming a purposively constructed organ or part of an organ.

The genetic system must therefore possess a receptor system (a mosaic) receiving biochemical or biophysical information about the concrete morphogenetic situation surrounding the cell (or nucleus) and, in conjunction with external signals, determining the type of work to be undertaken by the genome, the direction of specialization.

This approach to the problem of regulation of the switching on and off of various systems of genes enables us to understand the genetic significance of the phenomena of morphogenetic induction, hormonal induction, and other factors capable of modifying the character of specialization (expression of the genes) at the various stages of morphogenesis. Without a specific receptor system within the genome sensitive to extrachromosomal factors, external influences would be of no morphogenetic importance, and the question of the nature of their action on development cannot therefore be considered without regard to genetic determination of development.

Many of these factors act on the chromosome system via the cytoplasm, and the question of the influence of the cytoplasm on the character of functional differentiation of genetic systems of the cell must therefore be examined from the same standpoint.

The factual material concerning the influence of cytoplasmic factors on nuclear functions and, in particular, on nuclear morphogenetic functions is very extensive and varied and has been collected from a very large number of lower and higher forms of multicellular organisms studied by many different methods. There is no need to make a detailed analysis here of all the relevant material, because this task has already been carried out adequately in a number of recent surveys (Brachet, 1960; Neifakh, 1962, 1963; Shmal'gauzen, 1964; Ebert, 1965; Collier, 1965).

I shall therefore consider only some of the principal questions relating to the effect of the cytoplasm on chromosomal function and morphogenetic differentiation, and only insofar as this can help us to understand the mechanism of action of the main controlling system. It should be realized at once that this aspect of study of the role of the cytoplasm differs essentially from what has been called cytoplasmic inheritance.

a) Differentiation of the Cytoplasm during Cleavage of the Oocyte and Its Morphogenetic Importance (Ooplasmic Segregation). Earlier in the book we examined the facts and material showing how the nucleus has not begun to perform its active morphogenetic function in the period of primary cleavage of the fertilized oocyte, until about the late blastula stage, and differentiation of the cells takes place on account of differentiation of the cytoplasm of the oocyte and of the messenger RNA and ribosomes accumulated by the oocyte during oogenesis. As a result of intensive cleavage accompanied by a decrease in size of the cells, the cytoplasm of the oocyte is distributed among newly formed cells in such a way that different cells take different parts of the cytoplasm of the oocyte into their own system, so that if differentiation of this cytoplasm has occurred, it persists after cleavage and results in differences in composition of the cytoplasm of the blastomeres, and subsequently of different parts of the blastula, gastrula, etc., thus determining the divergence of subsequent pathways of specialization followed by different tissue zones of the primary embryo.

This phenomenon, which has been called ooplasmic segregation, may be demonstrated particularly clearly in the so-called mosaic oocytes in which physiological and biochemical differentiation of the cytoplasm is particularly clearly defined, being accompanied by other additional external characteristics: pigmentation, granule-formation, etc. This phenomenon has become a classical object of study in embryology, and the subject of much research, so that I shall mention only one or two typical examples by way of illustration.

Chemical and morphological differentiation of the cytoplasm of the oocyte commences during oogenesis, in the course of growth and maturation of the oocyte, i.e., it is an active morphogenetic process. It is during this period that a special functional form of chromosomes (lamp brush type) appears, to control this differentiation. Various axes of symmetry appear in the oocytes,

§1] EFFECT OF THE CYTOPLASM ON FUNCTIONAL DIFFERENTIATION 299

relative to which the cytoplasmic inclusions are arranged in a particular fashion.

Numerous examples of segregation of this type have been examined by Raven (1961) in a special monograph to the study of oogenesis from the standpoint of accumulation of morphogenetic information during this process. These areas of cytoplasm of the oocyte, having undergone differentiation in different ways, differ in their fate: they are used as primordial material for different anlagen. This has been examined in many different organisms using different markers, including classical experiments involving vital staining of different parts of the oocyte and studying the fate of the stained area. Two types of oocytes were thus distinguished: mosaic (with very clearly defined differentiation) and regulated

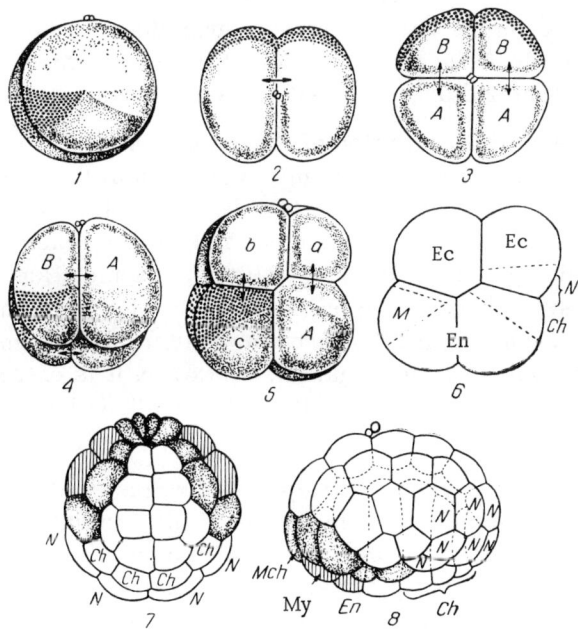

Fig. 94. Cleavage of ascidians. Distribution of material during cleavage of the Styela oocyte (Kühn, 1955). 1-5) First, second, and third cleavage divisions; 6) scheme of cleavage; 7) 64-cell stage; 8) 76-112-cell embryo, view from the left side; A, B, a, b, denote blastomeres; Ch, material of notochord; Ec, material of ectoderm; En, of endoderm; M, of mesoderm; Mch, of mesenchyme; My, of myoblasts; N, of nervous system.

(with less marked differentiation), although in the second case many of the characteristics of clearly defined cytoplasmic differentiation appeared during the first cleavage divisions, i.e., before the nucleus exerted its morphogenetic action on development, and this process could take place up to a certain period even after removal of the nucleus.

The pattern of the first cleavage divisions of an ascidian, Styela, is shown in Fig. 94 (Kühn, 1955). The fate of many groups of cells is already practically completely predetermined at this stage by cytoplasmic differentiation. The surface layer of cytoplasm in the oocytes of this animal contains an uneven (sickle-shaped) distribution of pigment granules and variously stained zones of cytoplasm, enabling better location of the various parts of the cytoplasm during subsequent stages of cleavage. In this mosaic type of development the removal of groups of cells, or even of individual cells, in the earliest stages (even the stages of 2 or 4 blastomeres) leads to the formation of larvae in which one or more strictly determined organs or systems are absent.

Primary differentiation of the cytoplasm of the oocyte of this type has now been studied not only in relation to the external characteristics of the membrane or incorporation of the cytoplasm, but also in relation to many biochemical and physico-chemical indices, such as redox potential, content of lipids, enzymes, and RNA, rate of protein synthesis, and so on, in many species of animals. It is impossible even to make a short examination of the many different variants of ooplasmic segregation. However, it is extremely important, because it determines the primary directions of differentiation.

This process is chiefly important because undifferentiated, equal, and totipotent nuclei (incapable of differentiation during continuous divisions) become surrounded as a result of a series of divisions by a different cytoplasm, which has a different action on them and activates different genetic loci and systems of differentiation.

The principles governing this process have been very well represented by Raven (1961) and are illustrated in Fig. 95. Ooplasmic differentiation does not determine the character of the final results, but only determines the pathway of differentiation.

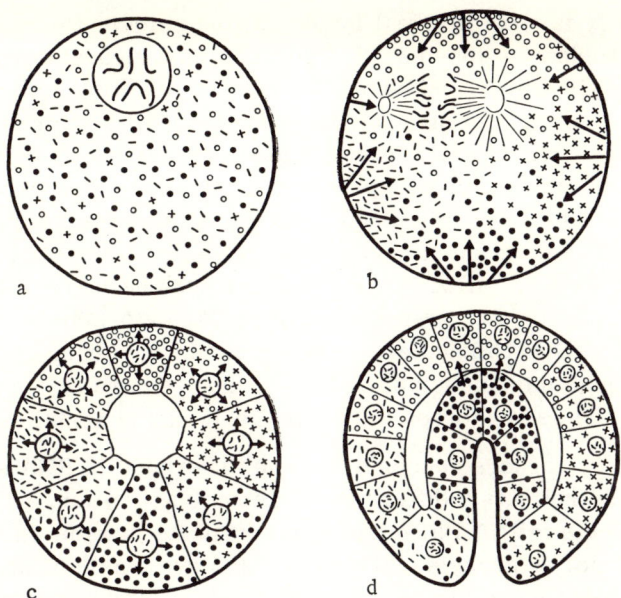

Fig. 95. Scheme of embryogenesis (Raven, 1961). a) Fertilized oocyte: nucleus with chromosomes, cytoplasm in which various substances are distributed at random, cortical layer; b) segregation of cytoplasmic substances, cortical layer regulates movement of these substances (continuous arrows) and localization of nuclei and division spindles (broken arrows); c) stage of cleavage: differences between cytoplasm of different cells due to preceding chemical differentiation, give rise to different forms of interaction with nuclear genes (arrows); d) topogenesis: changes in shape of cells and in their affinity lead to morphogenetic movements; new topographic relationships of cells cause induction (arrows).

It is based entirely on information of the maternal genotype, because paternal genes do not appear in the embryo until the gastrula stage, yet nevertheless the contribution of the paternal genome to formation of the characteristics is equal in importance to the maternal.

Some very interesting results to illustrate the mechanism of cytoplasmic control of primary differentiation were obtained by Davidson and co-workers (1965). They studied the early embryogenesis of the mollusk *Ilyanassa obsoleta*, an organism which was previously shown to offer exceptional opportunities for investigations of this type. During oogenesis the cytoplasm of the oocyte periodically forms spherical lobules at the poles. If one of these purely cytoplasmic lobules is removed during cleavage, embryogenesis at first proceeds normally, but the resulting embryo will not contain several organs (the heart, intestinal mucous membrane, and so on). Consequently, the cytoplasm of this lobule must contain certain factors for derepression (activation) of particular genes. By the use of this model, Davidson and co-workers found that definite activation of genes takes place in this animal in early embryogenesis and RNA synthesis in the nucleus increases. However, in embryos after removal of the polar cytoplasmic lobule, RNA synthesis in the nuclei is sharply inhibited during gastrulation, although the number of cells is unchanged and no other differences can be seen between the control and experimental embryos.

These workers conclude that the difference in RNA synthesis in this case reflects a difference in genome activation, and that it extends to the mRNA which has to perform its function in the later stages of morphogenesis. The part of the cytoplasm which was removed must therefore contain genetic determinants formed during oogenesis and determining the direction of nuclear function in the primary periods of development.

Ooplasmic segregation in the oocyte (polarity and gradient), as already mentioned, itself takes place under the direct control of the nucleus during maturation of the oocyte, i.e., as a morphogenetic process. It is unquestionably controlled by a genetic programming system whose elements can be clearly distinguished in the functional activity of giant lamp brush chromosomes of the oocyte.

This sequence of events is a sign that morphogenesis takes place in definite stages, for which the structure of the genetic system is evidently responsible. Initially, the genetic programming system active during oogenesis creates a special type of cell specialization: a cytoplasmic mosaic, unquestionably attributable to the structure and composition of the cytoplasm. These cytoplasmic differences later, as the cytoplasm breaks up into cells by extremely rapid cleavage (division of the nuclei and membrane formation with virtually no increase in man), become nucleo-cytoplasmic and are transmitted to the nuclei.

In this way receptor-activator systems are incorporated in

the nuclei and the character of functions of the genotype also undergo differentiation. This is the way and, evidently, the only way in which one cell, one nucleus, one genetic program can be translated into functional changes in tens and hundreds of other cells and other nuclei, into activation of tens and hundreds of different programs. Experiments involving transplantation of nuclei carried out in connection with the experimental study of the initial processes of embryonic development afford clear evidence that these cytoplasmic differences in fact exert a dominant influence at this period on functional differentiation of the nucleus.

b) Nuclear Transplantation Experiments in the Study of Primary Mechanisms of Differentiation. An extensive literature on transplantation of nuclei as a method of studying embryogenesis has now accumulated and this has been summarized and generalized in a series of surveys (Briggs and King, 1959; Brachet, 1960; Lehman, 1957; Stroeva and Nikitina, 1960; Rott, 1962; Moore, 1962; Gurdon, 1963, 1964; Hennen, 1963; Signoret, 1965; Gallien, 1966).

Preliminary experiments on nuclear transplantation in order to study mechanisms of embryonic development were started by Spemann (1928), more than 40 years ago. He tied a ligature around a triton oocyte before cleavage began, so that the nucleus was present in only one half of the oocyte. Cleavage began to take place in this half. After the number of blastomeres had reached 16, Spemann relaxed the ligature and introduced one of the 16 nuclei into the other half. He then tightened the ligature completely, but a normal embryo developed from the second half of the oocyte which had received the blastomere nucleus. The identical experiment was subsequently carried out on oocytes of insects (Seidel, 1932, 1938). The use of direct transplantation of the nucleus for studying morphogenetic properties of blastula nuclei was first suggested by Lopashov (1945). However, it was Briggs and King who used the method of nuclear transplantation most widely in order to study various genetic aspects of differentiation (Briggs and King, 1952, 1953, 1959; King and Briggs, 1956). These workers developed a very elegant method of direct transplantation of diploid nuclei from tissues of frog embryos into an activated frog oocyte from which its own nucleus had previously been removed. If the transplanted nucleus was equivalent in genetic capability to the oocyte nucleus, or to use their terminology, if it was totipotent, the host oocyte gave rise to a normal embryo. If, however, the genetic functions of the nucleus were modified, the embryo would be abnormal or its development would stop at a certain stage.

Originally Briggs and King used nuclei of blastula and early gastrula cells in their experiments as donor nuclei. After such transplantations the host oocyte (with its own nucleus removed) in most cases developed into a normal tadpole. This shows that such nuclei were totipotent or quickly became totipotent under the influence of the cytoplasm of the oocyte.

In later investigations (King and Briggs, 1954, 1955) these workers used nuclei from late gastrula cells (the yolk plug stage). They found that nuclei isolated from the chordomesoblast of the late gastrula cannot insure complete development of recipient oocytes. Besides normal tadpoles, blocked blastulas and gastrulas and also abnormal neurulas were obtained. King and Briggs concluded that incomplete development of embryos after transplantation of nuclei from the region of the chordomesoblasts of the late gastrula reflects changes in the state of differentiation of the nuclei in this part of the embryo, with the loss of their totipotency.

In another series of experiments, the same workers (Briggs and King, 1957, 1960) progressively limited the potency of the entodermal nuclei; this time the nuclei were taken from cells located at the base of the primary intestine, i.e., from the region of the presumptive midgut. As a result, some nuclei were found to be still totipotent and to enable normal and complete development of recipient oocytes to take place, but most entodermal nuclei proved to be differentiated, and after transplantation the most common result was the production of blocked blastulas and gastrulas or abnormal neurulas. In other words, the nuclei of this tissue showed mosaic properties. In this case the nuclei were members of a heterogeneous population. It is interesting to note that the degree of this mosaic pattern did not increase, but decreased toward the period of formation of a motile tadpole.

The next step in these experiments, providing an answer to the question of the reversibility or irreversibility of the changes, was studied by the method of nuclear clones (King and Briggs, 1956). In this method the donor nucleus is introduced into the oocyte which is allowed to proceed to the blastula stage (no disturbances can be detected at this period even if an abnormal nucleus is introduced), after which these nuclei descendants of the primarily transplanted nucleus are used for transplantation. Embryos developing as a result of this secondary transplantation form a clone in which each embryo contains nuclei derived from a common original nucleus used for the first transplantation. All embryos in this case are genetically identical. The procedure can be repeated several times and several such clones obtained.

These experiments showed that the changes undergone by the nucleus during differentiation are stable (irreversible) and are transmitted hereditarily. On the other hand, they also showed clearly that nuclei from zones of the embryo determining their path of development to some degree or other in some cases could have their program modified in the cytoplasm of the oocyte and give rise to an embryo of corresponding normality. It thus became clear that by returning the nucleus from cells whose fate had been determined back to the oocyte we give back to this nucleus its totipotency completely or partially.

The ability of nuclei from more or less specialized cells to revert to the program of totipotent nuclei was demonstrated particularly clearly by analogous experiments on embryos of the African toad *Xenopus*, an extremely convenient object for experiments of this type.

These investigations were begun by Fischberg (1958, 1959) and continued by Gurdon (1958, 1960, 1962a). After transplantation of ectodermal and entodermal nuclei cases of complete development of embryos occurred, even if these nuclei were taken at a much later stage than in the experiments of Briggs and King. For example, nuclei taken from somites 9 h before these became capable of contraction and from the intestine of feeding tadpoles were still capable of causing normal development when transplanted into an unfertilized oocyte. This could be either because the nuclei of *Xenopus* remain totipotent until the late stage of embryonic development, or they recover their totipotency when transplanted into an unfertilized oocyte. The latter hypothesis seems to be most probable.

These experiments acted as a stimulus to the transplantation of nuclei of completely differentiated cells into the oocyte, i.e., nuclei in which the spectrum of active and repressed loci have become stabilized and is responsible, not for the formation, but for maintenance of specialization.

In this series of experiments (King and McKinnell, 1960; McKinnell, 1962), the most unexpected results were obtained by transplantation of nuclei from cells of a frog renal adrenocarcinoma into the oocyte of the frog *Rana pipiens*, i.e., by the use of nuclei from tumor cells which had become tumor cells after they had been differentiated kidney cells. The first experiment showed that of 142 recipient oocytes into which fragmented tumor cells were injected, most stopped development at very early stages, although occasionally neurulas and abnormal late neurulas were formed.

In subsequent experiments (McKinnell, 1962), when the recipient oocyte received a diploid nucleus from an adrenocarcinoma cell, the tadpoles which were obtained, although possessing anomalies, were capable of swimming and possessed well-differentiated tissues: skin, brain, muscles, notochord, and so on. In this case a nucleus which had previously been a nucleus of a kidney cell, when transplanted into the cytoplasm of the oocyte, led to the formation of many types of specialization. Its total genetic material had not undergone irreversible inactivation.

However, in these experiments not just the somatic nucleus, but several fragmented cells with their nuclei, were injected into the enucleated oocyte. An improvement in the transplantation technique enabling just one nucleus to be injected revealed that early delay of development (mainly in the blastula stage) occurs after injection of nuclei of renal adrenocarcinoma cells. Retransplantation of the progeny of these nuclei led to more prolonged morphogenesis, as far as the neurula and sometimes the larval stage (King and Di Berardino, 1965). These findings were subsequently confirmed by investigations using a wide range of experimental material. Nikitina (1964) obtained normal tadpoles of the green toad by transplanting nuclei of cells of the primitive eye into the oocyte at a relatively late stage.

Gurdon (1962a) obtained fully developed *Xenopus* embryos by transplanting nuclei of epithelial cells of the intestinal mucous membrane taken from large tadpoles into the oocyte, using as nuclear donor entodermal cells of *Xenopus laevis* taken at various stages ranging from blastula to tadpole. In another series of experiments (1962b) he obtained a high percentage of normally developed embryos, of which 150 continued to the stage of adult animals.

Some interesting experiments on transplantation of cell nuclei of the frog embryo into the enucleated oocyte were carried out by Smith (1965). He compared nuclei of somatic entodermal cells and nuclei from cells of the genital tract lying next to them. In the latter case, development of embryos after nuclear transplantation took place quite normally in a higher proportion of transplantations, whereas if entodermal nuclei were transplanted, normal development never once occurred, particularly after the end of cleavage. Smith considers that entodermal cells, in contrast to cells of the genital tract, lose their totipotency.

Another method of studying the effect of cytoplasm on development and differentiation of the nucleus by nuclear transplantation was initiated with the work of Moore (1957, 1958), who used hybrid transplantation of nuclei into the oocyte of another species of frog. He showed that if the diploid nucleus of *Rana sylvatica* is transplanted into the enucleated and unfertilized oocyte of *R. pipiens*, development stops at the late blastula stage. The reverse combination (transplantation of the diploid nucleus of *R. pipiens* into the enucleated oocyte of *R. sylvatica*) gives the same result. In later experiments nuclei of the blocked blastula of *R. sylvatica*, having received a nucleus from *R. pipiens*, were transplanted back into the oocyte of *R. pipiens*. Development thereupon stopped at the early gastrula stage. This experiment clearly shows that under the influence of the cytoplasm of *R. sylvatica* the original chromosomes of *R. pipiens* underwent irreversible changes.

This method was still further perfected by Hennen (1963) who transplanted nuclei from normal blastula cells of *R. pipiens* into enucleated oocytes of *R. sylvatica*. After 10-12 divisions in the foreign cytoplasm, the nuclear progeny were retransplanted into enucleated oocytes of *R. pipiens*. Although the embryos (with one exception) had severe developmental anomalies, they developed to later stages than in Moore's experiments — from late blastula to the young tadpole. However, the harmful action of the foreign cytoplasm was definitely established and was accompanied by chromosomal aberrations. Similar results were obtained by Gurdon (1962c) who transplanted nuclei between two species of *Xenopus*.

During recent years experiments on transplantations of this type have been extended to new species: the axolotl, *Amblystoma mexicanum* (Briggs et al., 1964), and the frog *Rana palustris* (Hennen, 1965). Experiments have also been started with the object of elucidating the nature of the restricted and total potency of nuclei from differentiated cells causing developmental aberrations on transplantation (Subtelny, 1965).

The most important conclusion drawn from all this interesting series of investigations is that the cytoplasm contains substances affecting functional activity of the chromosomes and essentially determining the pattern of activity of genetic loci switching on and off the various systems of specialization.

It is interesting to note that a small nucleus taken from any somatic cell of a frog embryo, after transplantation into an oocyte, very quickly undergoes hypertrophy and grows almost to the size of the normal oocyte nucleus, i.e., almost 30 times bigger (Gurdon and Brown, 1965).

The nature of these substances is not yet known. Moore (1962) suggested that the principal role in reprogramming the transplanted nucleus is played by cytoplasmic (reserve) DNA present in large quantities in the cytoplasm of oocytes (see Chapter 5) and used for synthesis of chromosomal DNA during cleavage. Moore considers, in particular, that during interspecific nuclear transplantations of the type *R. pipiens* $\times$ *R. sylvatica* it is replication of chromosomal DNA at the expense of the "foreign" cytoplasmic DNA which causes structural aberrations of the DNA and chromosomes of the nucleus. However, there is no doubt that many other substances in the oocyte may have a similar effect on chromosomes of the transplanted nucleus, acting through an induction mechanism. In normal morphogenesis, changes in the program of specialization arise as a rule through induction, i.e., through the action of various substances exerting their action on the nucleus through the cytoplasm.

Another method of detecting the mutual importance of nucleus and cytoplasm, especially valuable for detecting their relative role in heredity, was developed by Astaurov, working with the silkworm (Astaurov, 1937, 1947, 1948; Astaurov and Ostryakova-Varshaver, 1957). The method consists of fertilizing oocytes with a killed nucleus and subsequently producing androgenetic development of these oocytes in which the nucleus is derived only from the nucleus of the sperm cell. In the experiments of Astaurov and Ostryakova-Varshaver (1957) with an interspecific combination of nucleus and cytoplasm, the complete dependence of inherited characteristics on the origin of the nucleus, but not of the cytoplasm, was demonstrated. The method of nuclear transplantation from the blastula of one readily distinguishable subspecies into enucleated oocytes of another was subsequently used with success in experiments of this type (Sambuichi, 1961; McKinnel, 1962b; Gurdon, 1961). The results of these experiments fully agreed with those obtained in the work with silkworms.

A special case of induction was described by Harris (1967). He transplanted nuclei from chicken erythrocytes into the cytoplasm of HeLa cells. The inert, completely repressed nucleus, became reactivated and brought about the resumption of RNA and DNA synthesis.

§ 2. Molecular-Genetic Aspects of the Phenomena of Embryonic Induction

The phenomenon of embryonic (morphogenetic) induction has been studied now for many decades, and a wealth of experimental material has accumulated. It is described in a series of monographs (Needham, 1942; Child, 1946; Saxon and Toivonen, 1962; Shmal'gauzen, 1964) and in numerous surveys (Lopashov, 1961; Lopashov and Stroeva, 1963; Toivonen, 1963; Grobstein, 1962; Yamada, 1962; Elsdale and Johns, 1965; Mitskevich, 1965; and others). Nevertheless, the biochemical and genetic aspects of morphogenetic induction have not yet been fully analyzed, the factual material is fragmentary and contradictory, and research into this interesting phenomenon is still carried out mainly by classical embryological, histological, and physiological methods.

Not very long ago attempts were actually made to contradict the principles of induction with the principles of genetic control of development, induction being presented as a purely epigenetic process. Nowadays, with the development of research in the field of regulation of gene activity through hormonal, chemical, and physico-chemical factors, including regulation by the principle of induction, it has become obvious that the study of the phenomena of embryonic induction can reveal new aspects and properties of the genetic apparatus of cells in their function of coordinating and programming development.

In the developing multicellular organism the genetic regulatory system consists of millions of individual nuclei, each of which nevertheless contains a total program of development for the whole organism. At the same time, the structure of the nucleus, the chromosomes, and the whole system of nucleo-chromosomal relationships is such that it bears the main responsibility for processes taking place directly in a given cell, but not in the organism as a whole. The nucleus is the control panel but only for events taking place in one cell. However, the very precise morphogenesis of the whole organism is built up of the processes taking place in millions of individual cells. This integration of individual cell processes into a common, goal-directed system of successive morphogenetic phenomena is certainly brought about through systems of highly sensitive adjustment to external conditions which exist in each nucleus. Because of the presence of the total program of development in each nucleus, purposive morphogenesis can take place without a supracellular regulatory system if this total program can be split up into groups of specialization with receptor loci capable of reacting in a definite manner to a particular external situation (whether chemical, physical, or physico-chemical).

In this case the cytoplasm of an oocyte differentiating under the influence of the specific program of its own nucleus gives during oogenesis the first stimulus toward differentiation of genetic systems, toward the activation of primary specialization in some cells in accordance with type A, in other cells with type B, in a third group with type C, and so on (Fig. 96). Each of these types of primary specialization can be differentiated further in several directions ($B \rightarrow b, d, i, \ldots A \rightarrow a, s, n, \ldots$) and so on, and each of the subtypes (b, d, i, ...) has a receptor system (a switch) reacting to a particular external situation, i.e., in this particular case to induction by a group of cells lying near to it or in contact with it and possessing a different type of primary specialization. Under the influence of induction normally one of the receptor systems, let us say i, is switched on, thereby regulating specialization of one general type, for example, nervous. In the process of formation of primary nerve tissue, further differentiation may take place in it with new systems of cross-induction.

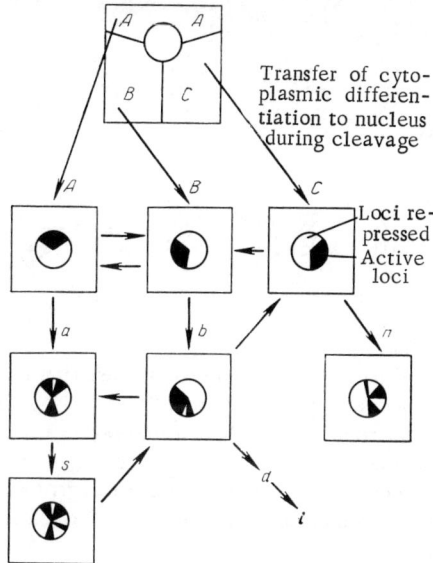

Fig. 96. Scheme of induction in early embryogenesis (explanation in text).

In this case the production of embryonic inducing agents may be programmed genetically and their qualitative composition is connected with a definite system of specialization. However, the inductor in some cases may be the nonspecific reflection of general biochemical specialization of a particular type.

Embryonic induction is thus definitely not the result of epigenetic elements of morphogenesis, but the necessary consequence of the existence of a strict genetic program, but a program represented in millions of cells by millions of systems on a local, intracellular scale.

This general view of the problem of embryonic induction is confirmed by most of the extensive factual material of experimental embryology, particularly if it is examined systematically and not chaotically. First, we know that the possibility of changes taking place in a group of cells in different directions in fact diminishes with each successive phase of development. Induction not only stimulates a group of loci, but as a result of this causes repression of the other systems of specialization. The activated

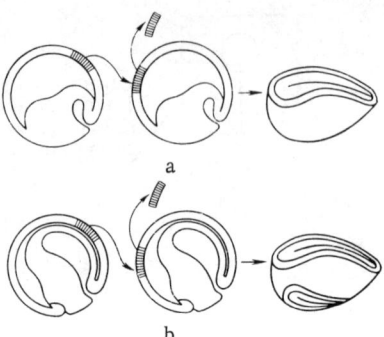

Fig. 97. Scheme of Spemann's experiments to establish times of determination of the neural plate in amphibian embryogenesis (from Saxen and Toivonen, 1962). a) Region of presumptive neural plate of triton embryo at the early gastrula stage, transplanted to ventral aspect of another embryo at the same stage; induction did not take place and graft develops in accordance with its new surroundings; b) analogous region but in the late gastrula stage, development now takes place in the new situation in accordance with its origin.

system evidently itself represses manifestation of activity of the other system. This was clearly demonstrated long ago by the classical experiments of Spemann (1916, 1918), undertaken to establish the times of determination of the neural plate in amphibian embryogenesis. These experiments are illustrated schematically in Fig. 97 (from Saxen and Toivonen, 1962).

These experiments showed that when, for example, a piece of ectoderm from the region of the future (presumptive) neural plate of the early gastrula was transplanted onto the ventral aspect of a new host of the same age, the graft developed in the same way as its new surroundings, and not in accordance with its origin (Fig. 97 a). Consequently, the neural plate tissue was not determined before the operation. This fact was confirmed by cultivating a piece of presumptive neuroepithelium as an explant; in this case the material of the presumptive neural plate formed only undifferentiated ectodermal cells. If, however, the experiment was performed at the late gastrula or early neurula stage, the graft developed in its new situation, on the ventral aspect, into part of the central nervous system in accordance with its origin (Fig. 97 b). The part of the presumptive neural plate now formed nerve tissue in the explant also. These results show that the neural plate acquires its differentiation during gastrulation, and at the same time loses its power of development in other directions, becoming capable only of differentiation in accordance with its origin. However, it still possesses many opportunities for differentiation in different ways.

The factor responsible for this determination of the piece of ectoderm was induction by substances present in the tissue of the upper lip of the blastopore, which invaginates within the embryo and lies beneath the presumptive neuroepithelium. Contact between these groups of cells brings about their mutual determination. If the tissue of the upper lip of the blastopore is grafted on the ventral side of another early gastrula, it induces a second central nervous system there, and along with it, forms a second embryo which may attain the same completeness of structure as the first.

Many examples of such investigations undertaken at various phases of development could be cited, showing that induction is connected as a rule with contact between tissues and that it implies the activation of systems restricting the morphogenetic potential of a particular area to within the boundaries of a definite system of specialization.

Another classical example of this type of process in the later stages of embryogenesis is formation of the eye, the experimental study of which was also started by Spemann, and subsequently continued by many other investigators. Mechanisms of development of the eye in vertebrate embryogenesis have been studied in particular detail by Lopashov and co-workers (Lopashov, 1960; Lopashov and Stroeva, 1963).

The anlage of the eye, when it appears in amphibians, possesses definite inducing properties. As soon as it comes into contact with the ectoderm, specialization of the ectoderm cells begins to take place at the point of contact: a lens is formed (Fig. 98). If contact is prevented, no lens develops.

If the eye anlage (the optic cup) is grafted into another situation, but beneath the ectoderm, a normal lens develops from the ectoderm at the new site. On the other hand, if the cranial ectoderm of the trunk is replaced in an early stage of development, after contact with the optic cup a lens is also formed from it. Differentiation within the lens under these circumstances also

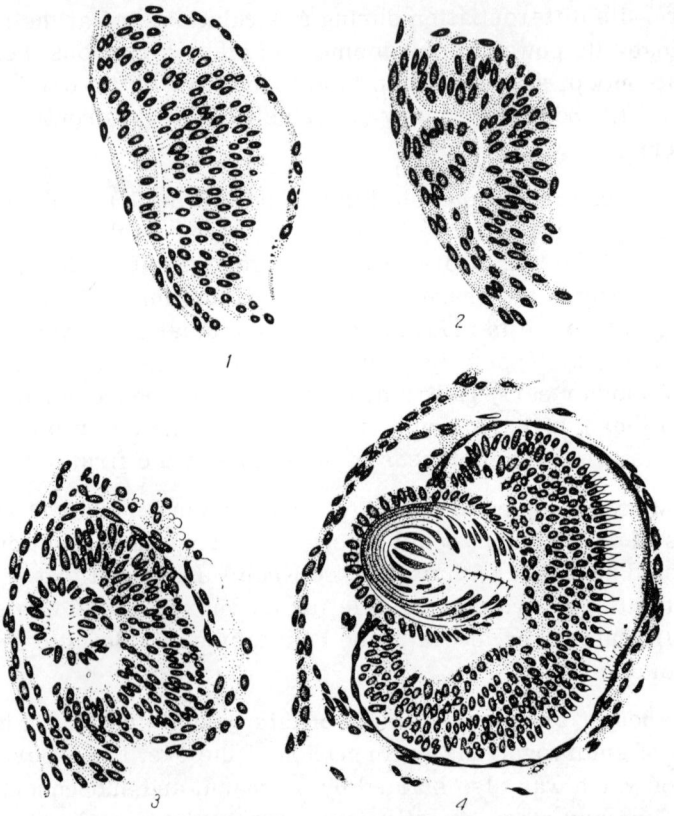

Fig. 98. Normal development of the eye in the axolotl (Kühn, 1955).

depends on the inductive effect of the inner wall of the optic cup, i.e., of the retinal anlage. In the lens, which to begin with has the appearance of a hollow vesicle, the inner wall facing the retina thickens, and the outer wall becomes thinner. The cells of the inner wall are converted into transparent fibers forming the fibrous nucleus of the lens, which acts as the refractory body of the eye. Cells of the outer wall form the thin transparent cover. Under the influence of the eye, the adjacent skin is also transformed into a transparent cornea. In this case also, transplantation experiments have demonstrated the inductive properties of the eye.

Mutual inducibility of this type by various systems in the course of formation is a normal feature of development, and by now many scores of different induction systems have been studied: the development of vertebral cartilage from mesenchyme under the influence of the neural tube and notochord, formation of the renal tubules by inductive interaction between mesenchyme and epithelium, formation of the olfactory organs under the influence of the anterior edge of the neural plate, formation of the otic vesicles from ectoderm under the influence of the hind brain, and so on.

Second, it follows from the theoretical argument presented at the beginning of this section that it is impossible for a tissue when specialized under the influence of induction to possess a wide range of possible directions of differentiation. Only a few receptor (trigger) groups, connected with a few systems of specialization, can be in a state of active adjustment. Accordingly, the inductor cannot be hyperspecific, especially in the case of the short-term contacts typical of early embryogenesis. The wide variety of chemical and biochemical agents which can simulate normal induction is therefore not unexpected.

Investigations in this field, which have been in progress for many years (Brachet, 1960; Saxen and Toivonen, 1962; Toivonen, 1963; Yamada, 1962), have shown that induction of the neural tube in amphibians, which normally takes place during gastrulation, can be brought about experimentally by many different products and agents.

Under normal conditions the inductor is a diffusible substance or group of substances, metabolic products of the neighboring tissue. The closer the contact, the higher the concentration of these products. For this reason the reaction of the receptor systems has not developed very narrow specificity in the course of evolution. A reaction has developed to a signal, which could be an individual compound. However, this does not mean that other compounds cannot act as such signals. Under normal conditions other compounds simply are not present, and the receptor systems have not therefore developed very narrow specificity of reaction such as occurs during the allosteric interaction of specific proteins in the phenomena of substrate induction in bacteria (see Chapter 3).

For this reason, investigations into the chemical nature of induction have split up into two directions. Some workers have used fractionation methods, microchemical reactions, and so on to study which concrete substances are in fact responsible for particular acts of induction under the conditions of normal development. Others have followed the less fruitful path of investigation of substances and agents which, in general, can simulate and reproduce normal induction or a particular act of induction.

It is only in the first direction that any significant number of investigations has yet been carried out, and interest in simulation of induction phenomena by various compounds has already begun to wane.

So far as natural inductors are concerned, their study has been handicapped for several reasons, mainly the negligible concentrations of active diffusible compounds and the difficulty in obtaining adequate quantities of embryonic tissues.

How real this difficulty is can be seen from the work of a group of biochemists from the Netherlands, who first isolated in a chemically purified form a natural embryonic inductor produced by the ventral half of the embryonic spinal cord of chick embryos. This inductor, penetrating into the region of contact of somite cells, induces the formation of vertebral cartilage. Similar induction is produced by the notochord. The nature of this chondrogenic factor was studied by these Dutch biochemists (Hommes et al., 1962; Lash et al., 1962).

Production of this inductor occurs only at a certain period of embryogenesis, the first few hours of contact between the tissues. The chemical isolation of the active factor involved considerable work. For each analysis batches of 1000 embryos were taken (1000 eggs, at stage 4-5 days of development). To isolate an extremely small quantity of the compound, these workers had to process altogether 70,000 incubated eggs, almost 5 tons of raw material. From each embryo they had to remove a portion of nerve tissue (the neural tube) with its inductor. The active factor was found in the fraction of free nucleotides, and after separation by electrophoresis and chromatography, it was preliminarily identified as a nucleotide—peptide compound (it contained cytidine and guanosine monophosphates and a number of amino acids: aspartic and glutamic acids, glycine, alanine, valine, serine). The compound also contained a hexosamine.

This investigation was virtually the first case of isolation of an embryonic inducing agent in a chemically pure form. However, it should be mentioned that a number of different nucleotide—peptide compounds have previously been found in a wide variety of biological subjects: bacteria, yeasts, animal and plant tissues. Their functions, however, still remain unknown. The first papers reporting the existence of this biochemical fraction only began to appear in 1957 and 1958. However, in our survey of nucleotide—peptide compounds (Medvedev and Khavkin, 1962) we could cite more than 100 publications on this subject. No fewer than 200 papers on this class of compounds have now appeared, yet their functions are still largely unexplained.

We cannot deduce any general conclusions regarding the nature of inductors on the grounds that one inductor has been shown to be of nucleotide—peptide nature. In each concrete case the nature of the inducing action may be different. It has been shown, in particular, that RNA is apparently the active substance secreted into the medium in the case of the chordomesoderm (Niu, 1956, 1958). An interesting investigation from this standpoint was carried out recently by Hillman and Niu (1965). Working with ex-planted chick embryos (in the fifth stage, early), these workers showed that RNA isolated from the brain of embryos aged 11-13 days induces development of the brain, while RNA isolated from the notochord induces formation of the notochord correspondingly. The RNA of these two organs thus carried information which started differentiation of the corresponding organ. Similar evidence of tissue-specific induction by RNA was obtained in an investigation of another induction system (Sanyal and Niu, 1966). It is difficult to decide at this stage, however, whether processes of this type take place in normal embryogenesis. On the other hand, differentiation of sympathetic nerve cells in embryogenesis of the mouse was found to depend on the presence of a specific factor of protein nature (Levi-Montalcini and Angeletti, 1961).

Some very interesting information on the mechanism of induction of muscle differentiation was obtained in experiments *in vitro* (Hauschka and Konigsberg, 1966). Earlier investigations in this laboratory had shown that the individual embryonic muscle cell from the chick embryo can form a differentiated macroscopic colony of multinuclear muscle cells in culture, including the formation of muscle (actomyosin) fibrils. However, differentiation of this type *in vitro* occurs only in cultures containing the solution previously used for keeping a dense population of cells. The investigation showed that the substance inducing and maintaining muscle differentiation *in vitro* is collagen, a protein normally intimately associated with muscle tissues. Collagen from rat tails was just as effective as collagen produced by fibroblasts of the embryo. Hauschka and Konigsberg suggest that in normal embryogenesis contact between myoblast and fibroblasts determines muscle specialization.

Third, inducing activity in normal embryogenesis, from the point of view of induction represented in Fig. 96, can be shown not only by a compound or group of compounds from the neighboring tissue, but in some cases simply by the general biochemical situation, arising locally but regularly for that particular tissue zone at a given period of development. In fact, certain concrete morphogenetic changes are associated with such general conditions as the oxygen concentration (aeration), the loose or compact structure of the tissue, and mechanical factors such as the conditions of stretching or relaxation of the tissues.

The importance of these factors was demonstrated very clearly by the work of Lopashov (1960) studying morphogenesis of eye structures in amphibians, notably differentiation of the eye anlage into retina and pigmented epithelium. These structures may be converted into one another for a very long time in the eye anlage under the influence of external conditions, mainly the oxygen supply.

Development of the pigmented epithelium takes place when the tissue of the eye anlage is stretched on a surface and has good contact with the environment. Metabolism can take place at high intensity also if the blood supply is good. If the eye anlage is transplanted into the wall of a blood vessel, that part of it which faces the lumen and is bathed with blood differentiates into pigmented epithelium. If, however, the anlage of the pigmented epithelium has formed folds and groups of cells deficient in blood supply, they differentiate into retinal tissue. Reactions of the same type can also be reproduced *in vitro*. If parts of the optic vesicle or optic cup are joined together into a continuous mass in saline, they develop only into retina. If, however, these parts are stretched out into a thin layer, they do not produce a retina, but in the presence of mesenchymal cells they are converted into sheets of vesicles of pigmented epithelium.

In normal development of the optic vesicle, its outer wall is rather thicker from the outset and lies in close contact with the covering ectoderm. In this solid mass of cells, accumulation of metabolic products naturally takes place easily, while the supply of oxygen is difficult. This determines development of the optic vesicle into the retina. Close connection with the ectoderm and the formation of a limiting membrane lead to stretching of the growing wall of the eye. This facilitates its conversion into a cup. At the same time, the lens is formed in the ectoderm and fills the hollow of the cup. All these close contacts maintain the intended development into a retina. As the outer wall of the vesicle turns to face inward it becomes the inner wall of the cup, and finds itself completely isolated from the external environment, since the inlet into the cup is closed by the lens. Conversely, the wall of the vesicle facing the brain is on the outside and it stretches as the cup is formed and grows. This produces a thin layer of cells, freely in contact with the external environment. Metabolic products of the cells are easily removed, and a free oxygen supply can be obtained. Just as in the saline cultures, this situation does not favor differentiation of the retina, but favors formation of the pigmented membrane. However, for this to take place, just as in cultures, mesenchymal cells must be present, and these surround the optic cup in a single layer forming a very thin membrane. Later, blood vessels penetrate into this layer of mesenchyme (to form the vascular coat of the eye), and only when this has happened and the oxygen supply is insured does the pigmented membrane of the eye attain complete development.

Induction taking place as a result of the influence of one anlage on another (heterotypic induction), although existing in many different forms and produced by many active factors, is evidently the dominant factor in formation of whole anlagen. With the transition to finer differences, to microdifferentiation, contacts and interactions between homologous cells of the anlagen (homotypic induction) become of great importance. This phenomenon of homotypic induction has been studied particularly closely in connection with morphogenesis of the vertebrate eye (Lopashov and Stroeva, 1963; Lopashov and Khoperskaya, 1967; Cahn and Cahn, 1966). The method of organ cultures, in which eye anlagen were cultivated as an unfolded layer of cells and thus isolated from the influence of other organs and tissues, proved particularly suitable for this purpose. In this case, for example, the thickness of the layer of cells becomes a differentiating factor, and under the influence of very simple homotypic contacts, subprograms of intensive and varied specialization of the cells forming the eye can evolve. Chemical and biochemical factors of homotypic inductions are at present virtually unknown.

This type of distribution of roles between heterotypic induction (a common plan of structure) and homotypic (fabrication of the details) is evidently a feature of the design of any complex structure. When a building is constructed, a basic plan is first produced, and then special drawings are made for the larger and smaller details of the rooms independently.

In his monograph, Shmal'gauzen (1964) analyzed the interactions between all the conditions present during formation of the eye which we have described above, together with other morphogenetic processes, and deduced the main conclusion that morphogenesis, with all its interactions, was established historically in accordance with principles of selection of the most expedient forms. Shmal'gauzen points out that in every case of embryonic development, with normal relationships between the parts, interactions within the system, whether between the cells of a particular tissue or between different anlagen and tissues, have the character of mutual stimulation of processes leading to the progressive development of that system.

However, it must not be forgotten that the purpose of this progressive development is the translation of genetic information of the chromosomes of individual cells, so that systems of interaction between cells are primarily systems of interaction between individual genetic systems and the developing organism, leading to an expression of activity of a type corresponding to the ultimate purpose of the morphogenetic process.

§3. Hormonal Control of Differentiation and the Influence of Hormones on Functional Activity of Genetic Systems

The importance of hormones for morphogenetic processes is brilliantly illustrated by the role of the sex hormones in development of the secondary sexual characteristics in animals at puberty, associated with many different morphogenetic acts. For a long time no attempt was made to probe deeply into the mechanism of action of most hormones, for until recently most studies of their action were concerned mainly with their effect on protein, carbohydrate, lipid, mineral, and other forms of metabolism familiar to biochemists, and on enzymes of the cytoplasm and nucleus. However, since all these processes, especially enzyme synthesis, are the result of functional activity of genetic systems, providing the cytoplasm with messenger RNA templates of specific protein synthesis, it was natural that until the link was established between current biochemical processes, and the continuous manifestation of intracellular genetic activity, workers studying the mechanism of action of hormones should virtually disregard the role of hormones in the manifestation of activity of the genes. The study of the genetic aspects of hormonal action only began quite recently and it quickly led to the conclusion that very often the genetic structures of organs are systems affecting the inductive functions of hormones through selective stimulation by the chromosomes of synthesis of corresponding forms of messenger RNA.

In this respect hormones were found to behave like embryonic inductors, also changing the spectra of active genes, but with the difference that inductors are substances with a shorter action on the tissue or system of tissues in a state of active morphogenesis, while hormones are substances with constant or, at least, prolonged action which not only induce but also constantly or periodically maintain differentiation of the tissues.

The changing pattern of hormonal action has become a means of functional regulation of metabolism through the genetic apparatus and has assumed the responsibility for many biological processes, notably certain forms of periodicity and rhythm, certain periodic phenomena of the sexual and reproductive cycles, a more labile connection between metabolism and behavior, on the one hand, and environmental conditions on the other. To illustrate the relationships between hormones and genes I shall mention only a few of the most typical and vivid examples, because any attempt to survey this problem fully would now take up considerable space, even though the systematic analysis of gene—hormone phenomena started only a few years ago.

Because the study of the problem of relationships between hormones and the genetic apparatus of cells is still in its infancy, no detailed surveys of it have yet been published. However, in two recent monographs covering certain aspects of the molecular biology of development (Bonner, 1965) and the influence of hormones on cell reproduction (Epifanova, 1965), sufficient attention has in fact been paid already to the question of hormonal action on genetic activity.

The sex hormones have their most marked action on certain tissues and organs which have been described as target organs (or tissues). A series of investigations (Mueller et al., 1958; Aizawa and Mueller, 1960, 1961) have shown that the primary action of estrogens on metabolic processes in the rat uterus is effected through an increase in activity of certain enzymes, which is associated with the synthesis of enzymes *de novo*. Later work (Gorski and Mueller, 1963; Noteboom and Gorski, 1963) showed that the primary response to the action of estrogen was not protein synthesis but synthesis of RNA and increased activity of RNA-polymerase. This synthesis and its stimulation by estrogens were suppressed by actinomycin D (Ui and Mueller, 1963), evidence that the point of application of the hormone is DNA-dependent (genetic) RNA synthesis, because actinomycin is a highly specific inhibitor of this process. Actinomycin, moreover, inhibited all histological and growth changes in the uterus associated with the action of estrogen. If actinomycin was given 1-2 h after injections of the hormone, its inhibitory action was considerably weakened: the required number of RNA templates had succeeded in being formed.

Later investigations by this group of investigators (Gorski and Nelson, 1965; Gorski, Noteboom, and Nicolette, 1965) showed that estrogen primarily stimulates synthesis of ribosomal and messenger RNA, although they were unable to find methods for determining which particular fraction of messenger RNA is synthesized preferentially. When discussing their findings they suggest that selectivity of action of the hormone on the uterus is due in fact to the presence of specific protein receptor molecules in chromosomes of the uterine cells which change their properties, in accordance with the principles of allosteric interaction, when acted upon by the estrogen, and this complex then has a specific activating effect on certain genes.

The existence of this type of genetic mechanism of action of estrogens on histological and biochemical processes in the uterus was confirmed by work in other laboratories which shed light on other aspects of this mechanism (Hamilton, 1963, 1964; Hamilton et al., 1965; Wilson, 1963; Moore and Hamilton, 1964; Greenman and Kenney, 1964).

It was shown (Wicks et al., 1965; Greenman et al., 1965) that steroid hormones stimulate synthesis of all fractions of RNA including the soluble. In the last case it is synthesis of sRNA which is stimulated and not exchange of the terminal group. However, this general stimulation is evidently a late effect. Analysis of the action of estrogen on the uterus after shorter periods of time showed (Notides and Gorski, 1966) that after 30 min synthesis of one particular protein is induced, and that this stimulation is insensitive to actinomycin D (the translation level). General stimulation of RNA and protein synthesis developed later.

Segal and co-workers (1965) showed that RNA isolated from estrogen-induced uterine tissue could induce a group of specific changes in the corresponding tissue of gonadectomized rats similar to those produced by the hormone directly. No significant changes in the histone/DNA ratio are produced in the uterine tissues by estrogen (Cohen et al., 1964).

The male sex hormone testosterone, as Williams-Ashman (1965) and Liao (1965) have shown, also acts on its target organs through stimulation of messenger RNA synthesis. It has been discovered no less clearly that the action of adrenal hormones, especially cortisone, is also associated with induction of certain changes in the genetic systems of cells.

Comparatively recently a sharp increase in activity of a number of liver enzymes in rats and other animals is produced by cortisone (Knox et al., 1956). This stimulation of enzyme (glutamate-tyrosine transaminase, tryptophan-pyrrolase, glucose-6-phosphatase, etc.) activity by cortisone in experiments on adrenalectomized rats was associated with the synthesis of these enzymes *de novo* (Kenney, 1962; Alievskaya, 1965). As might be expected, this increase in enzyme synthesis was preceded by a sharp increase in synthesis of RNA, particularly RNA of the cell nuclei (Fig. 99) (Kenney and Kull, 1963). The RNA fraction which was stimulated corresponded in its sedimentation characteristics to fast-labeled (messenger) RNA.

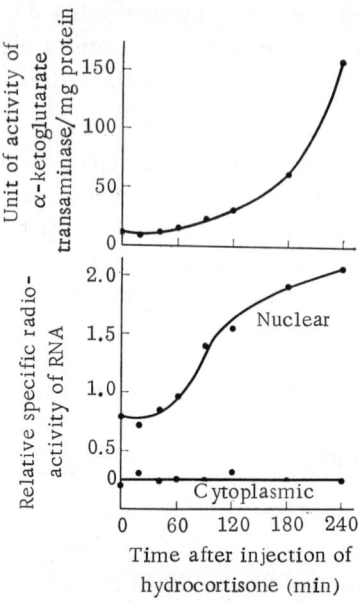

Fig. 99. Stimulation of nuclear RNA synthesis during induction of enzyme synthesis by administration of hydrocortisone (Kenney and Kull, 1963).

However, synthesis of ribosomal and soluble RNA is also stimulated by cortisone (Kenney et al., 1965). The action of actinomycin inhibited the effect of cortisone both on RNA synthesis and on activity of the enzymes.

The results of Shereshevskaya's (1962, 1963) experiments to study the effect of cortisone on RNA synthesis in the rat liver were somewhat unexpected. She observed inhibition of RNA synthesis in the liver after injections of cortisone except in old rats, in which these injections produced some stimulation of RNA synthesis. However, her experiments were performed on nonadrenalectomized rats, so that injections of the hormone evidently produced too high concentrations in the body. In old rats, the functions of the endocrine glands were weakened, and this led to the stimulatory effect.

Segal and Kim (1963) showed that cortisone stimulates synthesis of a number of enzymes very selectively. For instance, the content of glutamine-alanine transaminase, amounting to about

0.24% of the soluble liver proteins of adrenalectomized rats before receiving cortisone, was increased to 1.3% after injection of cortisone.

The action of cortisone on activity of certain enzymes (tryptophan-pyrrolase, tyrosine-α-ketoglutarate transaminase, etc.) through preliminary stimulation of DNA-dependent synthesis of certain fractions of messenger RNA has also been demonstrated by other investigations (Greengard et al., 1963; Garen, Howell, Tomkins, and Crocco, 1964; Garen, Howell, and Tomkins, 1964). Cortisone also stimulates activity of RNA-polymerase in the rat liver (Barnabei et al., 1965). Kenney, Wicks, and Greenman (1965) recently showed that synthesis of rRNA and metabolism of the terminal adenine in tRNA are also stimulated by hydrocortisone.

Important preliminary findings were reported by Lang (1965) when he spoke at the symposium in the discussion on the paper cited above (Kenney, Wick, and Greenman, 1965). Using tritium-labeled cortisone, Lang found that some of it is quickly and firmly bound with histones soon after injection.

Dahmus and Bonner (1965) showed that chromatin isolated from the liver possesses increased powers of DNA-dependent RNA synthesis after administration of cortisone to adrenalectomized rats and also showed that this difference between the control and experimental material disappears after removal of DNA-bound proteins from the chromatin.

These experiments thus confirm the possibility that the mechanism of cortisone action is through its link with chromosomal proteins. Further data on the mechanism of the RNA-stimulating action of hydrocortisone were obtained by Sluyser (1966) who showed that this hormone is bound with histones, selectively, in fact, with the histone fraction 3 characterized by a low lysine content.

Basically similar pathways of hormonal action on metabolism through their influence on selective stimulation of chromosomal synthesis of particular RNA fractions, with selective synthesis of corresponding enzymes in the target organ, have also been demonstrated by studies of other hormones, including pituitary hormones (Farese and Reddy, 1963; Korner, 1963, 1965), insulin (Wool and Munro, 1963; Weber et al., 1965), and thyroxin (Tata, 1963, 1964).

However, thyroxin was also found to stimulate protein synthesis at the ribosomal level, independently of its effect on mRNA synthesis (Sokoloff et al., 1964).

Hormonal effects on ribosomal protein synthesis (translation control) rather than at the gene level have been demonstrated in connection with hormonal stimulation (by luteinizing hormone) of the synthesis of steroid hormones by the ovaries (Gorski and Padnos, 1966). The possibility is not ruled out that in phenomena of this type activation may take place through the "unmasking" of messenger RNA as mentioned previously.

Comparison of the influence of different hormones on changes in the pattern of newly synthesized RNA clearly revealed the selectivity of this influence at the level of DNA-dependent RNA synthesis (Kidson and Kirby, 1964).

The direct action of hormones on transcription or translation of genetic information is proved by the exceptional rapidity of the reaction. In a number of cases the effect of hormones on synthesis of RNA and proteins was found 1-2 min after injection (Kidson, 1967; Means and Hamilton, 1966).

These experimental findings show clearly that the genetic apparatus, as the most stable, yet at the same time economic, system regulating metabolism, can evidently act also as a special type of intermediate link transforming external and internal influences into purposive biochemical reactions.

It is only because of genetic differentiation of the cells that the reaction to biologically active compounds can be so selective and specific, at many different levels, with respect to the type of organ (receptor organs or target organs), the type of tissue, and the type of locus (not all to the same extent) activated and, correspondingly, to the type of enzyme and biochemical reaction activated.

At the same time, some hormonal inductions of rapid action are evidently brought about by activation of preexisting but masked forms of messenger RNA.

§4. Mechanism of Rapid Induction at the Cytoplasmic Level. A Possible Model

I mentioned in Chapter 5 cases when a "masked" messenger RNA is formed, with a biochemical program for cell differentiation (specialized protein synthesis) which does not operate until sometime after its formation. According to Spirin's (1966) suggestion, this RNA is masked by the formation of protein capsules for its molecules (the formation of "informosomes"). In the case of oogenesis, the activator of this RNA, bringing about its participation in protein synthesis, is fertilization. However, Spirin also considers that many other embryonic inductions may take place in the same way. Some form of inducing agent acts, not on the level of gene or chromosome, but on preformed, masked RNA, thereby initiating its functional activity. Such a mechanism is evidently particularly useful for rapid reactions in which activation of the gene and formation of the required amount of mRNA act as a break in the process. In such cases the receptor for the inducing agent is not the regulator gene, but the protein membrane of mRNA, the protein component, and it evidently cannot be ruled out that this component may possess tissue specificity to account for selectivity of action of the inducing factor.

§5. Intrachromosomal Factors Determining the Level of Activity and Pattern of Function of Genes and Gene Systems

When examining regulation of the functions of the bacterial genome we had the occasion to analyze the work of regulator genes which, depending on the conditions, switch on various genes or entire operons of the bacterial chromosome. In multicellular organisms, the existence of a complex system of regulator gene is well established, and we have already examined a possible model of the working of one such regulator in connection with the regulation of the change from synthesis of fetal hemoglobin to synthesis of adult hemoglobin. Many inducing agents and hormones also exert their action on the spectrum of working genes through regulator genes which possess receptors for a particular inductor or hormone. However, besides phenomena of this type in which function of the genome bears a constant correlation with external extrachromosomal factors through special systems of regulator genes, there are other factors which indicate the existence of a genotypically programmed intrachromosomal regulation of the expression of particular genes, such as the genes which determine the group of cytoplasmic and other characteristics which, in turn, influence structural and regulator genes. Additionally, purely self-contained intrachromosomal interactions are possible, not emerging into the cytoplasm. The possibility of existence of such regulatory systems has been postulated in genetics for a long time, in the classical genetics of *Drosophila*, long before the emergence of bacterial genetics. The foundations of this trend in genetics were laid, or it would be more correct to say the problem first arose, when the position effect on gene activity was discovered by Sturtevant (1925) during the study of the Bar mutation (bar-shaped eyes), a dominant mutation in the X-chromosome of *Drosophila*. I mentioned this mutation in Chapter 4 in connection with its localization. The Bar mutation, as was stated there, is connected with the formation (by crossing over) of duplication of a group of genes in the Bar locus, while the double Bar mutation is associated with the tripling of this group.

By studying different variants of localization of the Bar segment and comparing them with the number of facets on each eye, Sturtevant showed that when the Bar segments lie in one chromosome their action (determined by a decrease in the number of facets) is manifested more strongly than when they lie in different chromosomes. This was called the position effect. Sturtevant's work marks the beginning of an extensive group of investigations, so that what was virtually a new trend in genetics arose, and very many different mutations were studied in which the mutant gene was placed (by translocation) in different surroundings. In most cases, in new surroundings, the phenotypic manifestation of the mutation is changed, usually weakened. The effect of this change in position sometimes extends not only to the mutant locus, but also to its new neighbors, the degree of disturbance studied from this locus to the next diminishing with distance. This variant of the position effect also belongs to the mosaic type. Dubinin and Sidorov (1935) first showed that the mosaic position effect is reversible, and if the gene is taken from its new environment (from the heterochromatin region) the wild-type characteristics are restored.

Although several attempts have been made to suggest a mechanism of the position effect (Serra, 1960), and the factual and theoretical material in this field has recently been analyzed from the standpoint of regulation of development (Sand, 1965), no acceptable explanation of this phenomenon has yet appeared and its molecular basis cannot be postulated.

Clear and convincing evidence that genes can influence each other in many different ways was obtained in a series of brilliant investigations on the corn plant, a classical genetic subject, by McClintock (1956a, b; 1961). Since these experiments, notwithstanding their genetic significance, are not directly concerned with molecular biology and the genetics of morphogenesis, I shall only mention here the main conclusions reached. A detailed account of McClintock's method and a discussion of her results can be found in several modern textbooks of genetics, notably that of Sager and Ryan (1961).

In her first experiments McClintock discovered extremely unusual genetic elements (controller genes) in corn which she called Ds and Ac (from the English words dissociation and activator). The Ds gene could produce ruptures and mutations in any part of the chromosome where it lay and could move from one locus to another. The second element Ac was essential for manifestation of Ds activity, and the time when the results of this activity become manifest varied with the number of Ac elements present. The Ac element could also change its position and cause ruptures and mutations, and was similar to the Ds in this respect. They differed, however, in that Ds activity could be exhibited only in the presence of a C, whereas Ac could act by itself.

Translocation of Ac and Ds elements into different parts of the chromosomes was determined by the linking of their action with different markers. Pigmentation of corn grains proved to be a convenient system for study of the phenotypic action of Ac and Ds elements.

The Ds factor caused ruptures (detected by loss of dominant markers lying side by side and connected with pigmentation), the timing of the ruptures (at the beginning, in the middle, or at the end of development of the grains) depending on the number of Ac alleles present in the chromosome. If one Ac allele was present, ruptures (somatic pigmentation mutations) took place in the early stages of development of the grain, while if two or three alleles were present they took place progressively later and later. Ac apparently has metabolic control over Ds. The Ds factor could migrate suddenly within the genome, and in the new position it no longer produced ruptures but changes in the phenotypic expression of the genes, resembling mutations. The number of mutations taking place in embryonic cells under the influence of Ds subsequently was stable, even in the presence of Ac. This gave rise to the hypothesis that in such mutations the Ds element disappeared from that particular locus, and in fact in many cases Ds activity could be found subsequently in a different place. McClintock interpreted these observations absolutely directly, as indicating the physical migration of the Ds element from one place to another.

In other experiments McClintock identified new elements similar to Ac and Ds, but with a rather different type of action and relationship to the conditions and she attempted to draw an analogy between the controlling elements in the corn genome and the regulatory system of the bacterial genome.

McClintock's experiments were confirmed and extended by the work of other investigators, notably Brink. In his recent survey (Brink, 1964), he summed up the importance of the whole series of investigations on unstable loci in plants as a series of generalizations linking this branch of genetics with general problems of regulation of gene activity in all its aspects, including the morphogenetic. Brink considers that the facts described and their analysis in the writings of McClintock and other authors confirmed the validity of the following observations.

1. Chromosomes consist of two clearly distinguished elements: a) genes in the classical sense of specific determinants (structural genes) and b) elements called controllers, additional to the ordinary genes. The controllers regulate the expression of the genes during development of the individuals. The existence of controllers migrating from one part of the genome into others, and also of untransportable controllers, has been proved.

2. The presence of unstable loci has been demonstrated, their high mutability, however, depending not on changes in themselves, but on the presence of the controller gene.

3. Controllers can change their position in the genome and the number of elements. A case of cyclic change in a controller

has been described, in which it can change from the active to the inactive state and vice versa (McClintock, 1961).

4. The action of controller genes is usually to inhibit completely or partially the gene associated with it (the target gene). Removal of the controller allows full expression of the gene.

5. Mutant phenotypes arising at a certain loci under the influence of controllers are often similar to corn mutations already known; it can therefore be assumed that many mutations are the result of migration of controllers.

6. Several controllers have been discovered in the corn genome and the number of those being found for the first time continues to increase. This suggests that in different variants they fill the genome, and many of them have not yet been detected in the appropriate systems.

7. Controllers may act independently but often their action is dependent on the presence of other controllers in the same nucleus. The association in this case is specific. This suggests that different controllers in fact differ from one another.

8. At the same time, the same controller can change the expression of different genes and, conversely, the same gene can be changed by the action of different controllers.

9. In some cases the controller has a spreading effect, i.e., it affects several neighboring genes with an activity that diminishes with increasing distance.

10. Some controllers can be transferred into very many loci in different chromosomes; however, wandering controllers are sometimes firmly fixed in one particular locus.

11. In some cases a correlation is found between the stage of development of the plants and the phenotypic expression of instability of the locus. McClintock postulates from this that controllers play an important role in regulating the time of action of genes in the course of individual development.

12. It has not yet proved possible to link the controlling elements with any definite zones in the structure of chromosomes.

Besides these facts indicating the existence of mechanisms of control of genetic activity, built into the genome, in multicellular organisms (although not revealing their molecular nature or mode of action), the case of genetic control of the change from synthesis of fetal hemoglobin to synthesis of adult hemoglobin by the hemopoietic systems of animals at the end of pregnancy, examined in detail in Chapter 5, can also be cited. Genetic analysis shows that this change also is produced by a special regulator gene.

To some extent the existence of suppressor mutations, in which a mutation in a new locus alters the degree of manifestation of another preexisting mutation, can also be taken as evidence of the activity of regulator systems. So far many suppressor mutations have been described at all levels of organization of living systems, although the mechanism of suppression may differ in many cases.

Apart from the system for the change from fetal to adult hemoglobin in animals, no other system has yet been found which is convenient for study of controller genes similar to the controlling elements studied in corn. However, fragments of such systems are beginning to be discovered. Paigen (1961) found that the time of appearance of β-glucuronidase in mice in ontogenesis is under genetic control, under the influence of a locus closely connected or perhaps even united with the locus determining the structure of this enzyme. Synthesis of this enzyme is one of the simple Mendelian characteristics, but by crossing mutants differing in relation to this enzyme it has been shown that the time of appearance and the pattern of increase in the content of this enzyme in the liver and other organs in time are also under genetic control, but the gene controlling the temporal pattern lies in the same chromosome.

It is not yet clear how controller genes act on other genes, but these mechanisms, in the case of mutual effects of one chromosome on another, may have outlets into the nuclear juice and cytoplasm. Several such genes have been discovered in the axolotl (Humphrey, 1964, 1966). These genes act on development of the axolotl from oogenesis until the larval stage. The action of these genes on the other genes takes place through the cytoplasm, sometimes in a very complex order. Attention has so far been concentrated on the mutant 0 gene, the presence of which causes a specific defect in the cytoplasm of the oocyte. The 0 gene is a simple recessive, while $+/0$ heterozygotes develop like the wild type, and 0/0 homozygotes give anomalies in the larval stage, expressed as loss of regenerating power, slow growth, and so on. The 0/0 combination arises as a variant during fertilization or crossing of heterozygotes, and oogenesis was thus normal in each of the original heterozygotes. However, in the developing 0/0 homozygote, the action of the 0 gene is manifested in oogenesis and the oocytes possess a defect, so that after fertilization with either 0 sperm or + sperm cleavage stops at the gastrula stage. The 0 gene changes something in the cytoplasm in oogenesis, and a deficiency of some cytoplasmic factor prevents development at a time when reprogramming of the nucleus normally takes place and differentiation begins. This interesting genetic phenomenon was used by Briggs and Cassens (1967) when they attempted to discover the nature of the cytoplasmic factor linking the 0 mutation with the arrest of cleavage. They found that injection of cytoplasm from a normal $+/+$ or even from a $+/0$ oocyte into a defective oocyte corrects the action of the 0 gene. Fertilized 0/0 oocytes injected with normal cytoplasm at the stage of 1-2 cells show significantly better development. The correcting component is present in highest concentration in nuclear juice. The nature of the component has not yet been established, although the possibility has not been ruled out that it is a special "masked" RNA.

As I mentioned above, there has as yet been no successful attempts to elucidate the molecular mechanism of the position effect, and no such mechanism has been suggested. An interesting thought in this direction was expressed by Taylor (1964). On the basis of differences in the time of replication along the chromosome he suggested that the same gene, but in different parts of the chromosome, and under different conditions, falls into a special type of repression which spreads to systems of combined genes larger than replicon operons. Each chromosome contains a series of replicons, each with its own regulation (principally of replication), and each replicon consists of a series of operons.

§6. Periodicity and Rhythm of Cell Biochemical Activity and Its Genetic Programming. The Genetic Clock

Biological rhythms of physiological and biochemical activity (diurnal, monthly, seasonal, and so on) are associated with several aspects of the genetic problem as a whole, and especially with the problem of temporal regulation of vital functions (the biological clock). The literature on the various aspects of biological rhythms is extensive, and a good introduction to the topics covered by investigations of this biological phenomenon can be obtained from several monographs (Bünning, 1958; Emme, 1962; Cloudsley-Thompson, 1961; Goodwin, 1963; Sollberger, 1962).

However, the only part of this extensive field which concerns us at present is the genetic regulation of rhythms and, consequently, the temporal, rhythmic regulation of activity of the genetic system itself.

The essence of a rhythm is that a process such as synthetic activity, the rate of oxidation, secretion, and so on, shows a well-defined temporal periodicity, usually coinciding with changes in the external environment: diurnal, monthly, seasonal. At one time or period (daytime, summer) these processes take place actively, but at another time or period (night, winter, and so on) they are strongly inhibited. However, in many cases of periodicity the link with changes in the environment is purely evolutionary. Periodicity is still present under constant environmental conditions, and both the type and amplitude of periodicity are inherited characters transmitted to the progeny and behave as genetic properties during crossing.

The usual explanation of this phenomenon is that organisms during their historical development have formed autonomous oscillator systems transmitted genetically, parallel in their action with natural periods, and capable of adjustment. These systems are somehow connected with the controlling systems of metabolism, i.e., with the chromosomes.

The problem to be faced in this case is: what is it which makes the genetic apparatus of the cells (usually almost synchronously) function at an increased rate and what causes it to stop or, more correctly, lowers the intensity of its activity after periodic intervals of time in the case of an inherited rhythm. I stress this type of rhythm, because by no means all rhythms are determined entirely by genetic factors. Many rhythms are directly connected with changes in the environment, information concerning which is transmitted to the genetic apparatus through a complex system of biochemical connections.

In the case of inherited rhythms (biological clocks) the problem is thus essentially to determine the nature of the oscillator system, the system switching on and off the complex group of processes leading to periodic stimulation and inhibition of metabolism.

No solution to this problem has yet been obtained experimentally. Many workers have tried to solve it by studying the action of various substances and specific inhibitors on the parameters of some of these rhythms, but no precise data have yet been obtained in this direction. The first important advance toward discovery of specific inhibition of circadian rhythms by means of chemical compounds was made by the use of actinomycin D (Karakashian and Hastings, 1962), a specific inhibitor of DNA-dependent RNA synthesis, and of puromycin. The circadian rhythms of bioluminescence and photosynthesis of the alga *Gonyaulax polyedra* were used in these experiments. These inhibitors of nucleic acid metabolism very characteristically inhibited the rhythmic nature of these processes, abolished the peaks and maintained the process at the minimal level. It is interesting to note that the action of actinomycin (as opposed to puromycin) was not manifested immediately, but after two or three normal cycles. The impression was created that this rhythm is associated with stable RNA templates of a certain type. Puromycin (inhibiting protein synthesis on RNA) acted immediately, but actinomycin acted only after the reserves of preexisting templates were exhausted. Chloramphenicol (acting at the level of liberation of protein from the templates) had no effect. The results of these experiments are illustrated in Fig. 100.

The link between periodicity and genetic systems of organisms has been examined by Kalmus (1962). It has also been shown

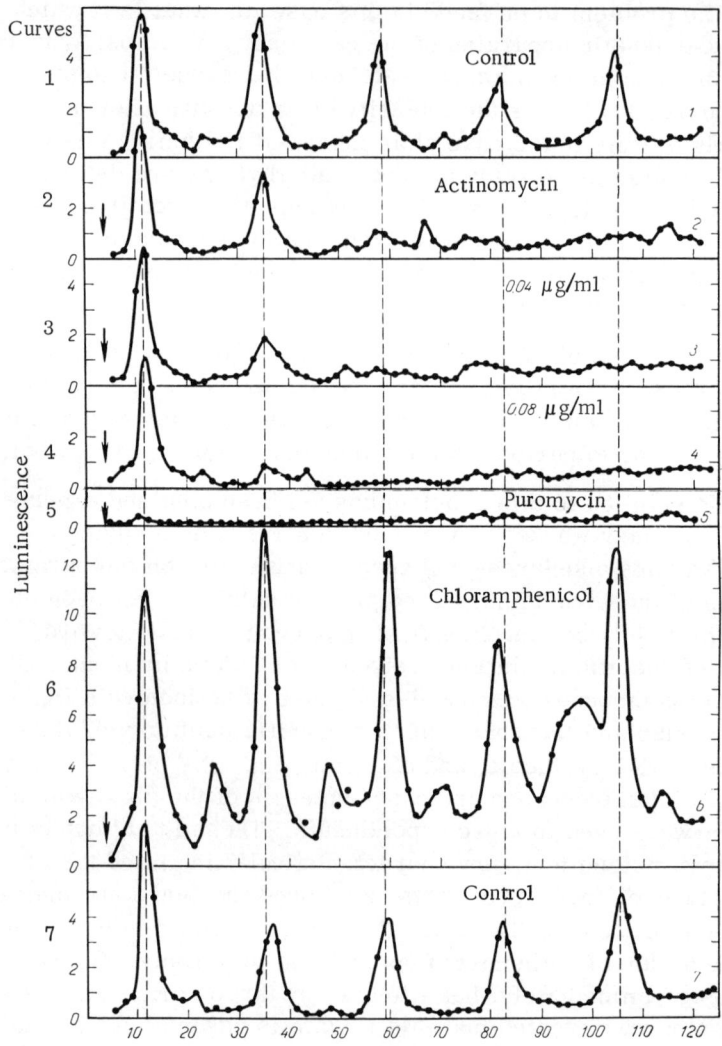

Fig. 100. Action of actinomycin D (concentrations given in figure), chloramphenicol (3×10^{-4} M), and puromycin (10^{-5} M) on the established rhythm of bioluminescence (of the alga *Gonyaulax polyedra*. Significance of curves: 1) Control; 2-4) different concentrations of actinomycin; 7) control for experiments with puromycin and chloramphenicol (5 and 6). Cells grown alternately in darkness and light for 12 h each, and then transferred to constant weak illumination and constant temperature. Beginning of experiments indicated by arrows. Luminescence in conventional units (Karakashian and Hastings, 1962).

(Schweiger et al., 1964) that circadian rhythms of respiration in the unicellular alga *Acetabularia* are associated with the nucleus. If nuclei are transplanted (exchanged) in lines with opposite rhythm, the rhythms of the cytoplasmic processes change to correspond to the nuclear program. However, after removal of the nucleus, some rhythms remain in this alga, and inhibition of some rhythms generally speaking does not necessarily influence others (Hastings, 1962).

New evidence of the connection between diurnal and other rhythms in algae and synthesis of RNA and proteins has recently been obtained by Driessche (1966) and Feldman (1967).

Literature Cited

Aizawa, Y., and Mueller, G. C., 1960, Federation Proc., 19:170.
Aizawa, Y., and Mueller, G. C., 1961, J. Biol. Chem., 236:381.
Alievskaya, L. L., 1965, Vopr. Med. Khimii, 9 (3):57.
Astaurov, B. L., 1937, Biol. Zh., 6:1.
Astaurov, B. L., 1947, Zh. Obshch. Biol., 8:421.
Astaurov, B. L., 1948, Usp. Sovr. Biol., 25:49.
Astaurov, B. L., and Ostryakova-Varshaver, V. P., 1957, Izv. Akad. Nauk SSSR, Ser. Biol., 2:154.
Barnabei, O., Romano, B., and Di Bitonto, G., 1965, Arch. Biochem. Biophys., 109:266.
Bonner, J., 1965, The Molecular Biology of Development, Oxford Univ. Press, New York—Oxford.
Brachet, J., 1960, The Biochemistry of Development, Pergamon Press, New York—London.
Briggs, R., and Cassens, G., 1966, Proc. Nat. Acad. Sci., USA, 55:1103.
Briggs, R., and King, T. J., 1952, Proc. Nat. Acad. Sci., USA, 38:455.
Briggs, R., and King, T. J., 1953, J. Exp. Zool., 122:485.
Briggs, R., and King, T. J., 1957, J. Morphol., 100:269.
Briggs, R., and King, T. J., 1960, Develop. Biol., 2:252.
Briggs, R., and King, T. J., 1959, In: The Cell, ed. J. Brachet and D. Mirsky, Vol. I. Academic Press, New York, p. 538.
Briggs, R., Signoret, J., and Humphrey, R. R., 1964, Develop. Biol., 10:233.
Brink, R. A., 1964, Am. Naturalist, 98:193.
Bünning, E., 1958, Die physiologische Uhr, Springer, Berlin.
Cahn, R. D., and Cahn, M. B., 1966, Proc. Nat. Acad. Sci., USA, 55:106.
Child, C. M., 1946, Organizers in Development and Organizer Concept.
Cloudsley-Thompson, J. F., 1961, Rhythmic Activity in Animal Physiology and Behaviour, Academic Press, New York—London.
Cohen, S. N., Spicer, S. S., and Yielding, K. L., 1964, Biochim. Biophys. Acta, 87:511.
Collier, J. R., 1965, In: The Biochemistry of Animal Development, Vol. 1, ed. R. Weber, Academic Press, New York, p. 203.

Dahmus, M. E., and Bonner, J., 1965, Proc. Nat. Acad. Sci., USA, 54:1370.
Davidson, E. H., Haslett, G. W., Finney, R. J., Allfrey, V. G., and Mirsky, A. E., 1965, Proc. Nat. Acad. Sci., USA, 54:696.
Driessche, Th. V., 1966, Biochim. Biophys. Acta, 126:456.
Dubinin, N. P., and Sidorov, B. N., 1935, Biol. Zh., 4:555.
Ebert, J., 1965, Interacting Systems in Development, Holt, Rinehart and Winston, New York.
Elsdale, T., and Johns, K., 1965, Usp. Sovr. Biol., 59:470.
Émme, A. M., 1962, Clocks in Living Nature, Izd. Sovetskaya Rossiya, Moscow.
Epifanova, O. I., 1965, Hormones and Cell Multiplication, Izd. Nauka, Moscow.
Farese, R. V., and Reddy, W. J., 1963, Biochim. Biophys. Acta, 76:145.
Feldman, J. F., 1967, Proc. Nat. Acad. Sci., USA, 57:1080.
Fischberg, M., Gurdon, J. B., and Elsdale, T. R., 1958, Nature 181:424.
Fischberg, M., Gurdon, J. B., and Elsdale, T. R., Exp. Cell Res., Suppl. 6:161.
Gallien, L., 1966, Ann. Biol., 5 (5-6):241.
Garren, L. D., Howell, R. R., and Tomkins, G. M., 1964, J. Mol. Biol., 9:100.
Garren, L. D., Howell, R. R., Tomkins, G. M., and Crocco, R. M., 1964, Proc. Nat. Acad. Sci., USA, 52:1121.
Gorski, J., and Mueller, G. C., 1963, Arch. Biochem. Biophys., 102:21.
Gorski, J., and Nelson, J., 1965, Arch. Biochem. Biophys., 110:284.
Gorski, J., Noteboom, W. D., and Nicolette, J. A., 1965, In: Symp. on Hormonal Control of Protein Biosynthesis, J. Cell Comp. Physiol., 66, Suppl. 1:91.
Gorski, J., and Padnos, D., 1966, Arch. Biochem. Biophys., 113:100.
Goodwin, B. C., 1963, Temporal Organization in Cells, Academic Press, New York—London.
Greengard, O., Smith, M. A., and Acs, G., 1963, J. Biol. Chem., 238:1548.
Greenman, D. L., and Kenney, F. T., 1964, Arch. Biochem. Biophys., 107:1.
Greenman, D. L., Wicks, W. D., and Kenney, F. T., 1965, J. Biol. Chem., 240:4420.
Grobstein, C., 1962, J. Cell. Comp. Physiol., Suppl. 1, Vol. 60(2):35.
Gurdon, J. B., 1960, J. Embryol. Exp. Morphol., 8:327.
Gurdon, J. B., 1961, Heredity, 16:305.
Gurdon, J. B., 1962a, J. Embryol. Exp. Morphol., 10:622.
Gurdon, J. B., 1962b, Devel. Biol., 4:256.
Gurdon, J. B., 1962c, Develop. Biol., 5:68.
Gurdon, J. B., 1963, Quart. Rev. Biol., 38:54.
Gurdon, J. B., 1964, Advan. Morphogenesis, 4:1.
Gurdon, J. B., and Brown, D. D., 1965, J. Mol. Biol., 12:27.
Gurdon, J. B., Elsdale, T. R., and Fischberg, M., 1958, Nature, 182:64.
Hamilton, T. H., 1963, Proc. Nat. Acad. Sci., USA, 49:373.
Hamilton, T. H., 1964, Proc. Nat. Acad. Sci., USA, 51:83.
Hamilton, T. H., Widnell, C. C., and Tata, J. R., 1965, Biochim. Biophys. Acta, 108:168.
Harris, H., 1967, J. Cell Sci., 2:23.
Hastings, J. W., 1962, In: Proc. XXII Intern. Congr. Physiol. Sci., Vol. I, p. 1, 37.
Hauschka, S. D., and Konigsberg, I. R., 1966, Proc. Nat. Acad. Sci., USA, 55:119.
Hennen, S., 1963, Develop. Biol., 6:1331.

Hennen, S., 1965, Develop. Biol., 11:243.
Hillmann, N. W., and Niu, M. C., 1965, Proc. Nat. Acad. Sci., USA, 50:486.
Hommes, F. A., van Leeuwen, G., and Zilliken, F., 1962, Biochim. Biophys. Acta, 56:320.
Humphrey, R. R., 1964, J. Exp. Zool., 155:139.
Humphrey, R. R., 1966, Develop. Biol., 13:57.
Kalmus, H., 1962, Ann. N. Y. Acad. Sci., 98:1083.
Karakashian, M. W., and Hastings, J. W., 1962, Proc. Nat. Acad. Sci., USA, 48:2130.
Kenney, F., 1962, J. Biol. Chem., 237:3495.
Kenney, F. T., Greenman, D. L., Wicks, W. D., and Albritton, W. L., 1965, In: Advances in Enzyme Regulation, Vol. 3, Pergamon Press, London, p. 1.
Kenney, F., and Kull, F. J., 1963, Proc. Nat. Acad. Sci., USA, 50:493.
Kenney, F. T., Wicks, W. D., and Greenman, D. L., 1965, J. Cell Comp. Physiol., 66, Suppl. 1:125.
Kidson, C., 1967, Nature, 213:779.
Kidson, C., and Kirby, K. S., 1964, Nature, 203:599.
King, T. J., and Briggs, R. J., 1954, J. Embryol. Exp. Morphol., 2:73.
King, T. J., and Briggs, R., 1955, Proc. Nat. Acad. Sci., USA, 41:321.
King, T. J., and Briggs, R., 1956, Cold Spring Harbor Symp. Quant. Biol., 21:271.
King, T., and Di Berardino, M. A., 1954, Ann. N. Y. Acad. Sci., 126:115.
King, T. J., and McKinnell, R. G., 1960, In: Cell Physiology and Neoplasia, University of Texas, Austin, p. 591.
Knox, W. E., Auerbach, U. H., and Lin, E. C., 1956, Physiol. Rev., 36:164.
Korner, E., 1963, Biochem. Biophys. Res. Comm., 13:386.
Korner, A., 1965, J. Cell Comp. Physiol., 66, Suppl. 1:153.
Kühn, A., 1955, Vorlesunger über Entwicklungsphysiologie, Berlin-Göttingen-Heidelberg.
Lang, N., 1965, J. Cell Comp. Physiol., 66, Suppl. 1:132.
Lash, J. W., Hommes, F. A., and Zilliken, F., 1962, Biochim. Biophys. Acta, 56:313.
Lehman, H. E., 1957, In: The Beginnings of Embryonic Development, Am. Assoc. Advan. Sci., Washington, D. C., p. 201-230.
Levi-Montalcini, R., and Angeletti, P. U., 1961, Quart. Rev. Biol., 36:99.
Liao, S., 1965, J. Biol. Chem., 240:1236.
Lopashov, G. V., 1945, In: Research Abstracts of Biological Departments of the Academy of Sciences of the USSR, Izd. AN SSSR, p. 88.
Lopashov, G. V., 1960, Mechanisms of Development of Anlagen of the Eyes in Vertebrate Embryogenesis, Izd. AN SSSR, Moscow.
Lopashov, G. V., 1961, Zh. Obshch. Biol., 22:241.
Lopashov, G. V., and Stroeva, O. G., 1963, Development of the Eye in the Light of Experimental Research, Izd. AN SSSR, Moscow.
Lopashov, G. V., and Khoperskaya, O. A., 1967, Dokl. Akad. Nauk SSSR, 175:962.
McClintock, B., 1956a, Brookhaven Symp. Biol., 8:58.
McClintock, B., 1956b, Cold Spring Harbor Symp. Quant. Biol., 21:197.
McClintock, B., 1961, Am. Naturalist, 95:265.
McKinnell, R. G., 1962a, Am. Zool., 2:430.
McKinnell, R. G., 1962b, J. Heredity, 53:199.

Means, A. R., and Hamilton, T. H., 1966, Proc. Nat. Acad. Sci., USA, 56:1594.
Medvedev, Zh. A., and Khavkin, E. E., 1962, Isvest. Timiryazevsk. Sel'skokhoz. Akad., No. 2:188.
Mitskevich, M. S., 1965, Usp. Sovr. Biol., 60:287.
Moore, J. A., 1957, Exp. Zool., 105:349.
Moore, J. A., 1958, Exp. Cell Res., 14:532.
Moore, J. A., 1962, J. Cell Comp. Physiol. Suppl. I, Vol. 60(2):19.
Moore, R. J., and Hamilton, T. H., 1964, Proc. Nat. Acad. Sci., USA, 52:439.
Mueller, G. C., Herranen, A. M., and Jervell, K. F., 1958, Recent Progr. Hormone Res., 14:95.
Needham, J., 1942, Biochemistry and Morphogenesis, Cambridge Univ. Press.
Neifakh, A. A., 1962, Relationships between Nucleus and Cytoplasm in Development, A. N. Severtsov Institute of Animal Morphology, Moscow.
Neifakh, A. A., 1963, Zh. Vses. Khim. Obshchestva im. D. I. Mendeleeva, 8:403.
Nikitina, L. A., 1964, Dokl. Akad. Nauk SSSR, 156:1468.
Niu, M. C., 1956, In: Cellular Mechanisms of Differentiation and Growth, ed. D. Rudnick, Princeton Univ. Press, Princeton, p. 155.
Niu, M. C., 1958, Proc. Nat. Acad. Sci., USA, 44:1264.
Noteboom, W., and Gorski, J., 1963, Proc. Nat. Acad. Sci., USA, 50:250.
Notides, A., and Gorski, J., 1966, Proc. Nat. Acad. Sci., USA, 56:230.
Paigen, K., 1961, Proc. Nat. Acad. Sci., USA, 47:1641.
Raven, C. P., 1961, Oogenesis, the Storage of Developmental Information, Pergamon Press, New York—London.
Rott, N. N., 1962, Usp. Sovr. Biol., 54:355.
Sager, R., and Ryan, F., 1961, Cell Heredity, Wiley and Sons, New York.
Sambuichi, H. J., 1961, Sci. Hiroshima Univ. B I. 20:1.
Sanyal, S., and Niu, M. C., 1966, Proc. Nat. Acad. Sci., USA, 55:743.
Sand, S. A., 1965, Am. Naturalist, 99:33.
Saxen, L., and Toivonen, S., 1962, Primary Embryonic Induction, Academic Press, New York.
Schweiger, E., Wallraff, H. G., and Schweiger, H. G., 1964, Science, 146:658.
Segal, H., and Kim, Y. S., 1963, Proc. Nat. Acad. Sci., USA, 50:912.
Segal, S. J., and Davidson, O. W., and Wada, K., 1965, Proc. Nat. Acad. Sci., USA, 54:782.
Seidel, F., 1932, Arch. Entwicklungsmech. Organ., 126:213.
Seidel, F., 1938, Arch. Entwicklungsmech. Organ., 138:345.
Serra, J. A., 1960, Rev. Portuguesa Zool. Biol. General., 2:153.
Shereshevskaya, T. M., 1962, Transactions of the Research Institute of Biology and Biological Faculty of Khar'kov University, Vol. 33-34, p. 95.
Shereshevskaya, T. M., 1963, Ukr. Biokhim. Zh., 35:656.
Shmal'gauzen, I. I., 1964, Regulation of Morphogenesis in Individual Development, Izd. Nauka, Moscow.
Signoret, J., 1965, Arch. Biol., 76:591.
Slyser, M., 1966, J. Mol. Biol., 19:591.
Smith, L. D., 1965, Proc. Nat. Acad. Sci., USA, 54:101.
Sokoloff, L., Francis, C. M., and Campbell, P. L., 1964, Proc. Nat. Acad. Sci., USA, 52:728.

Solberger, A., 1962, Biological Rhythm Research, Elsevier Publ. Company, Holland.
Spemann, H., 1916, S. B. Ges. Naturf. Fr. Berl., 9:306.
Spemann, H., 1918, Arch. Entwicklungsmech. Organ., 43:448.
Spemann, H., 1928, Z. Wiss. Zool., 132:105.
Spirin, A. S., 1966, Zh. Evol. Biokhim. Physiol., 2:285.
Stroeva, O. G., and Nikitina, L. A., 1960, Zh. Obshch. Biol., 21:335.
Sturtevant, A. H., 1925, Genetics, 10:117.
Subtelny, S., 1965, J. Exp. Zool., 159:47, 59.
Tata, J. R., 1963, Nature, 197:1167.
Tata, J. R., 1964, Biochim. Biophys. Acta, 87:528.
Taylor, J. H., 1964, In: Symp. Intern. Soc. Cell Biol., Vol. 3. Cytogenetics of Cells in Culture, Academic Press, New York, p. 175.
Toivonen, S., 1963, Usp. Sovr. Biol., 55:87.
Ui, H., and Mueller, G. C., 1963, Proc. Nat. Acad. Sci., USA, 50:256.
Weber, G., Singhal, R. L., and Srivastava, S. K., 1965, Proc. Nat. Acad. Sci., USA, 53:96.
Wicks, W. D., Greenman, D. L., and Kenney, F. T., 1965, J. Biol. Chem., 240:4414.
Williams-Ashman, H. G., 1965, J. Cell Comp. Physiol., 66, Suppl. 1:111.
Wilson, J., 1963, Proc. Nat. Acad. Sci., USA, 50:93.
Wool, I. G., and Munroe, A. J., 1963, Proc. Nat. Acad. Sci., USA, 50:918.
Yamada, T., 1962, J. Cell Comp. Physiol., Vol. 60, No. 2, Suppl. 49.

Chapter 8

Molecular Mechanisms Programming Morphogenesis and Differentiation. A Theoretical Analysis

Introduction

Although in previous chapters in some cases we have attempted a theoretical analysis of some aspects of morphogenesis, the main purpose of these chapters was to describe the up-to-date factual material, presented systematically and in logical sequence, and also to indicate the problems still requiring explanation and further experimental analysis. We examined many phenomena and processes concerned with morphogenesis on its molecular-genetic plane. However, we did not examine many more phenomena and processes because of absence or inadequacy of the available factual material.

In the study of most biological, biochemical, and molecular-genetic problems, just as in the study of many other scientific problems, a full analysis of the facts obtained by experimental investigation of actual phenomena can usually elucidate the mechanisms of the phenomenon, thereby eliminating the need for theoretical hypotheses. In this case such hypotheses as are put forward are simply a research tool enabling the best choice of productive trends of experimental investigation to be made, thereby eliminating chaos from experimental research and providing a plan for the regular conduct of the investigation. The following example will illustrate this point. Some years ago a series of hypotheses was developed to represent the possible structure of hemoglobin A and its difference from fetal hemoglobin, using indirect characteristics. At the present time, when the sequence of amino acids in all the subunits (α, β, and γ) composing these hemoglobins is known, and their tertiary and quaternary structures have been studied, the need for hypotheses of this type has vanished and the structure of the hemoglobins and the nature of their differences can be clearly demonstrated by accurate factual material.

Ultimately the same situation must arise at a given moment in relation to morphogenesis and differentiation. When we have precise facts to explain the molecular-genetic mechanism of programmed development of the organism and how it is manifested at different levels of morphogenesis, naturally there will be no need for hypothetical explanations and models.

In this chapter I shall indicate how far from this stage in the study of morphogenesis we still are, and I shall consider whether the available factual material can give us any precise idea of the mechanisms of purposive development of individual forms or whether the light they shed on these problems is still insufficient so that hypotheses and models are necessary to make up for the deficient factual material.

Anticipating the final answer to the question posed in this chapter, we must admit that the mechanisms of morphogenesis cannot yet be clearly elucidated from the facts, and hypotheses, ideas, and theoretical models in this field are still essential and potentially highly fruitful. However, in my examination of this theoretical material I shall not attempt to give a full historical survey of all theories and hypotheses of development, starting from the ideas of primitive preformation. I shall examine only the ideas and hypotheses which are based on current factual evidence, ideas and hypotheses which, when analyzed in conjunction with the facts described in earlier chapters, will enable us to create a modern and soundly based (allowing for the existing level of knowledge) conception of the mechanisms of programmed morphogenesis (which could provide a basis in the future for the planned analysis of this problem).

§ 1. Outlines of the Problem of Programmed Morphogenesis. Questions to Be Answered

The main problem in biology, the problem of adaptive expediency of the highly complex structure of living systems and of the increasing complexity and development of these systems from the phylogenetic standpoint, leading to the infinite variation of living organisms in nature, was solved more than a century ago by Charles Darwin's theory of natural selection and the wider theory of evolution.

However, a fuller understanding of the problem of evolution only became possible with the development of genetics and its discovery of the mechanisms of inheritance and variation, because the fundamental genetic discoveries and biochemical discoveries connected with them (the nature of the genetic code and mechanisms of synthesis of DNA, RNA, and proteins) were necessary before an explanation could be found for the nature of succession of generations and of the preservation and reproduction of the results of variations in living systems.

The problem of individual development (morphogenesis and differentiation) is the third principal stage along the path to the complete understanding of the fundamental laws of the animal world, just as important to a true understanding of the nature of life as the problems of evolution and heredity. It must be clearly emphasized that no true solution to this third fundamental problem in biology has yet been found, nor any productive hypothesis put forward, although many interesting ideas and models which have been suggested require our attention as representing some progress toward resolving the theoretical difficulties associated with the problem as a whole. Before turning to the theoretical analysis of these ideas and attempting to make generalizations, we must therefore make a schematic examination of the nature of the problems and tasks to be solved.

It is obvious from the factual and theoretical material described in earlier chapters that the main problem to be solved is the mechanism of functional differentiation of total genomes of individual cells, i.e., the problem of what regulates the selective activation of genes to correspond precisely with the tasks assigned by the overall plan of development in space and time. In other words, we must discover why at each given moment of morphogenesis only those genes insuring the required specialization of the cells are switched on. This leads us to the next problem: what is the molecular-genetic expression of the program (the overall plan) of development insuring that the genome reacts as it should to signals from the extragenetic environment — from other parts of the embryo as they are being formed?

The first point to mention is that differentiation and specialization, as a gradual transition from identical to varied cell forms, do not imply total distinction even in the ultimate forms, and from the biochemical point of view we might evidently distinguish three groups of characteristics and processes: the special, general, and universal. It is likewise evident that divergence in relation to a special characteristic usually accompanies differentiation.

Many of the functions of every cell are associated with the activity of certain enzymes, the regular spatial arrangement of enzymes on intracellular structures, and the purposive organization of these structures (ribosomes, mitochondria, plastids, and so on). In different types of differentiation and specialization, individual cells of different tissues may differ from each other in hundreds of micromorphological and biochemical characteristics: new assortments of enzymes, different proportions of enzyme groups common to them all, differences in the structure of membranes and other structures, in size and shape, in rate of growth and division, and so on. At the same time, many hundreds of characteristics and properties of the cell are qualitatively identical despite different types of specialization: all cells must have a system of protein synthesis, ribosomes, and mitochondria, all must carry out oxidation and glycolysis and many other metabolic reactions. Remembering that the synthesis of each protein, each enzyme, and each structural unit of the cell is effected by one gene or group of genes in the cell chromosomes, it will be obvious that whatever the type of specialization, several hundreds of identical genes and several hundreds or scores of different genes will be in an active state.

As a rule, processes and structural features characteristic of living systems at almost all levels of organization of life are identical for different types of specialization. For instance, ribosomes isolated from bacteria (for example, from *Escherichia coli*), and the system of amino acid activation in bacteria can synthesize typical mammalian hemoglobin *in vitro* if the remaining part of the synthetic system (messenger RNA) is taken from rat reticulocytes. Adenosine triphosphate and other triphosphates generated by any organism can be used by any other system, and so on.

§ 1] OUTLINES OF THE PROBLEM OF PROGRAMMED MORPHOGENESIS 347

These biochemical properties and processes occurring in most cells regardless of their specialization we can describe as universal. Many biochemical features are characteristics of many but by no means all types of cell specialization. We can call these characteristics, which are common to several types of differentiation, general. Finally, some characteristics appear as a rule as a result of differentiation. They are narrowly specific and they usually reflect the functions of an organ or tissue (the formation of hemoglobin, myosin, actin, collagen, elastin, pigment, surface structures, secretions, and so on). Correspondingly, from the genetic aspect, cell function as a whole is brought about by the selective activity of three main groups of genes, which may also be called universal, general, and special.

On the assumption that specialized cells have a potentially identical genome, clearly a different assortment of active and passive (repressed) genes of specialized functions in fact operates in the presence of a relatively identical assortment of active genes of universal function, without which there could be no living cell. This type of schematic picture of differentiation can be conveniently used to depict various forms of specialization (differences in space and time).

In particularly advanced types of specialization, notably in mammalian erythrocytes, the cells may lose their nucleus and their whole genetic system at the same time as they lose several of their universal functions. However, this type of specialization can be regarded only as an exception, for it is accompanied by simultaneous loss of the potential viability of these cells.

The simplest general scheme of morphogenesis of multicellular systems can thus be illustrated by the scheme shown in Fig. 101. This scheme is purely an illustrative, simplified model of differentiation. It is important to note, however, that processes similar to the patterns illustrated do in fact take place during actual development. I have in mind the ontogenesis of chromosomal puffs, i.e., the arrangement of zones of increased synthesis of RNA and proteins on cell chromosomes at the phases of development and during specialization along different lines.

This schematic picture nevertheless clearly shows that the problem which must first be solved is that of the mechanism of differential activation of the required genes at the required time,

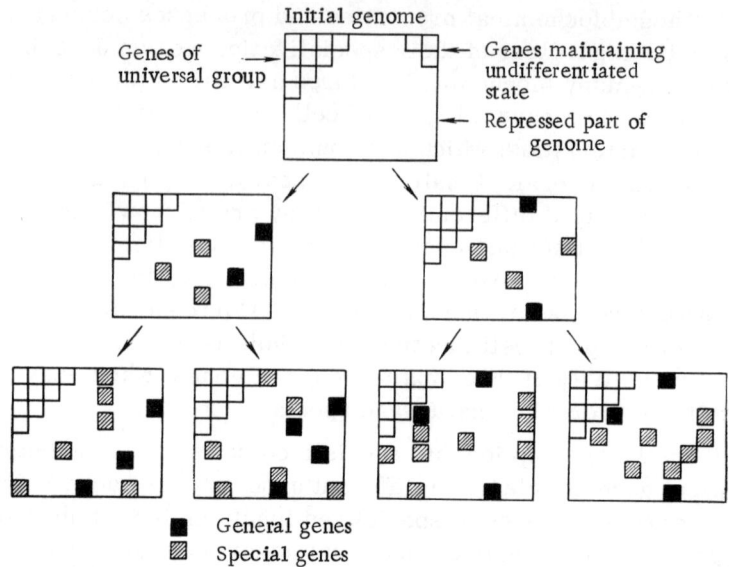

Fig. 101. Scheme of differentiation into four types of specialization (explanation in text).

in the required place, and at the required intensity, with the simultaneous repression of unnecessary genes.

In our examination of the factual material in previous chapters we saw that differentiation in embryogenesis usually is not absolutely independent, in its origin, of the conditions surrounding the cell or group of cells, but takes place under the influence of factors such as embryonic inductors, hormones, and physico-chemical or purely physical conditions. These conditions provide information concerning place and time, yet they are also the result of differentiation, of morphogenetic processes in other parts of the developing system, and themselves reflect a system of activation of a new group of genes. Accordingly, the second problem to be solved is that of the mechanism of interaction between the individual genome in the individual cell and other parts of the organism as a whole, between the individual program of differentiation of that cell and the general program of development.

As we have seen already, the inductor of differentiation is not narrowly specific, but often a comparatively simple compound, and even if it is a protein or RNA, it is not always a specific "activation" protein but in many cases it is something much more ordinary. Nevertheless, as a result of induction, complex processes take place in scores of genetic loci and cell function is modified. It is clear from this pattern of induction that in many cases the inducing agent simply switches on a completely prepared program of differentiation, with rigidly connected components. The low specificity of the inductor is evidently due to the limited choice of receptors with which it can interact.

This type of activation of preformed complex and purposive systems of genetic loci (operon systems) is evidence that the genome of every cell of a multicellular organism initially possesses latent differentiation, in the form of systems of intergenome or interoperon connections, into all the systems of specialization necessary for the fully formed organism. If this is so, then just as the purposive system of reaction of processing or synthesis of a compound is determined by a combination of genes (cistrons) into operons, switched on and off by a special operator gene, systems of these combinations creating functional specialization of the cell must also be combined into a single "federation" of operons, a suitable name for which could be "polyoperon," themselves switched on by an operator of their own.

In that case we can speak of the polyoperon of muscular specialization, the polyoperon of nervous specialization, the polyoperon of renal specialization, the polyoperon of specialization into intestinal epithelium, and so on. The third main problem of morphogenesis which we have to solve is thus the nature of the connection joining small groups of genes, functionally and often linearly linked together in operons, into the special polyoperon type of cooperation, extending to different chromosomes and capable, under the influence of a comparatively simple external agent, sometimes acting only once and for a short period, sometimes more permanently, of evolving in the manner of its transcription in the course of time.

Our ideas on the molecular and biochemical basis of life have mainly developed by the use of principles, ideas, and concepts common to organic and inorganic, physical, and colloid chemistry and physics. However, when these principles alone were used, for a long time it was impossible to understand and explain the basic molecular-biological phenomena such as protein synthesis, autoreplication of genetic material, and purposive coordination of a vast number of heterogeneous reactions, taking place at the same time and in the same cell space.

Initially, to explain these processes, some purely conventional, hypothetical and apparently unusual ideas were put forward, and in fact they were unusual from the point of view of the classical concepts of chemistry and physics. Nevertheless, as their mechanism was gradually revealed by theoretical and experimental analysis of these ideas, the real processes proved to be still more unusual, and their discovery marked a completely new era in biology, the discovery of the nature of fundamental biological processes. Three principal stages can be distinguished in this development.

Elaboration of the new principle of complementary autoreplication explained the mechanism of reproduction of genetic material (replication of DNA) and transmission of portions of this information for its realization (DNA-dependent RNA synthesis). Elaboration of the new principle of protein synthesis on RNA templates and the related principle of the nucleotide-amino acid code explained the mechanism of protein synthesis, a fundamental process of life.

Elaboration of the new principle of allosteric interactions between proteins and other biopolymers, a task which has only recently begun, together with ideas concerning operons — groups of functionally connected genes or cistrons — which may be switched either on or off, explained the phenomena of genetic regulation and purposive coordination of many different enzyme reactions taking place in the cell and their relationship to environmental conditions.

Elaboration of these three systems of interactions between molecules, completely new from the standpoint of classical chemistry, ultimately explained the basic properties and distinguishing features of living phenomena, but proved inadequate to explain the morphogenesis of multicellular organisms, to explain the phenomena of purposive, rapid, and faultless development of unusually complex systems, for explaining the nature and mechanism of realization of the genetic program of this development which undoubtedly existed. This suggests that the transition to this new stage in our knowledge of the nature of life demands a theoretical analysis and experimental verification of a new and, as yet unknown, principle of molecular interaction taking place within genetic systems and in connection with the temporal and spatial parameters of the developing embryo.

Obviously all three fundamental problems mentioned above require solution before we can understand the mechanisms of morphogenesis themselves require for their concrete analysis the discovery of this new biological principle, this new class of phenomena, this new system of interrelationships. In turn, analysis of the possible answers to these questions must certainly lead up to analysis of this new principle to serve as a basis for the understanding of mechanisms of development.

§ 2. Viruses and Bacteria; What Is Present and What Is Not in Their Genetic Systems Regulating and Programming Development

We have already examined in Chapters 2 and 3 the main facts and theories concerning molecular mechanisms of morphogenesis of viruses and bacteria. If we revert to this subject again it is not simply to summarize the results of investigations of these simplest forms of life. We are doing so at the end of the book so that, after examining descriptive and analytical aspects of morphogenesis of complex multicellular systems in the other chapters, we can try to understand more clearly and precisely the difference between the tasks and principles of individual development in lower and higher biological systems. The three new basic principles outlined above, discovery of which has greatly advanced our understanding of the nature of living phenomena — the principle of complementary autoreplication, the principle of template synthesis (with translation of the nucleotide alphabet into the amino acid alphabet), and the principle of allosteric interaction between polymers of activation and repression of operons — are universal and common to all forms of life, both higher and lower.

However, they are insufficient to explain the morphogenesis of complex systems, and the attempt to discover why these principles in their unicellular and virus forms do not work in accordance with this plan in multicellular organisms may be extremely useful for subsequent examination of the more important features distinguishing the controlling systems of morphogenesis of multicellular organisms.

a) Viruses. The simplest RNA-viruses with only two, three, or a few structural cistrons, control their morphogenesis with comparatively simple regulatory systems. Synthesis of virus-specific proteins is carried out in this case directly from the polycistron virus matrix. The individual velocity of synthesis of each protein can be satisfactorily explained by the hypothesis of nucleotide modulator triplets, i.e., by different ratios between the numbers of codon-anticodon interactions when templates are linked to acceptors (sRNA) by different codons for the same amino acids (Ames and Hartman, 1963). In this case the velocity of protein synthesis, being dependent on the frequency of repeated interactions of the sRNA carrying the amino acid with complementary zones of the template (codons) falls if the codon for the amino acid in one of the cistrons consists of a triplet variant which cannot provide an adequate concentration of anticodons in the transfer RNA fraction.

This principle of modulator triplets (modulators of the speed of translation) was put forward to explain changes in the velocity of protein synthesis within operons in bacteria (at the level of polycistron RNA). It is less suitable for viruses and phages, because in these forms the templates are virus-specific and sRNA fractions are provided by the host cell. However, the possibility is not ruled out that interaction between virus and host may also spread to this system of regulation, and that by selection the virus may become adapted to the conditions of synthesis present in the cell.

The boundaries between cistrons (zones for synthesis of individual proteins) within polycistron templates providing for synthesis of polypeptides of a particular length are satisfactorily explained (on the basis of theoretical and experimental data) by the hypothesis of the existence of nonsense triplets, i.e., codons to which there are no anticodons in the sRNA fraction and which do not code the position of any amino acid (Brenner et al., 1965). Of the 64 possible three-nucleotide codons, even allowing for degeneracy of the code, not all determine the position of an amino acid, but some, as has been shown, are signals for the end of the synthesized polypeptide chain.

In the more complex DNA-viruses, particularly in the group of T-even phages, in connection with the specificity of structure of certain nucleotides, the biochemical system of cell synthesis can no longer insure replication of the hereditary molecules, because two of the four nucleotides of their DNA differ in certain respects and cannot be synthesized in the cytoplasm of the infected bacterial cell. In this case, the DNA of the phage carries its own program for the formation of these nucleotides—a typical operon or group of linearly connected genes (cistrons), which, on entering the cells of the host bacterium, organize the synthesis of a series of enzymes for use in synthesizing the corresponding nucleotides. In this case, because of the increase in the total stock of hereditary information in the virus genome, with the synthesis of typical forms of messenger RNA as carriers of this information, with the precise temporal sequence of the syntheses (early and late proteins), and with differences in the amounts of the various proteins synthesized by different cistrons, the tasks of regulating the precise work of the genome are much more complex than in the case of the simplest viruses. The velocity of synthesis of individual proteins by different mRNA molecules in this case can be explained not only by modulator triplets, but also by differences in the life span of different mRNA molecules. Thus, different fractions of molecules of virus-specific messenger RNA are present in the infected cell in different concentrations, and different molecules of this RNA, because of their unequal stability, participate in the synthesis of different numbers of protein molecules. Some molecules of messenger RNA break up after synthesis of, for example, 15 protein molecules, others after 10 molecules, and so on. Furthermore, synthesis of different forms of messenger RNA on the DNA surface also may take place at different velocities, depending on many factors.

The order in which different genes are switched on in time can be satisfactorily explained by several mechanisms. First, it may be connected in some cases simply with the consecutive linear arrangement of genes in DNA corresponding (correlating) with the temporal sequence of their activation. In this case, consecutive or wavelike uncoiling of the DNA, creating conditions for

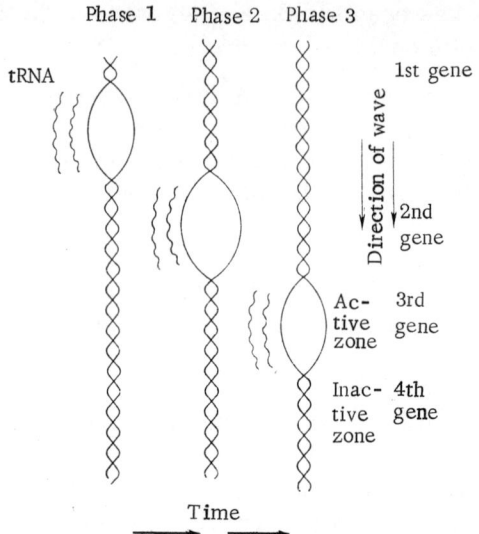

Fig. 102. Scheme of successive activation of DNA zones resembling the passage of a wave of uncoiling. Uncoiled areas contain single-stranded DNA capable of synthesizing RNA (Medvedev, 1963a).

successive RNA synthesis in different zones of this DNA, may be the simplest mechanism regulating the translation of information in accordance with a certain program (Fig. 102; Medvedev, 1963).

Finally, at this stage of evolutionary development of the genome, regulation of the repression—derepression type (allosteric interactions) is possible. Production by viruses of repressors of DNA and RNA synthesis in the host cells evidently takes place. Possibly, besides synthesis of early phage-specific proteins, a repressor of late phage operons with a definite life span (of the order of several minutes) is formed, while conversely, the late groups of genes, when they become activated, at the same time produce a repressor of the early genes.

In Chapter 2, I pointed out other possible ways in which the regulation and successive activation of virus genes could be organized. Although many details of this regulation have still to be filled in, the existence of correlation between the order of arrangement of the genes on the genetic DNA of the most complex

T-phages and on their circular genetic map and the order of their activation during phage replication, along with the grouping of genes in accordance with the character of their function, provide a real basis for understanding the molecular organization of the morphogenetic program of viruses.

The morphogenetic problems in reproduction of different forms of viruses are comparatively simple: this is reproduction or, more accurately, multiplication of a relatively simple system under relatively stable conditions of the cytoplasm or nucleus of the host cells. In this case the virus is practically independent of environmental conditions and receives no external inducing signals.

The whole process of its morphogenesis is carried out by the following systems controlling activity of the genes and products associated with it.

1. Regulation of the required and purposive order of activation of different genes in time (which may be determined by the linear order of the genes along the DNA macromolecule and by their grouping into functionally related systems, or operons, with mutual repression).

2. Regulation of the velocity of synthesis of expedient proteins in the necessary proportions for virus reproduction, which may be determined by variations in the half-life of different messenger RNA fractions resulting in different concentrations of these fractions at the site of synthesis and by a system of modulator codons for regulating protein synthesis at the template level.

3. Regulation of the correct assembly of the final virus particle from its structural protein components and the genetic macromolecule. In the simplest cases it is determined by the special structure of these protein components, enabling them to aggregate spontaneously into a particle with the required morphology. In more complex cases it may be determined by an additional structure (corresponding to the program) of special molecules linking the structural protein.

The genetic regulator system of the virus is not intended for, and is incapable of, solving the following problems:

a. Varying the development program and activity of loci in accordance with changing biochemical environmental conditions (the environment must be constant).

b. Forming different phenotypic variants with the same genotype (excluding the special case of lysogeny).

Because of the absence of these powers the possibilities of development of viruses are very limited and they are incapable of an independent existence.

b) Bacteria. As a comparison of the material in Chapters 2 and 3 has already shown, the system of regulation of gene activity in bacteria is much more complex than in viruses, principally because of the need, arising in the course of evolution, to perform new tasks as a result of interaction with changing external environmental conditions and with processes of assimilation and metabolism which enable this group of organisms to maintain their autonomy and independence. The volume of information contained in the genetic program of bacteria is far greater than in the case of viruses, and the elements of this program are therefore differentiated and subdivided into groups. In the simplest case these groups are operons, functionally connected groups of genes which together determine a given process: the stepwise synthesis of a certain substance, stepwise breakdown or stepwise transformation. The three principal systems (or problems) of regulation of activity of the genome which we distinguished for the solution of morphogenetic problems in viruses are also present in bacteria, but in this case they work at the level of loci on the genome and, in particular, at the operon level, coordinating events inside the operon, within a functionally connected group of genes. Specifically, we may note that:

1. Regulation of the required order of activation of the various genes (controlling, for example, the production of enzymes successively processing a certain substrate during either its synthesis or its breakdown), achieved by the linear order of gene distribution in DNA, is present also in some operons (but not all) in bacteria.

2. Regulation of the velocity of synthesis, and in the correct proportions, of proteins, such as enzymes, in bacteria as in viruses can be determined by the rate of renewal (by the half-life)

of different messenger RNA fractions, and in the case of polycistron forms of messenger RNA receiving information from the operon as a whole and carrying out the synthesis of several functionally connected proteins, it can be explained by the polarity of synthesis (as has been established experimentally) and by the presence of modulator codons, slowing synthetic activity of the templates.

3. Regulation of the proper assembling of the morphological elements, themselves equal in complexity to viruses (or more complex than viruses), on the basis of the same structural principles is found in bacteria during the formation of intracellular organelles, such as ribosomes and mitochondria.

This analogy between regulation problems of the element (fragment) of the bacterial genome and the entire virus genome is so close that theories of the secondary origin of viruses as an evolutionarily developed fragment of the bacterial genome are very widely held, and the possibility that the virus genome may be incorporated for a long time in a bacterial genome is well known.

However, in addition to these tasks, the bacterial genome also programs a series of new properties and phenomena. We can examine the most important of them by continuing our list of morphogenetic regulatory duties.

4. Unlike the genome of viruses, the bacterial genome is able to orient bacterial cells correctly (purposively) relative to chemical (assimilation) conditions of the external environment in two ways, to some extent opposite in direction.

a. If an assimilable product (an amino acid or nucleotide, for example) which, if absent, is synthesized from simple components by a series of enzymes through a chain of intermediate stages appears or accumulates in the intracellular medium, the appearing end-product represses the activity of the enzymes responsible for its synthesis (the principle of end-product inhibition). This repression also takes place in accordance with the principle of allosteric interaction, and in the modern view (Monod et al., 1965) it does so by interaction of the end-product, not with the gene and not with the template for synthesis of the enzyme, but directly with the first enzyme of the reaction cycle or the first members of this cycle, with a certain site (but not the active center) of the enzyme, modifying the steric properties of the enzyme molecule and inhibiting its activity. This type of inhibition has been called "allosteric," in contrast to "isosteric" inhibition, in which inhibitor and substrate are structural analogs and compete for the active group of the enzyme.

b. If a new substrate, which can be usefully assimilated or dissimilated, appears periodically in the medium, this substrate often induces the active formation of enzymes for its own processing, and these are usually grouped genetically into an appropriate operon. Information for synthesis of these enzymes is present in the genome in the form of a group of functionally, and sometimes linearly, connected genes, but in the absence of substrate it is not converted into the active form but is repressed. The substrate becomes an inducer activating this latent information by inactivating the repressor, evidently by the same mechanism of allosteric inhibition. This system of regulation (the model of Jacob and Monod) has already been considered earlier, and there is no need here to recapitulate its factual and theoretical aspects. All that need be said is that, while it cannot explain every case of induction and exceptions exist (presumably as a result of variations in mechanisms of regulation of different processes), which can be explained by other hypotheses (Stent, 1964; Brenner, 1965), nevertheless the existence of a regulatory system of substrate-induction type is clearly evident.

5. The bacterial genome correctly orients the bacterial cells relative to a number of general conditions created within the cells, and acting through a system of receptors reacting to these conditions, it activates and represses definite regions of the genome programming certain morphogenetic processes.

a. Processes of cell division and replication of bacterial chromosome are correlated with activation or repression, depending on certain parameters of increase in area of the membrane or volume of the cell.

b. The sexual process, associated with a comparatively long and complex series of morphogenetic reactions, is somehow activated by interaction between + and - forms and special internal conditions of sexualization.

Another special morphogenetic process, sporulation of bacteria, is evidently activated by a particular combination of external and internal conditions (exhaustion under the influence of unfavorable factors). These basic tasks of regulation apparently exhaust the morphogenetic and regulatory powers of the bacterial genome, sufficient to maintain the evolutionarily stable existence of the simplest unicellular forms and transmitting their adaptive achievements to new generations, primarily by direct divisions, by direct replication of the working genome and its cytoplasmic production.

Why is the bacterial genome, and the genome of unicellular forms in general, so restricted in its morphogenetic powers? This basic morphogenetic limitation has nothing to do with the total number of genes in the genome. As I pointed out in the introduction to Chapter 3, the simplest, smallest bacteria contain about half as much genetic DNA as the largest phages of the T group and the largest viruses of animals, such as vaccinia virus. Similarly, the genome of very complex unicellular organisms such as *Paramecium* has a much larger mass of DNA and many more genes than the genome of the simplest multicellular organisms, e.g., from the groups of the fungi and algae.

I consider that the main difference is that in the genome of unicellular organisms certain systems determining the power of specific, selective, permanent (or prolonged) repression and derepression (or activation) of genes and operons in response to various chemical, biological, or physico-chemical signals of short duration are absent. Inducible synthesis in bacteria requires the constant presence of inducing agents. If the inducer is removed, activity of the corresponding genes stops immediately, for the system of induction is based on constant linking of repressor with inducer. When, however, induction by factors external to the cell implies activation or, conversely, repression of genes for a long time, irrespective of the subsequent presence of the primary inducer in the medium, this also implies the appearance of a fundamentally new type of genome reaction — a reaction of differentiation.

Examination of the origin of genome reactions of this type to external impulses in the course of evolution is not a matter with which we are directly concerned. They evidently appeared gradually, through a series of stages, starting with a minor adaptive addition to the ordinary regulatory system of unicellular biosystems. However, the laws of biological evolution and natural selection are such that what is, in principle, a new system formed as a frequent but minor, perhaps even unnoticed adaptation, cannot disappear but must develop and exploit all its potentialities and must reveal, with the passage of time, all the forms and variants of which it is capable.

§ 3. Molecular-Genetic Mechanisms of Morphogenesis and Differentiation in Multicellular Organisms

a) Formulation of the General Outlines of a Basic Hypothesis. The regulatory system of adaptive substrate induction in bacteria and other types of cells can work only in the presence of an inducer and stops in its absence. This illustrates the main adaptive importance of the system: the substrate is the inducer, and activity of the corresponding operon controlling processing of the substrate should therefore be stopped if substrate is absent. Evolution and selection gave precision to this relationship. A similar combination of the role of genetic repressor with the products of a system of endogenous and interconnected reactions is responsible for precise regulation, purposive in nature, by a mechanism of feedback inhibition. In this case the substrate or end-product acts as a switch for activating certain regulatory systems, but a switch of the permanent action type, and not one acting on the trigger principle, switching on a process but not controlling it once it has started.

Activation by a trigger mechanism constitutes a new type of regulation, creating more lasting, yet reversible differences between cells, creating the differentiation of neighboring cells receiving different stimuli and existing under different conditions.

The possibilities of evolution of such a system of genome reactions are very wide, because on the one hand stimuli (triggers, inducers) which are not the direct substrate of the reaction may be infinitely varied; they may be biochemical, physico-chemical, physical, mechanical, and so on; they may be global, specific, nonspecific, and so on. On the other hand, the parts of the genome undergoing activation or repression may also be very varied, and may exist in any spatial or temporal combinations.

Here we see the basis of the functional combination, in the course of evolution, of genomes and cells sharing a common heredity into systems possessing differentiation, in which cells even though only temporarily under different conditions nevertheless differ in their pattern of activity gene. In this case evolution and selection may proceed along different lines: the creation of functional reactions of a particular genome which are adequate to the environmental conditions, or the acceptance of inadequacy to the direct external conditions but the creation of purposive differentiation, i.e., along the line of complementarity of differently modified genomes, providing them with a better and more perfect system of interaction with the environment.

In this case an individual genome is modified under the influence of the environment not for the cell which it directly serves, but for other cells which, together with that one, form a cell system.

Let us take the simplest example: a group of cells (Fig. 103) leading a stable existence under conditions when environmental factors X, essential to life (such as products for chemical assimilation) act on one side, factors Z (e.g., oxygen) act on another, and factors Y (e.g., harmful products of neighboring groups of cells)

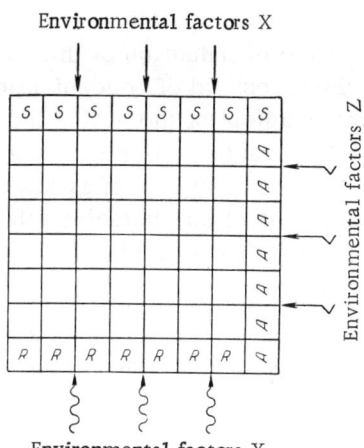

Fig. 103. Scheme of origin of differentiation (explanation in text).

act on a third side. Such a situation often arises in cell clones on a solid agar medium.

In this case it might be advantageous if cells first coming into contact with the group of factors X were to activate a system S (or to repress it) in order to allow vitally essential factors to penetrate into the whole group. This could also be advantageous for cells first coming into contact with factor Z. So far as the factors Y and the cells first coming into contact with them are concerned, the opposite would be advantageous: that they should activate the system R, inhibiting the penetration of the harmful factors Y into this cell system.

The regulatory system S, A, and R may be activated (triggered) in response, not to the sum total of the factors composing X, Z, or Y, but to one typical component, or to a typical companion of these factors.

This approach to the problem could thus be formulated as follows: a functional change in one region of the genome of a given cell in response to external induction of impulsive (trigger) or constant action, which is adaptive and advantageous not only for that cell but for a group of cells, in the rest of which this functional change has not taken place, is an elementary act of differentiation and morphogenesis. Typical morphogenesis is made up of a varied system of such acts.

In the simplest case of induction of this type the differentiating element may be the threshold of concentration of the inducing (triggering) agent, the gradient of the acting factors X, Y, and Z, invariably and inevitably creating effects of one system on another. It is easy to show that the importance of gradients and concentration thresholds in the action of any inducing effect is the basis of universality of the symmetrical organization of biological systems.

As soon as this type of elementary possibility of functional differentiation of the genome appeared as an adaptive phenomenon, evolution and selection then began to act in the direction of increasing the volume of information in those parts of the genome which were necessary, not for that cell alone, but for the system of cells. Gradually these parts of the genome became increasingly dominant both in their volume of information and in their mass, and the cell systems underwent increasingly complex differentiation.

Differentiating factors of the external environment and inducing agents became to an increasing extent factors of an internal environment which constantly became more and more complex, as a result of interaction between genomes, and this laid the foundations of typical morphogenesis.

However, the whole of this system of logical arguments, although possibly explaining the nature of morphogenesis from the standpoint of general evolution, still does not tell us anything about the molecular nature of the controlling systems of differentiation and development, or of the mechanism of the concrete intracellular events maintaining a given type of cell specialization. Hypotheses in this direction must take account of the facts which we examined in the previous chapters.

Since, as follows from the arguments given above, the cell genome can evidently be divided into two main systems: into a group of loci (genes) controlling processes essential to that cell only, and a group of special genes controlling processes catering to the whole differentiated multicellular system, clearly the regulatory mechanisms of genetic activity can also be divided into three separate types.

The first type, developing as a continuation of the unicellular form, controls stability of the basic, constant, vital activity of the cell under the constantly changing conditions of the internal and external environments, in accordance with the principle of a rapid increase or decrease in the intensity of processes characteristic of that particular cell (by induced synthesis and feedback inhibition, and by the action of many new functional influences typical of multicellular organisms).

The second type controls the constancy of maintenance of specialization (differentiation) through a mechanism of permanent or, more exactly, prolonged repression of definite groups of genes or activation of other definite groups of genes.

Finally, the third type controls the morphogenetic, programmed evolution of forms of specialization and differentiation brought about by the action of various inducers (triggers), with rapid, yet planned, changes in the spectrum of working and repressed loci, and with the establishment, after a series of phases, of a stable state (final activation of the second system).

Each of these systems of genetic control must have its own molecular mechanism, and our most difficult task is to elucidate the principle of operation of the third mechanism: the mechanism of active morphogenetic development.

Such are the general outlines of any hypothesis to explain the mechanisms of programmed morphogenesis which must take into account the up-to-date facts. These general outlines are essentially a generalized formulation of the hypothesis itself, for they provide a number of logical explanations of the mechanism of morphogenesis and they narrow the range of tasks for further elucidation.

Because of the complexity of the phenomenon, no explanation of the mechanisms of morphogenesis and differentiation in general can consist of any one hypothesis or any one model. It is important, therefore, to pick out the main systems, to break the problem up, to find the chief link, and then to try to attach the remaining links. As a result of the system of logical arguments given above, this chief link was reduced to demonstration of the need for the existence in the genome of multicellular organisms of latent repressed differentiations, under the control of interconnected regulator genes, and potentially determining various temporary (morphogenetic) and permanent specializations. These specializations, in turn, must be provided with receptor-activator mechanisms, tuned to receive various signals from outside sources, and they must correspond to the purposively successive change of these signals generated by the developing system itself and including different forms of differentiation, each of which, acting through a system of new signals, activates new latent differentiations, and so on until the complete formation of the organism with the minimal quota of mistakes.

After all these mostly general arguments, we must naturally turn now to the analysis of possible molecular models, and to the construction of a composite model from those molecular components of the functioning genome (DNA, RNA, genes, repressors, inducers, polymerases, replicases, operons, operators, modulators, nonsense triplets, etc.) and its physico-chemical changes (coiling and uncoiling of the chromosomes, transfer of information, coiling and uncoiling of DNA and of parts of it, and so on) with which we are familiar, to which the mutal arrangement and the system of connections between the components would enable the basic tasks of morphogenesis to be completed. It will be profitable to examine, selectively and briefly, some of the models which have already been proposed for this purpose.

b) Hypotheses of Mechanisms of Programmed Differentiation. Recently many attempts at the theoretical generalization of facts relating to manifestation of genetic systems of development control and at the identification of basic principles of this control have been published (Waddington, 1962; Shmal'gauzen, 1964; Shapot, 1965; Lopashov, 1965; Grobstein, 1964; Flickinger, 1963; Sand, 1965; Apter and Wolpert, 1965; Bonner, 1965; Ebert and Kaighn, 1967).

Examination of these writings shows that, despite the absence of radical achievements, whether experimental or theoretical, the general situation of our understanding of the genetic basis of morphogenesis of multicellular organisms is far from stagnant. Progress in both theoretical and experimental analysis, while not startling, has nevertheless been noted, the connecting links between the two have become more clearly understood, the range of problems requiring solution has been defined, the number of discovered elements of the basic controlling system (such as the operon, replicon, systems of repression and induction, and so on) has risen sharply, and as the factual and theoretical material described in this book shows, a basis for generalization undoubtedly exists. This must stimulate the attempt to create models, and even if no model has yet been suggested which satisfies the requirements of the main problems, it is apparent that the conditions for creation of such a model are close at hand. It is also certain that elements of a future model can already be detected in modern theoretical investigations of the problem of morphogenesis, and an examination of the ideas and tendencies which they contain from this point of view is therefore most important.

I shall examine this group of theoretical surveys which have recently appeared very selectively and briefly for two reasons. First, because I have already discussed the facts related to the various aspects of the problem, cited in these publications, earlier in this book, in most cases more fully and in greater detail, and together with an appropriate analysis. Second, in many of these publications numerous repetitions of the same schemes and models are to be found (variations of the regulatory scheme of Jacob and Monod, data on the genetic position effect, somatic mosaicism, etc.). In my analysis of these works I shall therefore try to pick out any new ideas and suggestions which they contain, in order to assess the richness of our total logical store of parts from which an integral concept can be built.

In this section I shall examine only those models in which the principles of modern molecular genetics and information theory have been used. I shall not be bound by chronology or by any other methods of formal classification, but I shall try to draw a general theoretical picture which will not be just a simple survey of explanations, but a synthesized, general picture, a general explanation derived from several different models.

Solution of the problem of the mechanisms of genetic programming of morphogenesis, as will be clearly understood from all I have written before, does not rest on the surface of our knowledge of the nature of the phenomena of life waiting to be picked up, it does not emerge as a simple consequence from the facts we already know. At this stage it requires a wealth of logical argument, and the creation of models on the basis of every new original principle. Boldness in the development of these principles may lead to success, as was the case in the preliminary theoretical analysis of the principles governing the coding of biological information, the principles of complementarity, the target principle in radiobiology, and the principles of template activity, allosteric interaction, and so on, which have given biology its modern guise.

I shall therefore start my analysis of models of morphogenesis with one of the most recent ideas, one developed by Bonner (1965) to explain the programming of morphogenetic processes in plants. Bonner chose plants in this case because of their simpler differentiation, but the principles he suggests are of more general interest.

In previous chapters I have often had occasion to cite material from Bonner's recent book "The Molecular Biology of Development" (1965). Although this small book was written in a popular manner for a wide range of readers, and although it lays emphasis on an examination of work done in his own laboratory rather than a general analysis of the problem, nevertheless it contains many novel thoughts and ideas not hitherto formulated in experimental papers published by Bonner and his collaborators. The last very short chapter of this book, entitled "Switching networks for developmental processes," contains an analysis of a series of logical and theoretical postulates which we shall do well to examine here as an example of a fundamentally novel, yet at the same time realistic, although hypothetical, approach to the analysis of the organization of the control of differentiation.

Bonner points out that from the standpoint of molecular biology the mechanism of development, in its elementary form, consists of the correctly programmed, successive, and orderly repression and derepression of individual structural elements and parts of the genome, expressed primarily as the appearance of specific genetic products in specific groups of cells, at strictly definite periods of development.

The appearance of a new pattern of active genes in the course of development is considered by Bonner to be the result of a system of "switchings on" in accordance with an existing latent program, connected in a hidden manner with the "switch," operated by several factors. These factors thus simply set in motion a ready-made mechanism unfolding a chain of purposive events, not directly connected with the switching on operation. For example, a resting vegetative bud, awakened by a single, brief treatment with ethylene chlorohydrin (a nonbiological agent), nevertheless develops into a shoot with all the parts present in a normal shoot. On the other hand, if the same bud is treated with flowering hormone, it develops in accordance with a completely different (but also normal and ready-made) program, producing step by step a flower bud, a flower, and a fruit.

To give this program its final form, very many genes must have been switched on and off in the period of its realization, in the proper places and at the proper time, and the entire interconnected system of switchings on and off can have been set in motion only as a result of the initial induction. After induction, as Bonner observes, a predetermined group of acts of development takes place automatically, like a song of which the words have already been inscribed in the genetic book. In this sense induction simply means choosing the correct and necessary song.

The whole development of the plant could be split up into several such "songs" in the sense that the total program of morphogenesis of plants could be compiled from a group of independent subprograms. For monocarpic plants, in Bonner's opinion, eight such subprograms can be identified: 1) the program of the cell cycle; 2) the program of embryonic development; 3) the program of seed development; 4) development of the bud; 5) leaf development; 6) stem development; 7) root development; 8) reproductive development (flowering).

Each of these subprograms consists of different processes and cell differentiations, and to choose, for example, the simplest of the eight — development of the bud, its formation into a definite morphological system requires very strict determination of the order of division and differentiation of the cells.

The principle of subdivision of the general program of morphogenesis into groups of subprograms, predetermined and associated with certain genetic receptors, is not new; it follows logically from the existence of different forms of differentiation. In particular, I elaborated it in my earlier book (Medvedev, 1963) as the principle of mutual exclusion of predetermined subprograms (activation of one subprogram inhibits all the rest). In my opinion, at the time of development the system of switching could be corrected by localization or by any other signal. This principle of grouping of genes and of gene systems (operons) into larger systems, constituting parts of the total program, facilitates our understanding of the planned formation of interconnected systems of specialization, but it does not explain the expediency of elementary acts within each system of differentiation.

To explain this purposiveness, Bonner suggested a new morphogenetic principle, that of testing the situation by the cells and reaction of each cell and tissue on the basis of this test. In his opinion, the meristematic tissue in the growing bud continually tests itself for size and number of cells, comparing the results of the test with the genetic program. When the correct value is attained, the next developmental stage begins (is switched on). In the initial period the bud can test for the presence or absence of flowering hormone, and its cells can test their neighbors for similarity or strangeness, i.e., the cells somehow or other can receive information about where they are and, depending on this information, they can switch on a particular internal mechanism. This, according to Bonner, constitutes the logic of differentiation.

Testing the situation in the sense of a definite reaction to hormones is a well-known fact, and the unique feature of Bonner's hypothesis is the broadening of the testing principle and its application to reactions to the most widely different conditions surrounding the cell, and the manner of switching on particular processes (commands) when the results of testing the situation are obtained. For a bud to develop, the following morphogenetic commands might be used:

1. Divide tangentially with growth.
2. Divide transversely with growth.
3. Grow but do not divide.
4. Test for size or cell number which must achieve a specified value.
5. Test for apical or not apical.
6. Test for inside or not inside.
7. Test for outside or not outside.
8. Call Max ⎫ stored sub-subroutines for differentiation
9. Call Map ⎭ into xylem and phloem respectively.
10. Call Epidermis sub-subroutine.
11. Stop.

There are thus three categories of instructions or commands for bud development. First, there are commands concerning cell division and growth, second there are commands for the execution of various tests, and third, commands for the selection (bringing into play) of a system of differentiation. One way in which these commands might be sequenced to guide the bud in its differentiation is indicated in Fig. 104.

The logic of the arrangements in this program is simple. The single apical cell divides transversely, and each of the daughter cells then tests itself to find out whether it is still the apical one or not. For the apical cell the result of the test is positive, and the resulting command is to behave as an apical cell should, i.e., to divide again. The other cell, which has discovered that it is not apical, receives a different command — namely to test for the size of the group of cells in which it lies and to compare this with the final size specified by the program. If the nonapical cell finds, as

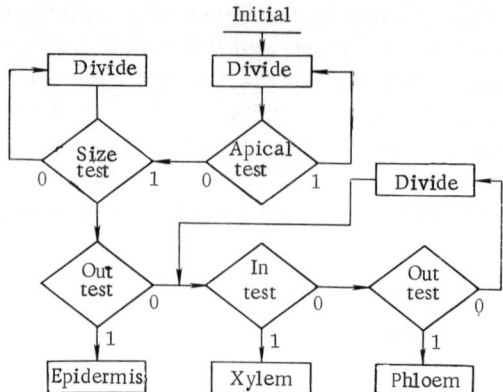

Fig. 104. An example of one way in which a set of genetic switching units might be interconnected to bring about the development of a single cell into an organ. In this example, a single apical cell generates a stem containing a variety of specialized cell types (Bonner, 1965).

a result of this test, that the group size is smaller than required, it receives the following command: divide again and again in alternating planes, testing after each division until the required size is attained. When the required size has been reached, each cell is required to test its own position in the group.

    Cells which find that they are on the surface of the group are instructed to follow the subprogram of differentiation into epidermis. If the result of the test is negative, however, the cell is required to test itself to find out whether it is in the center of the mass of cells. If the result of this test shows that the cell is in the center of the mass it receives the command to develop in accordance with the subprogram of differentiation into xylem. If the test shows nearness to the epidermis, the command is different: develop into phloem. Finally, the layer of cells between xylem and phloem does not differentiate but stays as cambium, consisting of dividing cells.

    Bonner further accepts the possibility that these tests are themselves programmed in time so that the time scale is subdivided into small time units during which the cell executes a single command — "it divides, tests, or what-not."

According to this hypothesis, the planned execution of the program, in the sense of selection of the path of development and the path of differentiation, is thus based on the principle of testing the situation. Because of their structure, in the tests (the possibility of which was discussed above) one "input" determines one of only two alternative "outputs." Life may use more complicated tests requiring more than one input.

What is the possible material basis of the cell tests? Bonner considers that the ability of the genome to obtain information about different situations may vary. For example, the presence or absence in the nucleus of a particular hormone or other substance in a definite concentration (depending on the position of the cell and its distance from the source of the hormone) may provide the material, physico-chemical basis of a variety of tests.

He cites the results of an investigation by Moscona and Moscona (1963), which have been interpreted as showing that certain cells of the chicken embryo continuously test adjacent cells for likeness or nonlikeness. If neighboring cells are alike, they stick together to form an organ. If neighboring cells are different, they do not stick together. As their preliminary results showed, this test is carried out by means of a specific protein secreted by the cell surface. This sensitive protein is continuously degraded (with a half-life of only a few minutes), but it is being continuously replenished through a special messenger RNA with a half-life longer than that of the protein, namely about 2 h. This system is thus connected with the genome through messenger RNA. Although these suggestions are still unconfirmed, they are nevertheless interesting from the standpoint of the search for mechanisms of cell interrelationships.

This continuous testing of neighbors is perhaps connected with ability to dedifferentiate and to regenerate up to a number when, on account of the changed conditions, the results of the test inform the genome about the nature of the changes which have occurred.

As Bonner postulates, the test for apicalness might rest on the physical basis of cell contact. The cell genome might possess a recognition system to tell whether or not it is in contact with other cells over an angle of 180° (apicality), 270°, or 360°. This recognition might also be connected with the concentration of a particular substance S (a substance of neighboring cells) in the nucleus, which could vary with the area of the surface through which it enters. The switching on of a new genetic system dependent on the concentration threshold could in this case be a response to localization of the cell.

The idea of testing the situation is very interesting, but so far it can be used only for the formal, logical analysis of differentiation and morphogenesis. This idea essentially indicates the logical possibility of development on the basis of application of an arbitrary logical combination of different reactions of genome and cell, all theoretically possible and with analogies in other processes. However, it is essentially still very far from experimental confirmation, although there is no doubt that experiments to verify it could now be carried out on the basis of a purely logical model.

Translating some of the premises of the model we have just considered into the language of concrete processes which we used to analyze the factual material in preceding chapters, they assume the following form:

a) Bonner's model requires the presence of a variety of molecular receptors in the cells sensitive to a variety of conditions external to the cell. These receptors are somehow connected with various genetic subprograms of differentiation, and when the receptor gives a positive reaction to a given external condition (created by the cell's position, by its neighbors, by an inducing agent, etc.), various subprograms are brought into play.

b) Receptor systems are not in a state of constant preparedness (or activity), but they are themselves switched on and off in accordance with a definite time schedule and are subdivided into groups. In this case a group of receptors (testing for apicalness, testing neighbors, etc.) is switched on sequentially "by schedule," by automatic commands "from within," and that is why the impression of active testing of the situation and collection of information by the cell is created. Testing of conditions by the cell constitutes a new logical principle calling for further verification. How far this principle facilitates, or, on the contrary, complicates analysis of the mechanisms of morphogenetic development will be clear after scrutiny of other schemes, primarily those based on the principles of cybernetics.

Comparatively many attempts have been made to find a logical explanation of the mechanisms of morphogenesis on the basis of the methods and principles of cybernetics, expressed through biological phenomena. The very concept of a "genetic program of development" is itself a cybernetic concept, and usually many concepts and terms of cybernetics are employed in the analysis of morphogenetic problems. However, I do not propose to describe the history and give a formal survey of the various attempts which have been made to analyze acts of development by cybernetic methods. I shall confine my attention to the scrutiny of the paper written by Lopashov (1965) entitled "Embryology and Cybernetics," which summarizes the possibilities of the cybernetic analysis of development.

Lopashov's analysis of the cybernetic aspects of development is distinguished by a much deeper understanding of the difficulties and problems, of their specific features, and of the productivity of various approaches to their logical elucidation than is usually achieved in the writings of specialists in theoretical biocybernetics, who as a rule employ a predominantly abstract approach to the problem and attempt to solve it as they would an exercise in cybernetics. He notes that information in development, because of its adaptive character, differs essentially from hereditary information. Whereas the latter must be transmitted from generation to generation with maximal stability, not significantly vulnerable to changes occurring with mutation, in the course of development, on the other hand, information concerning differentiation must differ at every point and in every cell of the body. In contrast to the problem of replication of information, the basic problem in the explanation of morphogenesis is therefore that of the distribution of different types of information among cells of the developing organism.

In Lopashov's opinion the theoretically most probable way of determining this differential information is its regulation by the states reached by the cells in the course of previous development, through feedbacks acting on the state of the nuclei or on the information issuing from them. This information must differ correspondingly in cells differentiating in different ways.

As examples of the existence of feedbacks of this type I could mention the facts of induction, the influence of hormones on differentiation, experiments on the reprogramming of translatable nuclear information when the nuclei are transplanted into a strange cytoplasm, and other phenomena we have discussed previously.

It will be evident, however, that a morphogenetically purposive reaction of cell nuclei to external stimuli of feedback type requires the existence of receptors and a state of preparedness to react to a definite stimulus in a strictly definite manner. Factors acting on nuclei in the manner of feedbacks have an influence whose results, as Lopashov points out, is limited by the information already recorded in the chromosomes. That is why the "choice of translatable nuclear information depends on morphogenetic connections joined together into definite systems." This choice of translatable information depends on the nature of the inducing agent. In normal embryogenesis, during its main events, inductive influences are precisely differentiated, and although they are not narrowly specific, Lopashov nevertheless considers that a number of groups of substances, associated with the direction of differentiation of parts of the embryo, can be distinguished on the basis of experiments to study different inductions.

In all these cases the agent is produced by a neighboring tissue; it comes from outside. However, at a certain stage of development the inducer may be a product of the tissue on which it acts. This has been shown for development of the retina, the mesoderm of the gastrula in amphibians, and for several other tissues whose fate varies with the degree of accumulation of inducing products of their own metabolism, but whose degree of morphological specificity has not yet been discovered. The switching mechanism of differentiation during such interactions between homogeneous cells ("homotypic induction") (Grobstein, 1964) may be the spatial character of the connection between the cells (arranged in one or several layers, etc.), modifying the character of distribution of the inducing agents and, hence, their influence on subsequent differentiation. This type of induction, in Lopashov's opinion, is in most cases more primary than external ("heterotypic") induction, in which the external influences of other parts of the embryo cannot operate directly on molecular-genetic syntheses in the cells, but can only regulate them through homotypic action.

In his evaluation of the cybernetic aspects of the facts of induction as an example of feedbacks, Lopashov points out that the sources of the feedback effects determining the issuing of qualitatively different items of information may vary. They can be regarded as a manifestation of the genetic code itself. In such a case, however, it must be considered that the code itself determines the differential issuing of information in different parts of the body. This point of view, in Lopashov's opinion, is unlikely to be true, for such a system would be self-contained and would have no source of movement.

According to this view, the sources of feedback influences must differ somehow in their origin from genetic information. They may be conditions or structures unable to reproduce the genetic code themselves and which must accordingly be different in their origin.

Lopashov sees this origin in the continuous organizing role of actual preexisting structures in the creation of new, without which the new structures could not arise. Such organizing influences are essential, regardless of the code, and the presence of the code is no substitute for them. Yet these influences do not lead to repetition of the previous state, i.e., they do not have the character of self-reproduction. The previous structure, the distribution of anlagen and of agents secreted by them, is thus essential for creation of the spatial organization of the latest stages. Without the preceding spatial distribution, further regular changes in the arrangement of differentiations is possible only to a limited degree, to the extent to which it corresponds to the relationship between cell groups and the environment.

There is considerable experimental evidence to show that this in fact occurs, and that it is interaction between different, existing differentiations at different levels which determines the direction of further development and its divergence into a more complex system, by a unique type of branching based on the regulatory value of particular interactions between differentiations at each stage of development. This point of view, with its several variants, is widely held although, as can easily be seen, it differs to some extent from Bonner's hypothesis in its approach to the identification of the main component in the explanation of development. We have here a new expression of the old dilemma of the importance of preformational and epigenetic principles, in which, at the new level, both principles lie within the bounds of genetics and the theory of information, and they are so close that the difference between them is almost imperceptible. However, this is not so. In Bonner's hypothesis the individual cell and its genome actively test the environment and react to external conditions in accordance with the results of the test. In this case the active principle of the morphogenetic process is the individual genome of the individual cell, and it ultimately controls the direction of movement.

In Lopashov's hypothesis genome and cell take part in a cycle of changes during morphogenesis, while the individual cell to some extent remains passive: information of the whole, composed of a system of connections between the differentiations of the preceding stage, plays the active role. A "self-development" of the system takes place, in which the fate of the individual cell is limited by the volume of its genetic information while the fate of the whole embryo depends on the more complex and higher information of the interacting anlagen and differentiations and genome as a whole, on their links with one another, and on their influences on the genomes and cytoplasm of individual cells.

Ratner (1966) raises a number of interesting points on cybernetic (controlling) systems of development in his book.

In our examination of morphogenesis in bacteria we saw that radical changes in the understanding of phenomena taking place at this stage of development of life occurred after genetic methods had demonstrated the existence of special regulator genes, responsible for activating and repressing the operons induced in the presence of the inducing substrate. It was this discovery, with the accurate mapping of a number of regulator genes, which prompted subsequent hypotheses concerning repressors, interactions between repressors and inducers, and so on, which were put forward initially simply to explain a possible mechanism of working of the regulator gene.

Regulator genes are essentially genes which switch other genes, or their functional groups (operons) on and off under specified conditions.

It is evident that the scheme of action of these genes when forming a repressor, acting on an operator located next to a functional group of genes and linked with the substrate, as developed by Jacob and Monod is an alternative and does not rule out the possibility that several different mechanisms of regulation may exist in bacteria and, in particular, in multicellular organisms. I would draw attention to the work of Lindegren (1963), who developed a hypothesis of activation of genes in accordance with a different principle and used it to explain induced synthesis of α-glucosidases in yeasts, which he had studied. He called this the "receptor hypothesis." He postulates that the inducer is not linked with the repressor, but with a special receptor group lying close to the gene which it regulates. The formation of a link between inducer and receptor converts the gene into an active state.

The existence of regulator genes with different powers and specificity is now established beyond all doubt, and on it is based the genome system that regulates development and differentiation of multicellular organisms. The existence of regulator genes in multicellular organisms can be presumed, and the abundant direct and indirect evidence of interaction between genes, the modifying influence of some genes on others, the effect of position of the gene in the chromosome on its expression (the position effect), somatic mosaicism, and many other aspects are examined in the standard works on genetics. The organization of morphogenesis is unquestionably associated with the complex, yet rigid, cooperation between these regulator genes to form definite mutually influential systems differentiated by groups.

The idea of aggregation of operons into an autonomous system, smaller than the chromosome, was formulated by Stern (1963), who called such a system a macrooperon. He considers that functional connections may be established between the operons themselves.

We can detect in these ideas the elements of molecular models of a system controlling the development of multicellular organisms, i.e., attempts to determine a system of interaction between all genetic elements which can serve as the basis for program determination.

We must take into consideration also works which will add to our total stock of genetic elements having a role in the regulation of development. This is a very important step, because it is only by venturing outside the limits of the customary categories common to both bacteria and multicellular organisms (the cistron, operon, regulator gene, and so on) and by discovering the additional elements found only in multicellular organisms, that progress can be made.

Another important matter to be mentioned is the existence of various concentration microgradients of active substances such as inducers, repressors, and so on, with an important role in differentiation. The significance of concentration gradients of inducers in the creation of polarity and symmetry in the case of embryonic induction is well known (Flickinger, 1963). Taylor (1964) has pointed out that concentration gradients may also be important at the chromosomal level, especially in connection with the degree of inhibition of various chromosomal regions by histones.

It can thus be concluded from the material examined that, besides receiving external signals playing an inductive role in differentiation (the role of cytoplasmic differences during cleavage, the role of induction by neighboring tissues, and so on), the cell genome of multicellular organisms also evidently possesses its own inbuilt regulating and programming systems, sometimes autonomous and sometimes synchronized with external stimuli. A clear picture of this "ladder" of morphogenetic stimuli influencing the genome was given by Kroeger (1963) in the form of the scheme in Fig. 105.

To conclude our examination of current conceptions of development, all that can be said is that the total number of factors

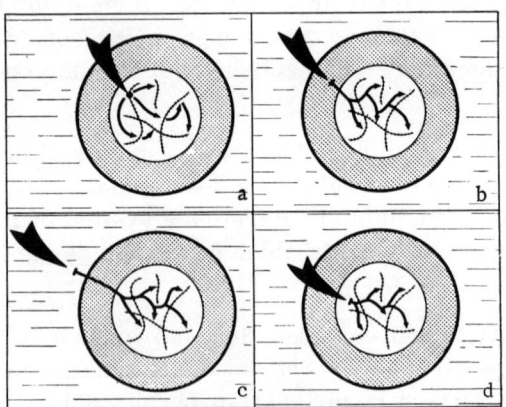

Fig. 105. Scheme illustrating activation of loci: a) During interaction between parts of the genome; b) effect of stimuli from the cytoplasm; c) stimuli from the external environment; d) stimuli from the nuclear juice (Kroeger, 1963).

or elements for the control and programming of morphogenesis is certainly very large. We have indicated some new aspects and features of the problem. Yet the control system itself, the character of interaction between its elements, so precise and so exactly reproduced in successive generations, still remain unexplained. We have not yet created a logical model of morphogenesis and differentiation which could bring all these factors into play, even if only temporarily, for the development of both experimental and theoretical research has followed the lines of discovery of individual elements playing a role in morphogenesis.

c) Theoretical Simulation of the Molecular Mechanisms of Morphogenesis and Differentiation. Concluding Remarks. When describing the factual material of all the present chapters, I tried to present it, not in the form of a descriptive survey of the modern data, but in the form of a theoretical analysis of these data with a discussion of the ideas, models, and hypotheses which have been suggested to explain the different phenomena of morphogenesis in living systems at different levels of complexity.

In concluding the book, I shall not, therefore, repeat this theoretical material in precise form; indeed, this has already been done to a large extent in the preceding sections. In this final section I shall endeavor to present a logical (theoretical) model of morphogenesis at the level of our knowledge of this phenomenon as it stood at the beginning of 1967. The problem is to try to bring together all the known elements of morphogenesis into a system of interconnections and interactions which can give a logical explanation of the precise programming of development, its automatism its purposiveness, and its reproducibility in each new generation. In addition, this explanation must conform to the requirements (outlines) formulated at the beginning of this section.

A few years ago I proposed an explanation (Medvedev, 1963a, b; 1964) of this type in the form of a hypothetical model (Fig. 106). This model was based on the suggestion that a synthetic apparatus is incorporated in the structure of the gene, producing intranuclear inducers. These selectively block certain groups of repressors, thereby activating the structural loci of the genome controlling synthesis of functional proteins. By analogy with the

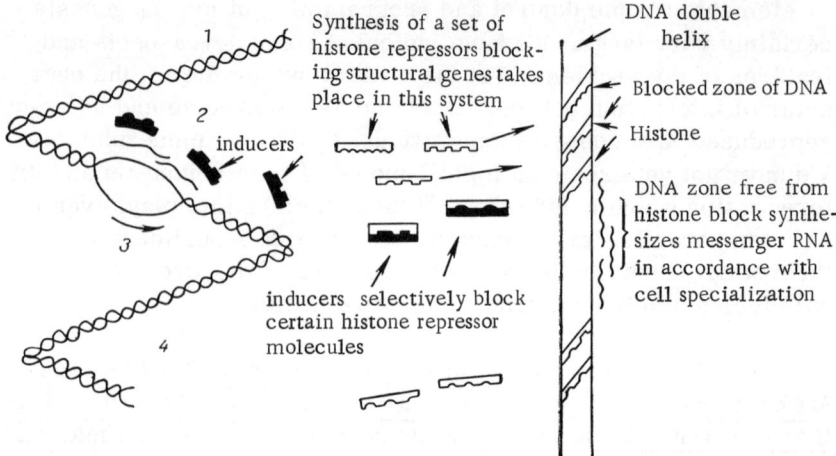

Fig. 106. General scheme of regulation of differential genetic activity in accordance with the localization of the cell during ontogenesis (Medvedev, 1963a). On the left is an aggregation of linearly joined DNA molecules providing the genetic time and localization service. Activation of zones 1, 2, 3, 4, ... depends on position; movement of the loop along the zone changes the character of the inducers with time.

giant DNA-phage or the moving loop of oocyte chromosomes (of lamp brush type), I designed the model in the form of a DNA system in which the zone of synthesis of messenger RNA molecules advances, in the course of time, along the chain (in the form of a zone of uncoiling), and at the corresponding time this molecule determines the synthesis of different forms of messenger RNA and different inducers. By virtue of this process, at different time intervals different structural genes are derepressed, and this enables the cell to change from one form of functional activity to another at the appropriate time.

To explain the difference in differentiation of cells in connection with their localization in space and in relation to other groups of cells, I postulated that this type of change in the assortment of inductors within the genome as a function of time is characteristic of the process of increase in, or primary formation of specialization in, time. In my view, many different molecules could exist in the genome for the time-controlled synthesis of inducer molecules, each for a different type of specialization. These inducer synthesizing molecules must have different receptors for different cytoplasmic conditions and for different external inductions, and they are brought into play by a system of mutual exclusion: activation of one (A, for example) implies repression of the others (B, C, D, etc.).

Such a model could provide a formal explanation of determination and the outwardly visible automatism of differentiation. Since, in connection with the ooplasmic segregation of the oocyte, its cytoplasm is differentiated and polarized, after a series of primary cleavage divisions identical nuclei find themselves in a cytoplasm with different properties. This may lead to the activation of a variety of inducer generators, each with a differently timed program. For this reason, derepression of different groups of loci begins in different zones of the blastula (or gastrula). In association with morphological events, contacts develop on this primary basis between parts of the developing system, new and additional inducer generators, varying from one part of the body to another, are brought into play, and in this way the total program of development is executed in each cell on the basis of interaction between individual genomes and the entire developing biosystem.

If the more recent material is taken into consideration, this hypothetical model has many shortcomings, but some of its basic premises can be brought up to date. This is true, in particular, of the principle of initial concealed differentiation of the genome into groups with functional specialization, in which each group of specialization (muscular, nervous, renal, etc.) is activated as a whole and is then formed in accordance with its own subprogram.

This principle of division of the total program of development into ready-made, predetermined subprograms, corrected for the conditions, is also accepted as an essential condition in other morphogenetic models (as in Bonner's hypothesis which we have already considered).

To obtain some idea of what additional elements and principles we must use, consider, or develop in order to be able to construct a more universal and realistic model, let us briefly recall some of the special features of morphogenetic development requiring explanation.

1. Morphogenesis of complex multicellular organisms is clearly divided into two periods: unicellular and multicellular. In the unicellular period maturing of the gametes takes place. Maturation of the oocyte (a very prolonged process), moreover, is associated with complex phenomena of special cytoplasmic differentiation (ooplasmic segregation) and often (but not always) with a special state of the chromosomes.

2. Morphogenesis after fertilization is also clearly divided into structural and temporal periods with a certain specificity of their events (primary cleavage, blastula, gastrula, neurula, etc.) and with critical phases of reprogramming which, in the early stages of development (blastula, gastrula, neurula), occur almost synchronously for all cells of the embryo (Neifakh, 1962, 1965; Spirin and Belitsina, 1965). It is possible that these phases of synchronous reprogramming may lie at the basis of what are called the critical periods of morphogenesis, revealed by numerous investigations principally by their effect on the degree to which environmental conditions influence the organism at different stages of its development (Svetlov, 1960; Trifonova, 1963).

3. The systems of orientation of morphogenetic processes in time unquestionably cannot extend to the whole of morphogenesis or have a single basis, but they cover only discrete phases. The system of orientation relative to the situation, to neighboring cells, and to external induction must be more important. The principle of testing the situation by the cells, suggested by Bonner, must certainly rest on a material, physiological, or biochemical basis, and it must itself form the basis of the morphogenetic reactions of the genome and selection of the proper program by the genome. However, it must be supplemented by the "principle of signaling," implying that every type of cell differentiation, besides providing for synthesis of the substances and enzymes necessary for that function, must produce a substance to act as a signal or as a visiting card, and this substance often becomes an inducer for neighboring cells.

Although they do not develop very high selectivity, under the conditions of induction receptor systems can nevertheless be tuned to distinguish very general differences, or global conditions. This would require a highly complex receptor system, and for that reason the possibility of signaling is an essential addition to the system of receptors.

4. The genome of multicellular organisms, consisting of several chromosomes or several tens of chromosomes, is undoubtedly subdivided into functional groups of genes which are larger than operons. In the past I called them polyoperons. However, differentiation of the genome may be more complicated still. Any system consisting of interacting elements (in this case, genes) must differentiate, as these elements grow quantitatively, into subgroups of more closely connected elements. The genes within one operon are connected with each other functionally, for they control a single process. If they are to be controlled jointly, they must therefore be connected both (in many cases) purely linearly and through a common operator, or by some other means. All the operons and other groups of genes and individual genes determining a particular specialization also are functionally connected, and some form of combination in the genome with the common regulator (subprogram) must exist for them also. It is perhaps not absolutely true to call this system a "polyoperon" or "macrooperon" because it may include other types of gene combinations besides operons, but for the sake of simplicity I shall use the term polyoperon. The genome of multicellular organisms, as we know, is subdivided into chromosomes and "linkage groups" of genes connected with them. This differentiation of the genome into chromosomes in the case of the X and Y sex chromosomes is of definite morphogenetic importance. If somatic chromosomes also had functional specialization (chromosome 1 determined muscular specialization; chromosome 2, nervous; and so on), in that case the problem of regulation of specialization would be more easily solved. However this is not so, and every type of specialization is built up from loci situated in various chromosomes. The chromosomes are therefore mainly a product of evolution and a structure for replication.

The existence of early and late replicating chromosomes shows that in the case of differentiation into chromosomes also this differentiation has functional significance.

Taking all these circumstances into consideration, together with the general outlines and premises for the hypothesis of programmed morphogenesis which we examined above, and the scheme of differentiation which I gave in connection with discussion of the material on morphogenetic induction (Fig. 96), we can indicate the following as a possible system of interaction between various factors and elements to explain the nature of the molecular-genetic system controlling morphogenesis.

a) Oogenesis, whose program contained in the genome is activated by conditions created for maturation of the oocyte. Besides the universal group of genes, working in all cells, a special program is activated (derepressed) for the formation of reserve materials, and a program of primary cytoplasmic gradients is also brought into play. The oocyte changes considerably with time (it grows larger). With time, therefore, the tasks of the genome also change sharply. Time orientation takes place either relative to the cell volume or to its surface area, or by a genetic clock such as a DNA molecule, emitting a variety of signals over a period of time in connection with movement of the zone of messenger RNA synthesis along this molecule in the manner shown in Fig. 106. Under the influence of sequential inductions within the genome the essential morphogenetic changes take place in the oocyte. Maturation of the oocyte is connected with inactivation of these programs of oogenesis.

b) Cleavage in the presence of segregation of the ooplasm brings totipotent and identically repressed genomes (only the universal genes are derepressed) into contact with different regions of the cytoplasm. Let us designate the different forms of cytoplasm by (+) and (-). Let us suppose that there is a special receptor group, consisting of two gene groups AB and CD, in one of the chromosomes in the genome of each cell (Fig. 107). In group AB, A is the receptor gene which, in response to the cytoplasmic stimulus (+), switches on production of the intragenome activator B. In the group CD, C is also a receptor gene which, under the influence of cytoplasmic conditions (-), switches on the synthesis of intragenome activator D.

Intragenome activator B activates (derepresses) gene loci (operons) x, y, z, and i, each of which has an operator (receptor) reacting with activator B or linking with repressors (or sites of their synthesis) for loci x, y, z, and i, if they have two allosteric groups: one different for all the loci x, y, z, and i, and the other identical in all of them for attachment to the activator B.

On the other hand, the intragenome activator D activates (derepresses) gene loci (operons) n, s, e, r, each of which has an operator (receptor) reacting to locus D or linking with repressors (or inhibiting their synthesis) for loci n, s, e, and r.

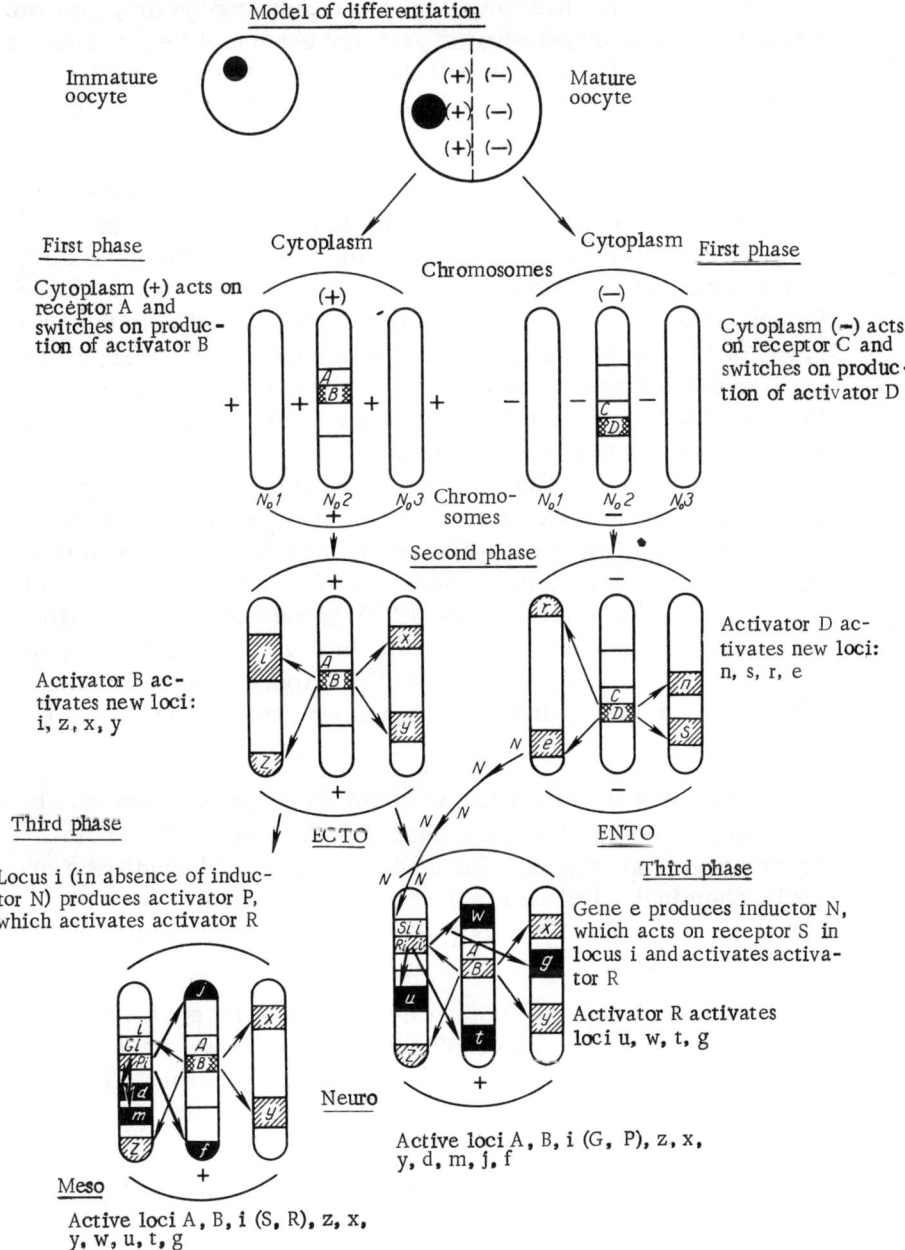

Fig. 107. General model of programmed differentiation of chromosomes during morphogenesis and activation of subprograms (explanation in text).

Cytoplasmic differentiation, the result of gene program No. 1 for the oocyte, thus activates primary genome differentiation in two groups of cells (we can call these differentiations ECTO and ENTO). In ENTO one group of loci (n, s, e, r) works, and in ECTO another group (x, y, z, i).

c) Subsequent development takes place in accordance with the same principles. Let us assume that the group of loci ENTO (n, s, e, r) contains a program according to which the cells of this group form different products, among them the substance N (formed, for example, by locus e). Let us further assume that in the ECTO group of loci (x, y, z, i) the locus i is a receptor-inducer, containing two gene groups SR and GP, one of which (SR) has a receptor locus S which, under the influences of substance N, activates the synthesis of a new intragenome activator R, derepressing new loci (operons) g, w, t, u. The other group (GR) contains a receptor group G, which activates the synthesis of intragenome activator P in the absence of external induction by substance N. Activator P derepresses genes d, m, j, and f. A further intragenome differentiation thus arises in the ECTO group of primary differentiation, and in some of the cells loci g, w, t, and u are activated (we shall call this the NEURO specialization), while in others loci d, m, j, and f are activated (the MESO specialization).

This process of activation of subprograms can continue further, and it is clear that such subprograms may contain both receptor-activator groups for new substances and conditions (physical, chemical, etc.) and loci whose activity is not concerned with any of the usual functions, but with the repression of previously functioning loci.

d) The group of loci activated by the intragenome activator (or activators if the receptor derepresses not one, but several activator genes) constitutes in this case a polyoperon. The following features combine it into a polyoperon (and distinguish it from other polyoperons).

First, the polyoperon consists of operons or loci, each of which has a derepression system reacting to the product or products of a strictly definite receptor-activator locus receiving information from outside.

Second, each polyoperon is represented in the external environment by one signaling compound or a globally specific group of compounds acting as an inductor or simply as an informer for neighboring systems of specialization.

Third, the activity of the sum of the genes of each polyoperon can organize the production of substances or stimuli with intragenome action, which inhibit other polyoperon systems (other subprograms), blocking their receptor-activator zones.

Fourth, each polyoperon of intermediate embryonic (temporary) specialization contains a free sensory receptor-activator locus, open to receive signals from outside and for further differentiation of that group of cells into new and different subprograms. The polyoperon of final specialization does not contain such a group, and dedifferentiation is required for reprogramming the cell genome in this case.

e) The work of the receptor-activator locus may, and in some cases must, be time-oriented. This may be organized as indicated, for example, in Fig. 106. Although this earlier model of mine is unsuitable as a model of morphogenesis, in some respects it can serve as a model for subprograms. In this case the receptor locus (S, for example) is the one which switches on the activator. However, activator R can be represented not by one gene, but by a group of sequentially arranged genes, participating in the synthesis of messenger RNA molecules (or activators) in a time sequence on the wave principle (displacement or widening of the zone of synthesis along the DNA molecule). In this case the operons of the polyoperon are switched on in a time sequence, under the influence of sequentially formed activators.

f) Once it has been started by the action of an inducer, even one whose action is of very short duration, the polyoperon subprogram must be executed, and during the period of its execution it must be isolated from constantly changing external signals. This can be achieved, for example, by the irreversibility (trigger mechanism) of the operation of switching on the receptor-activator locus. In each individual cell the possibility of a change of program can arise only after the messenger RNA molecules of, let us say, period No. 1 have become degraded on completion of their function, and new forms of messenger RNA may be "grafted" onto the ribosomes. The life span of the messenger RNA molecules thus serves as biological clock of morphogenetic stability for a variety of cells.

Precision of the morphogenetic process is evidently based on this system of very exact and rigid interactions and transitions from one state to another. Of course, real phenomena are much more complex and diverse, although, provided that everything is not reduced to a mass of unending interactions, they must be based on comparatively simple, yet capable of evolving, systems of interconnections.

It will be clear from everything that is written in the preceding chapters that an essential feature distinguishing the morphogenesis of higher multicellular organisms is the stepwise structure of differentiations, their special time-sequential hierarchy — an essential condition for the creation of higher complexity. Initially the embryo differentiates into a series of anlagen. For example, the neural anlage is activated, and it then differentiates into a series of subdifferentiations, each of which differentiates again. When the process has reached the stage of anlagen of definite organs, such as the eye, particularly delicate and far-reaching processes of final specialization, often irreversible, of groups of cells begin to take place in different directions. This stepwise sequence of specialization also reflects phylogenetic development. From the genetic point of view, it is associated with a continuous change in the spectrum of genes, accompanied by departure from the initial type and by divergence of the forms of differentiation from one another. Because of the nature of the process, accompanied by repression and derepression of the genes, we can speak of early repression, subsequent supplementary repression, triple repression, and so on.

In other words, repression may vary in its embryologic depth. For example, repression of the fetal hemoglobin system which takes place not long before birth is superficial repression of late type. In the same stem cells of the erythroid series ability to synthesize collagen or antibodies, for example, is repressed more and earlier. Syntheses characteristic of, for example, neural specialization possibly present in the genome of these cells are repressed still earlier (in the period of gastrulation) and more so.

It can thus be postulated that if a new, but intermediate subprogram contains the molecular conditions of repression (inactivation) of other subprograms, the new subprogram of narrower specialization which follows it superimposes on the previous repression new factors of inactivation and buries the residual potentialities of the genome ever deeper and more completely.

Repression of fetal hemoglobin synthesis (late repression) in cells of the erythroid series, as we have seen, is not absolute. However, it is nearly absolute in cells of a different specialization, in which it took place earlier. Although in both cases the same gene is repressed, it is undoubtedly repressed differently, by different repressors, by different methods. In cells of the erythroid series the repressor of the HbF gene is connected somehow with the genome activating synthesis of the β-chain of HbA; in leucocytes, for example, it acts in connection with activation of the subprogram of leucocytic specialization.

Every particular gene can thus be repressed selectively and individually through an associated receptor, and it can also, perhaps, be repressed additionally together with the polyoperon (through a polyoperon receptor), together with the complete subprogram, and so on. This overlapping, stratified repression insures the "purity" of operation of each subprogram, whereas in the case of individual late repression, or more especially in that of functional repression, it does not reach completeness.

The study of this stratification of intragenome morphogenetic correlations and interconnections during morphogenesis is exceptionally important not only from the standpoint of understanding its progressively broadening character, but also from that of analyzing the processes of involution in the definite organism which gradually destroy the remarkable perfection of its infinitely complex biological systems.

Literature Cited

Ames, B. N., and Hartman, P. E., 1963, Cold Spring Harbor Symp. Quant. Biol., 28:349.
Apter, M. J., and Wolpert, T. L., 1965, J. Theoret. Biol., 8:244.

Bonner, J., 1965, The Molecular Biology of Development, Oxford. Univ. Press, New York and Oxford.
Brenner, S., 1965, Brit. Med. Bull., 21:244.
Brenner, S., Stretton, A. O. W., and Kaplan, S., 1965, Nature, 206:994.
Ebert, J. D., and Kaighn, M. E., 1967, In: Major Problems in Developmental Biology, Academic Press, New York, p. 29.
Flickinger, R. A., 1963, Science, 141:608.
Grobstein, C., 1964, Science, 143:643.
Kroeger, H. J., 1963, J. Cell Comp. Physiol., 62, Suppl. I, No. 2:45.
Lopashov, G. V., 1965, In: Cell Differentiation and Induction Mechanisms, ed. G. V. Lopashov et al., Izd. Nauka, Moscow, p. 242.
Medvedev, Zh. A., 1963, Protein Biosynthesis and Problems in Ontogenesis, Medgiz, Moscow.
Medvedev, Zh. A., 1963a, Zh. Vses. Khim. Obshchestva im. D. I. Mendeleeva, 8:384.
Medvedev, Zh. A., 1964, Adv. Gerontol. Res., 1:181.
Monod, J., Wyman, J., and Changeux, J. P., 1965, J. Mol. Biol., 12:88.
Moscona, M. N., and Moscona, A. A., 1963, Science, 142:1070.
Neifakh, A. A., 1962, The Problem of Nucleo-Cytoplasmic Relationships in Development, A. N. Severtsev Inst. Animal Morphol., Moscow.
Neifakh, A. A., 1965, In: Cell Differentiation and Induction Mechanisms, ed. G. V. Lopashov et al., Izd. Nauka, Moscow, p. 38.
Ratner, V. A., 1966, Genetic Controlling Systems, Izd. Nauka Press, Siberian Branch, Novosibirsk.
Shapot, V. S., 1965, In: Biosynthesis of Proteins and Nucleic Acids, ed. A. S. Spirin, Series: Basis of Molecular Biology, Izd. Nauka, Moscow, p. 171.
Shmal'gauzen, I. I., 1964, Regulation of Morphogenesis in Individual Development, Izd. Nauka, Moscow.
Spirin, A. S., and Belitsina, N. V., 1965, Usp. Sovr. Biol., 59:187.
Stent, G., 1964, Science, 144:816.
Svetlov, P. G., 1960, In: Problems in Cytology and General Physiology, Izd. AN SSSR, Moscow-Leningrad.
Taylor, J. H., 1964, In: Symp. Internat. Soc. Cell Biol., Vol. 3. Cytogenetics of Cells in Culture, Academic Press, New York, p. 175.
Trifonova, A. N., 1963, Usp. Sovr. Biol., 56:381.
Waddington, C. H., 1962, New Patterns in Genetics and Development, Columbia Univ. Press, New York.

Chapter 9

Recent Advances in the Study of Molecular-Genetic Mechanisms of Development*

The preceding chapter have given a survey and theoretical analysis of the literature up to and including the first half of 1967. Preparation of the English edition of the book has presented the opportunity, not to be missed, of outlining briefly the most important advances made in the study of development in the last 18 months. In this additional chapter, just as in the rest of the book, the principle of evolutin has been followed in the presentation of the factual material and each section can be regarded as a short supplement to the corresponding chapter in the main text.

§ 1. New Material on the Mechanism of Transfer of Genetic Information

In the last two years the task of deciphering of the nucleotide—amino acid genetic code has been completed. The table of the genetic code in the original Russian edition of this book still included a number of undeciphered codons. In this edition the table has been brought up to date with the completed version of the code (Caskey et al., 1968) as given in the paper by Nirenberg et al. at a special symposium on the genetic code held at the Twelfth International Genetic Congress in Tokyo in August 1968.

---
*Supplement to the English Edition.

The final unraveling of the genetic code, however, has resulted in many new problems in connection with the study of the transcription and translation of genetic information. Some of these are closely connected with molecular mechanisms of differentiation and morphogenesis. A matter of particular interest is the study of the morphogenetic and tissue specificity of relationships between different codons and anticodons with identical meaning in messenger and transfer RNA molecules, i.e., the morphogenetic use of degeneracy of the code (Garen, 1968). Recent investigations have revealed many new details of the interaction of transfer RNA molecules with ribosomes and polysomes, the structure of ribosomes and polysomes, and the synthesis of messenger, transfer, and ribosomal forms of RNA (Spirin and Gavrilova, 1968). One of the most interesting achievements in this field has been the chemical synthesis of a gene by Khorana and his collaborators (Khorana, 1968).

§ 2. New Data Concerning Mechanisms of Morphogenesis of Viruses

The intensive accumulation of facts has continued during the study of molecular mechanisms of reproduction and morphogenesis of plant, animal, and bacterial viruses. The finer details of phenomena discovered previously have been filled in, and new viruses have been used as experimental objects.

In the study of RNA-viruses, attention has been concentrated on the action of virus-specific RNA-polymerase (replicase) (Spiegelman et al., 1967; Felix et al., 1967). The RNA-replicase of viruses MS2 and Q^{β} has been shown to possess the ability not only to recognize virus RNA, distinguishing it from cell forms of RNA, but also to react only with intact, whole, undamaged molecules of virus RNA, so that only a perfectly normal virus progeny is reproduced. New work has been carried out to determine in detail how polycistron RNA of viruses functions as a template for the synthesis of several proteins (Vinuela, Salas, and Ochoa, 1967; Sugiyama and Nakada, 1967, 1968). These investigations also disclosed an interesting case of selective repression. As soon as the RNA of virus MS2 begins to synthesize virus coat protein, the complex formed by this RNA with the coat protein acts at the same time as a repressor of the synthesis of all other virus-specific proteins. In this way reproduction of the virus passes perfectly and smoothly into its terminal stage. A similar repression mechanism has been formed in another RNA-phage, f2 (Zinder, 1968; Robertson et al., 1968). The morphogenesis of this phage consists of the successive synthesis of three proteins: replicase, particle assembly protein, and coat protein. As soon as the polycistron RNA template starts to synthesize envelope proteins, translating activity in the direction of replicase and particle assembly protein is sharply reduced under the influence of envelope protein.

The study of molecular mechanisms of morphogenesis of the more complex DNA phages has also continued to develop along lines established previously. New facts have been obtained concerning changes in the transfer RNA population for leucine during reproduction of phage T4 in bacteria. The use of more modern methods of fractionation has shown (Waters and Novelli, 1967; Kano-Sueoka et al., 1968; Kano-Sueoka and Sueoka, 1968) that after infection of *Escherichia coli* with phage T4 not two, but five different forms of leucine sRNA are found. Three of them appear at the beginning of infection and the other two at its end. It is not yet clear whether synthesis of these new forms of leucine RNA takes place on account of the genes of phage T4, or whether modification of transfer RNA molecules takes place in the host (*Escherichia coli*) cells.

Weiss and co-workers (1968) have shown by the method of RNA/DNA hybridization that part of the leucyl and prolyl sRNA formed in *E. coli* cells after infection with phage T4 is phage-specific. It may thus be assumed that these phages have genes for the synthesis of a number of phage-specific transfer RNA molecules which are the factors switching cell synthesis over to phage reproduction.

The study of the mechanism of regulation of time-sequential transcription of phage DNA (switching on the "early," "middle," and "late" genes) has continued to attract considerable interest. An important discovery here has been that transcription usually does not begin simply from any denatured site of the DNA molecule, but from special starting zones, usually distinguished by an increased number of adenine-thymine (AT) pairs. DNA-dependent RNA polymerase recognizes these areas as the place of onset of transcription in different sectors of the DNA molecule (Szybalski et al., 1966; Cohen et al., 1967).

Confirmation has been obtained of findings indicating that the synthesis of "late" forms of RNA requires preliminary replication of DNA, and the original DNA for the most part forms only "early" forms of messenger RNA (Zograf et al., 1967). However, early phage RNA continues to be formed in the late periods, although it is not translated and does not act as template for protein synthesis (Friesen et al., 1967). A full analysis of the various aspects of development of the T4 phages has been given by Bolle and co-workers (1968).

Some interesting facts have been discovered concerning the molecular biology of development of phage λ. Lysogenic forms of this phage can be reproduced for a long time synchronously with DNA of the bacterial chromosome in the form of p r o p h a g e. It has been shown (Ptashne, 1967; Echols et al., 1968) that lysogenic cells contain a r e p r e s s o r of this phage. This repressor has been isolated, partially purified, and tested *in vitro* . These workers postulate that this repressor inhibits DNA-dependent RNA synthesis. Conversely, a phage-specific protein i n d u c e r of DNA transcription of phage λ has also been found, with a very short half-life of less than 3 minutes (Konrad, 1968).

The genetic significance and molecular structure of the circular character of DNA found in many phages has been studied in detail (Krylov et al., 1967; Thomas, 1967; Ritchie et al., 1967).

The molecular biology of viruses has recently been the subject of a comprehensive analysis by Fenner (1968) in his monograph, and of a number of special symposia.

§ 3. New Data on the Molecular Mechanisms of Regulation of the Morphogenesis of Bacteria

The study of the molecular and genetic mechanisms of morphogenesis of bacteria in the last 18 months has followed previously established lines. The main result has been to fill in the details of the principles and phenomena which were examined in Chapter 3.

New confirmation has been obtained of the universality of the o p e r o n principle for biochemically connected groups of reactions, and in addition to the study of previously established operons, several new ones have been described, forming polycistronic messenger RNAs for the synthesis of several proteins at once (Giles et al., 1967). Of the previously known operons, as before the lactose operon of *E. coli* and the histidine operon of *Salmonella typhimurium* have received most attention. For example, the orientation of the nonsense codons UGA, UAG, and UAA, separating individual cistrons in the lactose operon, has been established by the method of intracodon recombination (Zipser, 1967), the temporal parameters of transcription and translation of individual cistrons of the operon have been measured precisely (Leive and Kollin, 1967), and many other details of the regulation of this operon have been discovered (Beckwith, 1967). Mechanisms of regulation of the lactose operon were discussed at a special symposium of the Twelfth International Genetic Congress.

It was stated in Chapter 3 that the tryptophan operon of *S. typhimurium* consists of five cistrons joined into two suboperons, transcribing independently. It has subsequently been shown (Bauerle and Margolin, 1967) that these two operons (trpA−trpB) (trpE−trpD−trpC) have a common o p e r a t o r , but independent promotor-like in i t i a t o r elements.

As might have been expected, the first successful attempt to isolate and identify the genetic repressor for the lactose operon (Chapter 3, p. 73) stimulated a whole series of new investigations. The repressor of the lactose operon has been more thoroughly purified and studied biochemically (Riggs and Bourgeois, 1968). These workers conclude from their observations that several nucleotides are attached to each molecule of the repressor. The large repressor molecule (mol. wt. 150,000) consists of subunits with molecular weights of 40,000 and 50,000. The highly purified repressor is bound specifically both with substrate and with DNA of *E. coli* (Riggs, Bourgeois, et al., 1968). Mutants of *E. coli* capable of forming lac-repressor in large quantities have been isolated (Müller-Hill et al., 1968).

However, formation of the specific repressor by the regulator gene is evidently only one of the possible methods of repression of transcription of operons. Other methods of repression of the structural genes in bacteria have recently been found (Richmond, 1967; Bretscher, 1968) and the repressor function of aminoacyl-transfer RNAs in repression of the enzymes of amino acid synthesis has been studied in more detail (Freundlich, 1967).

§ 4. Functional Activity of the Chromosomes in Morphogenesis and Differentiation

The problems examined in Chapter 4 have continued to undergo active development in all directions. More detailed dynamic structural models of chromosomes have been suggested (Mosolov, 1968). The pattern of DNA replication in organ cells changes in embryogenesis and morphogenesis (Hill and Yunis, 1967; Lima-de-Faria and Jaworska, 1968; Rodman, 1968). Many new investigations of the biochemical functional activity of polytene chromosomes in connection with differentiation and morphogenesis of insect larvae and of the action of various factors on chromosome puffs have been published (Ashburner, 1967; Lezzi, 1967a, b, c; Berendes, 1967, 1968; Edström and Daneholt, 1967; Goodman, Coidl, and Richart, 1967; Mattingly and Parker, 1968). However, these studies revealed nothing fundamentally new as regards the phenomena or mechanisms associated with chromosomal functions, and they have mainly served to fill in the details of the earlier picture.

A basic discovery in the field of biochemistry of chromosomes, or more correctly in the field of the biochemical structure of the genome of higher organisms, was the discovery that particular nucleotide sequences are repeated many times over in the DNA of chromosomes. This fact suggests that certain genes or gene loci of DNA in cells of animals and plants exist in hundreds and thousands of copies.

The phenomenon of repetition of nucleotide sequences in DNA was first described as long ago as 1964. However, information about this work was published only in the Carnegie Institution Year Book (Bolton et al., 1965), and for that reason it did not receive adequate attention. More extensive and detailed investigations of this phenomenon were published in 1967 and 1968 (Britten and Kohne, 1967, 1968a). The same phenomenon has been studied by somewhat different methods by Georgiev and co-workers (Anan'eva et al., 1968).

The occurrence of repetition of DNA segments carrying the same information many times over (conjectural of repeated genes or cistrons) was discovered by DNA/DNA molecular hybridization. When heated in solutions with low ionic strength, DNA dissociates (denatures), and its polynucleotides, which were joined together into a double helix, separate after rupture of the hydrogen bonds. If such a system is cooled slowly, renaturation takes place: the double helical structure is restored through random interactions between complementary segments. Renaturation of long polynucleotides into a double helix is usually not ideal, and numerous loops are formed. If, however, the very long DNA molecules are broken by mechanical shearing into comparatively short segments, the character of the denaturation and renaturation processes is the same but more regular complementary short double helices are formed. Naturally the rate of renaturation is directly proportional to the number of possible collisions (coincidences) of the complementary segments, and inversely proportional to the total volume of information contained in the DNA or the genome, and to the number of informationally unique segments into which this DNA can be dissociated. If, for example, renaturation in a system

$$\begin{matrix} A-B \\ A-B \end{matrix} \rightarrow {}_A B^B A \rightarrow \begin{matrix} A-B \\ A-B \end{matrix}$$

takes place at a velocity of x, the same process in the system

$$\begin{matrix} A-B \\ C-D \end{matrix} \rightarrow {}_A{}^B_D \; {}^B_C \rightarrow \begin{matrix} A-B \\ C-D \end{matrix}$$

will take place only half as fast, at a velocity of x/2, because complementary replicas will collide with each other in such a system only half as frequently. In a system

$$\begin{matrix} A-B \\ C-D \\ E-F \end{matrix} \rightarrow {}_A{}^D C{}^F_B \; {}_E \rightarrow \begin{matrix} A-B \\ C-D \\ E-F \end{matrix}$$

the renaturation process will be three times slower, its velocity will be x/3, and so on.

In the group of DNA-viruses and during the study of DNA of bacteria in experiments on renaturation of DNA cut into equal segments, the velocity of reassociation of the double helices corresponded strictly to the theoretical volume of the genome. For example, when the DNA of the small virus SV-40 was cut into ten pieces, and the DNA of the large T4 virus was cut into pieces of the same size (350 pieces), if the concentration of DNA in the solution was the same the process of reassociation of T4 DNA took about 35 times longer (until the stage when the double helical structure of half the DNA segments was restored).

Britten and Kohne naturally supposed that when they turned to similar experiments on higher animals, in which the nuclei contain much larger quantities of DNA and tens, hundreds, or even thousands of times more genetic information in the form of the nucleotide text, the process of reassociation of denatured DNA would take place the same number of times more slowly. However, this was not observed. In all higher multicellular organisms which they studied, their DNA contained a fraction (about 30-50% of the total DNA) which reassociated many times faster than bacterial DNA in the same concentration. This indicated that the DNA of higher organisms contains a large proportion of frequently repeated identical sequences and their complementary replicas. The probability of collisions for such pieces was proportional to the number of their repetitions in the DNA molecule. Some areas had a repetition quota of about 1 million, others 100,000, a third group about 1000, and so on. The other half of the DNA genome consisted of unique nucleotide sequences, i.e., individual genes (cistrons) not repeated over and over again.

By extrapolation of these figures to the genetic level, it can be assumed that for a particular tissue whose genome contains, let us say, 100,000 genes, 98,000 genes in the DNA of this tissue may be represented in a haploid set of chromosomes by individual genes (alleles), while 2000 genes are present as dozens, hundreds, or thousands of copies.

The individuality and uniqueness of genes as units of recombination have long been established in genetics. It could therefore be supposed that this repetitive part of the genome is what is known as the reserve DNA, not carrying genetic information. However, Georgiev and collaborators (Anan'eva et al., 1968) came to similar conclusions regarding the presence of repeated segments in the DNA in the genome of a number of animals as a result of their experiment on hybridization of DNA with natural messenger RNA isolated from nuclei of the same cells. They thus showed that the repetitive segments of the DNA molecule are as active as the nonrepetitive segments and they produce messenger RNA.

I shall not discuss here the possible evolutionary aspects of this phenomenon, which have been fully examined by Britten and

Kohne (1967; 1968a,b). From the point of view of morphogenesis and the genome structure responsible for this, in the discussion of these findings the idea suggests itself that frequent repetition of certain genes may be connected with cell specialization and may be morphogenetic in nature. It might be supposed that during differentiation, when the cell becomes specialized for synthesis of large quantities of certain proteins: hemoglobins, collagen, keratin, etc., or for the constant production and secretion of proteins or other products, one gene in the composition of the DNA is insufficient for such massive synthesis, and to increase the capability of the genome a special linear polytenization (polymerization) of certain genes corresponding quantitatively to the intensity with which these genes work in cells of this particular type takes place. Britten and Kohne (1968a) found, however, that repeated sequences are present not only in somatic tissues, but also in sperm cells of many species. Nevertheless, this does not contradict the hypothesis put forward above. Spermatogenesis and, in particular, oogenesis are also highly specialized processes, and during their course large quantities of surplus DNA are formed in the cells. The presence of the surplus DNA in reproductive cells led to the hypothesis of the existence of "master" genes, concerned with genetic recombination, and "slave" sequences of genes, not participating in recombinations, and of a possible structural basis of this subdivision (Callan, 1967).

To verify the morphogenetic hypothesis, the possibility of cross hybridization must be determined: are the repeated sequences in the genomes of somatic tissues tissue-specific, i.e., can a situation exist in which in the liver, for example, genes A B C are repeated frequently, while in the thyroid genes K N M are repeated, and are these genes connected with products of the function of these organs?

The suggestion that selective gene amplification may be the mechanism of specialization of a type in which cells must rapidly synthesize a particular product in large quantities was put forward by a group of speakers at a symposium on cell biology in November 1968 (Cahn, Solursh, and Lasher, 1968). These workers point out that DNA-DNA and DNA-RNA hybridization reveals a difference in this respect between differentiated cells, although in the abstract which they publish they give no precise details to illustrate their point.

Bonner and Widholm (1967) carried out interesting experiments on hybridization of chromosomal RNA from organs of the pea plant with nuclear pea RNA. They found that RNA from different organs is partially complementary to different segments of the nuclear DNA. This was to be expected. However, in their experiments these workers did not divide the DNA into repetitive and nonrepetitive sequences, so that their findings still cannot be accepted as proof of the hypothesis put forward above.

§ 5. New Facts Concerning Changes in Proteins and Nucleic Acids during Morphogenetic Processes

An exceptionally wide range of new facts has been accumulated in the last 18 months on the matters discussed in Chapter 5. Since it is impossible to examine these facts fully in a short supplementary chapter, I shall pick out only a few of the most interesting.

Genetic control of protein synthesis has been studied recently in many protein systems. Development has been particularly intensive in the genetics of enzymes and isoenzymes (see the surveys by Serov, 1968; Alimova, 1968; Latner and Skillen, 1968).

In the course of the study of genetic control of synthesis of the polypeptide chains of the hemoglobins and their polymorphism, the suggestion was made (von Ehrenstein, 1966; Rifkin et al., 1967; Popp, 1967) that one gene controls the synthesis, not of a single polypeptide, but of a family of polypeptides, as the result of specific "planned" mistakes at the translation level. However, the possibility is not ruled out that these facts of polymorphism would be more easily explained by the influence of a special mutant gene, linked with the structural gene.

Changes in the protein pattern during morphogenetic processes have also been studied in numerous protein systems. As before, considerable attention has been paid to the changeover from the synthesis of embryonic and fetal hemoglobins to the synthesis of adult hemoglobins (Kovach et al., 1967; Fantoni et al., 1968; Dan and Hagiwara, 1967; Hall and Motulsky, 1968; Moss and Ingram, 1968).

Embryonic and postembryonic morphogenetic changes in the activity and spectrum of the following enzymes and isoenzyme systems have been investigated: lactate dehydrogenase (Genis-Galvez and Maisel, 1967; Auerbach and Brinster, 1967), alkaline phosphatase (Moog and Grey, 1967), isocitrate dehydrogenase (Mintz and Baker, 1967), desoxyribonuclease and DNA-polymerase (Muhammed et al., 1967), tyrosine-aminotransferase (Holt and Oliver, 1968), and many other enzymes. The morphogenetic polymorphism of lactate dehydrogenase has been shown to be controlled by a special regulator gene (Shows and Ruddle, 1968).

Changes in the spectrum of ribonucleic acids during morphogenesis and differentiation have also been studied intensively in recent times. Work in this field has been concerned with determination of changes in the composition of ribosomal RNA (Nemer and Infante, 1967), mRNA-polysome complexes (Infante and Nemer, 1967, 1968) in early sea urchin embryos, the characteristics of different types of RNA in the embryonic organs of various animals (Bresnick et al., 1967; Wicks, 1967), and other matters.

Two lines of research must be picked out as particularly interesting: the study of cases of the provision of stable RNA templates when specialization is particularly high (formation of the crystalline lens) and morphogenetic changes in the pattern of transfer RNAs.

Specialization of protein syntheses during formation of the lens goes further and deeper than during the formation of anuclear erythrocytes. Ultimately in this case synthesis is predominantly confined to the crystallins (α, β, γ), mainly of the γ-fraction. The dynamics of synthesis of these proteins has been studied along with the synthesis of special stable long-living forms of RNA (Papaconstantinou, 1967; Stewart and Papaconstantinou, 1967; Reader and Bell, 1967). These investigations showed that two types of RNA can be formed for the same proteins (crystallins): short-living — at a time when the epithelial cells, although specializing, have not yet lost their ability to divide, and stable, long-living RNA, as soon as cell division stops.

Different crystallins are formed at different phases of specialization. Formation of γ-crystallin begins before the cells have lost their ability to divide; α-crystallin is formed only in the concluding stages of specialization (McDevitt, 1968).

The study of changes in the pattern of transfer RNAs before long will unquestionably occupy an important place in the molecular biology of morphogenesis. Because of degeneracy of the genetic code, several types of transfer sRNAs with different anticodons exist or may exist for the same amino acid. The appearance of a new variant of sRNA may signify the possibility of translation of new forms of messenger RNA which have not been translated by transfer sRNAs hitherto in existence. In Chapter 3 I pointed out that a change in the composition of lysyl- and valyl-sRNAs is observed in bacteria during sporulation. Changes in the lysyl-sRNA pattern have also been found during differentiation of wheat shoots (Vold and Sypherd, 1968). It is interesting to note that the same sort of change in the ratio between two different forms of lysyl-sRNA has been found during early embryogenesis of the sea urchin (Yang and Comb, 1968). Changes in the transfer RNA pattern have also been found during morphogenetic processes in erythrocytes during embryogenesis of the chicken (Lee and Ingram, 1967).

On the question of biochemical determination, during oogenesis, of the processes of early embryogenesis the following new facts should be noted. First, the earlier hypothesis that the surplus DNA of the oocyte nuclei may contain extrachromosomal replicas of several genes has been confirmed experimentally. This extrachromosomal DNA has been shown to code the formation of large reserves of ribosomal RNA (Brown and Dawid, 1967).

During oogenesis large reserves of messenger RNA are also formed. They remain in an inactive state and start to perform their function of protein synthesis only after fertilization, in the period of early embryogenesis (Crippa et al., 1967; Humphreys, 1968). This preformed messenger RNA is kept in an inactive state in special capsules which also contain free reserve ribosomes. The nature of these capsules has been thoroughly studied (Stavy and Gross, 1967). Qualitatively new forms of messenger RNA and qualitatively new proteins (activation of new genes) do not begin to appear until the blastula period (Davidson et al., 1968).

§ 6. Chromosomal Proteins as Genetic Regulators

All aspects of the biochemistry of nuclear proteins have

been fully covered in recent surveys (Hnilica, 1967; Bonner, Dahmus, et al., 1968).

Histones and nonhistone (acid) nuclear proteins have continued to attract attention, mainly from the standpoint of their possible role in the regulation of genetic activity of the nucleus and chromosome, i.e., in the phenomena of selective repression and de-repression.

Attempts to demonstrate tissue and morphogenetic specificity of the histone as before have yielded very contradictory results. Comings (1967) compared the electrophoretic spectrum of histones and the ratio between lysine-rich, lysine-moderate, and arginine-rich histones in genetically active and inert chromatins (comparison of euchromatin with heterochromatin, metaphase cells with interphase, adult with dividing lymphocytes, after treatment with phytohemagglutinin). No appreciable differences in histone composition were found during these comparisons. By cytochemical methods Gorovsky and Woodard (1967) showed that the DNA/histone ratio is approximately the same in the puffs and condensed segments of chromosome No. 2 of the salivary glands of *Drosophila virilis*. No tissue differences were found in the composition of electrophoretically separable histone fractions in the embryonic tissues of the chicken (brain, liver, skin) (Kischer and Hnilica, 1967).

On the other hand, slight tissue differences in the ratio between fractions of β-histones isolated from chicken erythrocytes, liver, and spleen were observed by Bellair and Mauritzen (1967). Gutierrez and Hnilica (1967) studied phosphorylation of various histone fractions in rat tissues (by determining incorporation of P^{32}-labeled phosphates into histones) and observed definite fraction and tissue specificity in histone phosphorylation activity.

The hypothesis that it is not the presence or absence of histones, but their chemical modification (acetylation, methylation, phosphorylation) which determines the repressive or active (derepressed) state of DNA has often been put forward. Phosphorylation of histones takes place through the action of the enzyme histone phosphokinase after synthesis of the histone protein molecule (with the formation of phosphoserine). Phosphorylation is most perceptible in the lysine-rich fraction. Conversely, acetylation of histones, determined by incorporation of labeled acetate, takes place most actively in arginine-rich histone, and a relationship exists between the level of acetylation of histones and the ability of chromatin to synthesize nuclear RNA (Pogo et al., 1966; Allfrey et al., 1968; Pogo et al., 1968). The findings of Ellgaard (1967), which conflict with this view, were not confirmed by Allfrey and co-workers (1968) with the use of a more meticulous procedure.

Observations revealing changes in the pattern of histone fractions obtained by electrophoresis on polyacrylamide gel were made during the study of a morphogenetic process taking place in plants, the formation of lily pollen (Sheridan and Stern, 1967).

The ability of histones to repress the process of DNA transcription was confirmed in model experiments to study the effect of histone−DNA interaction on RNA synthesis (Bonner et al., 1968; Paul and Gilmour, 1968). A messenger RNA with features of specificity in relation to histone synthesis has been isolated (Borun et al., 1967). In addition, experiments have shown that histones can stimulate DNA synthesis (Holoubek and Hnilica, 1967) and that a definite correlation exists between histone synthesis and DNA synthesis in the cell cycle (Robins and Borun, 1967). However, the asynchronism of synthesis (during division of the nucleus), both along the chromosomes and between different chromosomes, characteristic of chromosomal DNA replication does not extend to the histones and other nuclear proteins (Cave, 1967). It is interesting to note that in mature frog erythrocytes, which have virtually ceased to synthesize RNA and hemoglobin, incorporation of leucine-C^{14} into the histones of the nucleus and into the remaining protein of the nucleus is observed (Medvedev et al., 1970).

As well as the study of histones, work on the examination of the role of nonhistone proteins of the nucleus in the regulation of its functions have continued to develop (Himes, 1967; Wang, 1967; Holoubek and Crocker, 1968). Interest has revived in the possible regulatory function of DNA-bound amino acids and peptides (Salser and Balis, 1967).

§ 7. Induction and Repression of Genetic Loci During Differentiation

From the molecular-genetic point of view, differentiation in the period of embryonic development is a regular change, occurring at a certain time and depending on localization, between processes of induction (de-repression) of certain groups of genes and inactivation (repression) of other groups of genes. It was pointed out in Chapter 7 that processes of repression and de-repression differ in character at different periods of ontogenesis. In the period of early cleavage of the oocytes and primary differentiation, the principal role in regulation of the character of genetic activity of the nucleus is played by the cytoplasm of the oocyte and the fine differentiation takes place in it in accordance with the program of oogenesis. In the period of early embryonic development and appearance of the organ anlagen, interaction between tissues and between cells and various forms of induction begin to take over the leading role. In late embryogenesis and in postembryonic development the role of specific genetic activators and repressors is transferred to hormones. Repression and derepression, apart from morphogenesis, also acquire the character of a constant or periodic function.

The presence of products predetermining the character of primary differentiation or cleavage in the cytoplasm of the oocyte has naturally led to attempts to isolate and identify these products. The first attempts of this nature were made in 1967 (Gurdon, 1967; Hörstadius et al., 1967). The first of these investigations showed that the factor inducing DNA synthesis in the gametes in connection with fertilization or transplantation of the nuclei is liberated through the action of pituitary hormones. In the second investigation, this factor was partially purified chromatographically.

The presence of factors in the cell cytoplasm capable of causing de-repression of the inactive nucleus was confirmed by the work of Harris (1967), showing that after transplantation of totally repressed hen erythrocyte nuclei into the cytoplasm of HeLa cells, RNA synthesis was resumed in the erythrocyte nuclei.

The present state of the problem of embryonic induction in connection with cell and tissue interaction in the period of active morphogenesis has been analyzed in several surveys (Gustafson and Wolpert, 1967; Ignat'eva, 1967; Lopashov, 1968). Increasingly, research in the field of embryonic morphogenesis is moving toward the study of the role of physico-chemical and chemical factors of tissue interaction during differentiation processes (Loewenstein, 1967; Pfohl and Giudice, 1967; Eagle and Levine, 1967; Toivonen and Saxen, 1968; Lopashov and Khoperskaya, 1967). New data have been obtained to indicate that special forms of RNA may play a role in repression and de-repression processes in embryogenesis (Moscona et al., 1968). The problem of interaction between genes and hormones has been studied particularly intensively in the last 18 months, with emphasis on the role of hormones as specific repressors and derepressors of particular genetic programs, and on the function of hormones at the level of translation of genetic information.

Additional material has been obtained concerning the action of estrogen (Hamilton et al., 1968), hydrocortisone and cortisone (Arbuzova et al., 1968; Kidson, 1967), glucagon (Greengard and Dewey, 1967), pituitary growth hormone (Sells and Takahashi, 1967), testosterone (Liao and Lin, 1967), and other hormones at the genetic level (on de-repression and synthesis of new messenger RNA). In some cases hormones act as specific inhibitors of synthesis of RNA and proteins (Peck et al., 1967). Hormones apparently cannot act directly on gene DNA, but interact with repressors within the genome. Cortisone has its effect principally on activity of RNA-polymerase, and thereby stimulates the synthesis of messenger RNA (Lukács and Sekeris, 1967; Turner et al., 1967). The action of hormones sometimes may be to initiate and stimulate the formation of ribosomes and other components of the system of protein synthesis, and thus to enable the growth effect of induction to take place (Tata, 1967, 1968).

Finally, turning to the molecular regulation of periodic functional phenomena (biological clocks), it may be mentioned that the link between biorhythms and the synthesis of proteins and RNA has been confirmed in a number of investigations (Driessche, 1966; Feldman, 1967; Tschudy et al., 1967).

Literature Cited

Alimova, M. M., 1968, Genetika, 4(8):165.
Allfrey, V. G., Pogo, B. G. T., Littau, V. C., Gershey, and Mirsky, A. E., 1968, Science, 159:314.
Anan'eva, L. N., Kozlov, Yu. V., Ryskov, A. P., and Georgiev, G. P., 1968, Molekul. Biol., 2:736.
Arbuznova, G. S., Gryaznova, I. M., Morozova, T. M., and Salganik, R. I., 1968, Molekul. Biol., 2:308.
Ashburner, M., 1967, Chromosoma (Berlin), 21:398.

LITERATURE CITED

Auerbach, S., and Brinster, R. L., 1968, Exptl. Cell Res., 46:89.
Bauerle, R. H., and Margolin, P., 1967, J. Mol. Biol., 26:423.
Beckwith, J. R., 1967, Science, 156:597.
Bellair, J. T., and Mauritzen, C. M., 1967, Biochim. Biophys. Acta, 133:263.
Berendes, H. D., 1967, Chromosoma (Berlin), 22:274.
Berendes, H. D., 1968, Chromosoma (Berlin), 24:418.
Bolle, A., Epstein, R. H., Salser, W., and Geiduschek, E. P., 1968, J. Mol. Biol., 31:325.
Bolton, E. T., Britten, R. J., Cowie, D. B., Roberts, R. B., Szafrasky, P., and Waring, M. J., 1965, In: Carnegie Institution Year Book 64, Annual Report of the Director of Dept. of Terrestrial Magnetism 1964-1965, Washington, D. C., p. 316.
Bonner, J., Dahmus, M. E., Fambrough, D., Huang, R. C., Marushige, K., and Tuan, D. Y., 1968, Science, 159:47.
Borun, T. W., Scharff, M. D., and Robbins, E., 1967, Proc. Nat. Acad. Sci., USA, 58:1977.
Bresnick, E., Eckles, S. G., and Lanclos, K. D., 1967, Biochemistry, 6:2481.
Bretscher, M. S., 1968, Nature, 217:509.
Britten, R. J., and Kohne, D. E., 1967, In: Carnegie Institution Year Book 65, Annual Report of the Director of Dept. of Terrestrial Magnetism, 1965-1966, Washington, D. C., p. 78.
Britten, R. J., and Kohne, D. E., 1968b, Science, 161:529.
Brown, D. D., and Dawid, J. B., 1967, Science, 160:272.
Cahn, R. D., Solursh, M., and Lasher, R., 1968, J. Cell Biol., Vol. 39, Abstr. 20a.
Callan, H. G., 1967, J. Cell Sci., 2:1.
Caskey, T., Wilcox, M., Anderson, F., Scolnick, E., Tompkins, R., and Nirenberg, M., 1968, Proc. XII Internat. Congress of Genetics, Vol. II, Tokyo, p. 37.
Cave, M. D., 1967, Exptl. Cell Res., 45:631.
Cohen, S. N., Maitra, U., and Hurwitz, J., 1967, J. Mol. Biol., 26:19.
Comings, D. E., 1967, J. Cell Biol., 35:699
Crippa, M., Davidson, E. H., and Mirsky, A. E., 1968, Proc. Nat. Acad. Sci., USA, 57:885.
Dan, M., and Hagiwara, A., 1967, Exptl. Cell Res., 46:596.
Davidson, E. H., Crippa, M., and Mirsky, A. E., 1968, Proc. Nat. Acad. Sci., USA, 60:152.
Driessche, T. V., 1967, Biochim. Biophys. Acta, 126:456.
Eagle, H., and Levine, E. M., 1967, Nature, Vol. 213.
Echols, H., Pilarski, L., and Cheng, P. Y., 1968, Proc. Nat. Acad. Sci., USA, 59:1016.
Edström, J. E., and Daneholt, B., 1967, J. Mol. Biol., 28:331.
von Ehrenstein, G., 1966, J. Cell Physiol., 67 (Suppl. I):47.
Ellgaard, E. G., 1967, Science, 157:1070.
Fantoni, A., de la Chapelle, A., Fifkind, R. A., and Marks, P. A., 1968, J. Mol. Biol., 33:79.
Feix, G., Schneider, M. C., Weissmann, C., and Ochoa, S., 1967, Science, 157:701.
Feldman, J. F., 1967, Proc. Nat. Acad. Sci., USA, 57:1080.
Fenner, F., 1968, The Biology of Animal Viruses, Vol. I, Molecular and Cellular Biology, Academic Press, New York.

Freundlich, M., 1967, Science, 157:823.
Friesen, J. D., Dale, B., and Bode, W., 1967, J. Mol. Biol., 28:413.
Garen, A., 1968, Science, 160:149.
Genis-Galvez, J. M., and Maisel, J., 1967, Nature, 213:283.
Giles, N. H., Case, M. E., Partridge, C. W. H., and Ahmed, S. J., 1967, Proc. Nat. Acad. Sci., USA, 58:1453.
Goodman, R. M., Coidl, J., and Richart, R. M., 1968, Proc. Nat. Acad. Sci., USA, 58:553.
Gorovsky, M. A., and Woodard, J., 1967, J. Cell Biol., 33:723.
Greengard, O., and Dewey, H. K., 1967, J. Biol. Chem., 242:2986.
Gurdon, J. B., 1967, Proc. Nat. Acad. Sci., USA, 58:545.
Gustafson, T., and Wolpert, L., 1967, Biol. Rev., 42:442.
Gutierez, R. M., and Hnilica, L. S., 1967, Science, 157:1324.
Hall, J. C., and Motulsky, A. G., 1968, Nature (London), 217:569.
Hamilton, T. H., Teng, Ch. S., and Means, A., 1968, Proc. Nat. Acad. Sci., USA, 59:1265.
Harris, H., 1967, J. Cell Sci., 2:23.
Hill, R. N., and Yunis, J. J., 1967, Science, 155:1120.
Himes, M., 1967, J. Cell Biol., 34:77.
Hnilica, L. S., 1967, In: Progress in Nucleic Acids Research and Molecular Biology, Vol. 7, eds. J. N. Davidson and W. E. Cohn, Academic Press, New York, p. 25.
Holoubek, V., and Crocker, T. T., 1968, Biochim. Biophys. Acta, 157:352.
Holoubek, V., and Hnilica, L. S., 1967, J. Nat. Cancer Inst., 39:187.
Holt, P. G., and Oliver, J. T., 1968, Biochem. J., 108:333.
Hörstadius, S., Hosefsson, L., and Runnstrom, J., 1967, Develop. Biol., 16:189.
Humphreys, T., 1968, J. Cell Biol., Vol. 39, Abstract 63a.
Ignat'eva, G. M., 1967, In: Embryology, ed. E. M. Vermel', Series "Itogi Nauki," Izd. Nauka, Moscow.
Infante, A. A., and Nemer, M., 1967, Proc. Nat. Acad. Sci., USA, 58:681.
Infante, A. A., and Nemer, M., 1968, J. Mol. Biol., 32:543.
Kano-Sueoka, T., Nirenberg, M., and Sueoka, M., 1968, J. Mol. Biol., 35:1.
Kano-Sueoka, T., and Sueoka, M., 1968, J. Mol. Biol., 37:475.
Kidson, C., 1967, Nature, 213:779.
Kischer, C., and Hnilica, L. S., 1967, Exp. Cell Res., 48:424.
Khorana, H. G., 1969, In: Cold Spring Harbor Symp. Quant. Biol. (in press).
Konrad, M. W., 1968, Proc. Nat. Acad. Sci., USA, 59:171.
Kovach, J. S., Marks, P. A., Russel, E. S., and Epler, H., 1967, J. Mol. Biol., 25:131.
Krylov, V. N., Alikhanyan, S. I., and Morozova, E. S., 1967, Genetika, No 5:128.
Latner, A. L., and Skillen, A. W., 1968, Isoenzymes in Biology and Medicine, Academic Press, New York.
Lee, J. C., and Ingram, V. M., 1967, Science, 158:1330.
Leive, L., and Kollin, V., 1967, J. Mol. Biol., 24:247.
Lezzi, M., 1967a, Chromosoma (Berlin), 21:72.
Lezzi, M., 1967b, Chromosoma (Berlin), 21:89
Lezzi, M., 1967c, Chromosoma (Berlin), 21:109.
Liao, Sh., and Lin, A. H., 1967, Proc. Nat. Acad. Sci., USA, 57:379.

LITERATURE CITED

Lima-de-Faria, A., and Jaworska, H., 1968, Nature, 217:138.
Loewenstein, W. R., 1967, Develop. Biol., 15:503.
Lopashov, G. V., 1968, What Lies at the Basis of Development of the Organism, Izd. Znanie, Moscow.
Lopashov, G. V., and Khoperskaya, O. A., 1967, Dokl. Akad. Nauk SSSR, 175:962.
Lukács, J., and Sekeris, C. E., 1967, Biochim. Biophys. Acta, 134:85.
McDevitt, D. S., 1968, J. Cell Biol., Vol. 39, Abstract 88a.
Mattingly, E., and Parker, C., 1968, Chromosoma (Berlin), 23:255.
Medvedev, Zh. A., Medvedeva, M. N., and Chaban, I. A., 1970, In: The Cell Nucleus and Its Ultrastructure, Proceedings of a Symposium, Izd. AN SSSR (in press).
Mintz, B., and Baker, W. W., 1967, Proc. Nat. Acad. Sci., USA, 58:592.
Moog, F., and Grey, R. D., 1967, J. Cell Biol., 32(2.CI.).
Moscona, A. A., Moscona, M. N., and Saenz, N., 1968, Proc. Nat. Acad. Sci., USA, 61:160.
Mosolov, A. N, 1968, Genetika, 4(12):135.
Moss, B., and Ingram, V. M., 1968, J. Mol. Biol., 32:493.
Muhammed, A., Conclaves, J. M., and Trosko, J. E., 1967, Develop. Biol., 15:23.
Müller-Hill, B., Crapo, L., and Gilbert, W., 1968, Proc. Nat. Acad. Sci., USA, 59:1259.
Nemer, M., and Infante, A. A., 1967, J. Mol. Biol., 27:73.
Papaconstantinou, J., 1967, Science, 156:338.
Paul, J., and Gilmour, R. S., 1968, J. Mol. Biol., 34:305.
Peck, W. A., Brand, T., and Miller, J., 1967, Proc. Nat. Acad. Sci., USA, 57:1599.
Pfohl, R. J., and Giudice, G., 1967, Biochim. Biophys. Acta, 142:263.
Pogo, B. G. T., Allfrey, V. G., and Mirsky, A. E., 1966, Proc. Nat. Acad. Sci., USA, 55:805.
Pogo, B. G. T., Pogo, A. O., Allfrey, V. G., and Mirsky, A. E., 1968, Proc. Nat. Acad. Sci., USA, 59:1337.
Popp, R. A., 1967, J. Mol. Biol., 27:9.
Ptashne, M., 1967, Nature, 214:232.
Richmond, M. H., 1967, Nature, 216:1191.
Reeder, R., and Bell, E., 1967, J. Mol. Biol., 23:577.
Rifkin, D. B., Hirsch, D. J., Rifkin, M. P., and Konigsberg, W., 1967, Cold Spring Harbor Symp. Quant. Biol., 31:715.
Riggs, A. D., and Bourgeois, S., 1968, J. Mol. Biol., 34:361.
Riggs, A. D., Bourgeois, S., Newby, R., and Cohn, M., 1968, J. Mol. Biol., 34:365.
Ritchie, D. A., Thomas, C. A., McHattie, L. A., and Wensink, P. C., 1967, J. Mol. Biol., 23:365.
Robbins, E., and Borun, T. W., 1967, Proc. Nat. Acad. Sci., USA, 57:409.
Robertson, H., Webster, R. E., and Zinder, N. D., 1968, Nature (London), 218:533.
Rodman, T. C., 1968, Chromosoma (Berlin), 23:271.
Salser, J. S., and Balis, M. E., 1967, Biochim. Biophys. Acta, 149:220.
Sells, B. H., and Takahashi, T., 1967, Biochim. Biophys. Acta, 134:69.
Serov, O. L., 1968, Genetika, 4(10):134.
Sheridan, W. F., and Stern, H., 1967, Exp. Cell Res., 45:323.

Spiegelman, S., Haruna, J., Pace, N. R., Mills, D. R., Bishop, D. H. L., Claybrook, J. R., and Peterson, R., 1967, Cell Physiol., Suppl. I to Vol. 70:35.
Spirin, A. S., and Gavrilova, L. P., 1968, The Ribosome, Izd. AN SSSR.
Stavy, L., and Gross, P. R., 1967, Proc. Nat. Acad. Sci., USA, 57:735.
Stewart, J. A., and Papaconstantinou, J., 1967, J. Mol. Biol., 29:357.
Sugiyama, T., and Nakada, D., 1967, Proc. Nat. Acad. Sci., USA, 57:1744.
Sugiyama, T., and Nakada, D., 1968, Proc. XII Intern. Congress of Genetics, Tokyo, 1:38.
Szybalski, W., Kobinski, H., and Scheldrick, P., 1966, Cold Spring Harbor Symp. Quant. Biol., Vol. 123.
Tata, J. R., 1967, Biochem. J., 104:1.
Tata, J. R., 1968, Nature, 219:331.
Terner, J. R., Goodman, R. M., and Spiro, D., 1967, Exp. Cell Res., 45:550.
Thomas, C. A., 1967, J. Cell Physiol., Suppl. I. to Vol. 79 (2):13.
Toivonen, S., and Saxen, L., 1968, Science, 159:539.
Tschudy, D., Waxman, A., and Collins, A., 1967, Proc. Nat. Acad. Sci., USA, 58:1944.
Vinuela, E., Salas, M., and Ochoa, S., 1957, Proc. Nat. Acad. Sci., USA, 57:729.
Vold, B. S., and Sypherd, P. S., 1968, Proc. Nat. Acad. Sci., USA, 59:453.
Wang, T. Y., 1967, J. Biol. Chem., 242:1220.
Waters, L. C., and Novelli, G. D., 1967, Proc. Nat. Acad. Sci., USA, 57:979.
Weiss, S. B., Hsu, W. T., Foft, J. W., and Scherberg, N. H., 1968, Proc. Nat. Acad. Sci, USA, 41:114.
Wicks, W. D., 1967, Arch. Biochem. Biophys., 121:55.
Yang, S. S., and Comb, D. G., 1968, J. Mol. Biol., 31:139.
Zinder, N. D., 1968, Proc. XII Internat Congress of Genetics, Tokyo, 2:34.
Zipser, D., 1967, Science, 157:1176.
Zograf, Yu. M., Nikiforov, V. G., and Shemyakin, M. F., 1967, Molekul. Biol., 1:94.

Index

Acetylcholinesterase 224, 226
Actin 193
Actinomycin D 31, 38-39, 165-166, 177, 239, 244, 322-324, 335-336
Actomyosin 193, 195, 317
Acyladenylate 10-11
Adaptation 345
Adapter molecules 11
Adenine 11, 167
Adenosine 12
Adenosine triphosphate (ATP) 10-11, 194-195, 346
Adenosine triphosphatase 193-195
Adrenalectomized rats 323-325
Adrenocarcinoma 306
Adrenocortisone 232
Alanine dehydrogenase 120
Albumin 197-199
Alcaligenes faecalis 107
Algae 79
Allosteric effect 101, 350
Allosteric groups 11, 59
Amblystoma mexicanum 308
Amino acid 6, 187
    activating enzymes 10-11
    activation 10-14, 220
    analogs 58
    dictionary 18
    genetic code *see* Genetic code
    residue 4-5
    synthesis 97
Ammonia 231
Amphibians 176

Amylase 226
    pancreatic 190
    salivary 190
Anemia 208-217
    high HBF syndrome 208-217
    sickle cell 185
    thalassemia 208-214
Animal viruses *see* Viruses
Antibiotics 165-166
    see individual entries
Anticodon 7-8, 12-13, 34
Antigen 36
Anthranilate synthease 91
Arabinose operon 92
Arginase 223
Arginine operon 98
    synthesis 97
Aspartate transaminase 226
Aspartate transcarbamylase 120
Autoreplication 3, 28, 38, 42, 44, 48, 52, 67, 350
Axolotl 308, 314, 333
Azotobacter vinelandii 121

Bacillus cereus, 118
Bacillus subtilis 70, 105, 112, 114, 118-120
Bacteria 2, 4, 7, 9, 79-125
    chromosomes 81, 104-121, 295, 327
    enzymes 82, 87, 119
    E. coli   *Escherichia coli*
    F-factors 113

Bacteria (Continued)
  genes 81-103, 260, 356-360
  genetic loci 96-99
  genetic map 97, 104-121
  genome 327-328
  membrane 112-113
  morphogenesis 80-87, 356-365
  operon 83-103
  reproduction 80, 117
  spore formation 80, 118-120
  substrate induction 315
Bacteriophage 3-5, 23-24, 36-46, 79-80
  assembly 71-72
  circular gene 62-63
  evolution 41-42
  hybridization 62
  replication 44
  synchronized infection 39
  β 37
  f2 37, 41
  fr 37
  λ 54, 60, 62, 68, 107, 394
  MS2 37-40
  ϕR 46
  1ϕ7 46
  R17 37
  S13 46
  SP8 69
  T-even 53-54, 57, 60-62, 69, 353
  T1 62
  T2 57, 59-62, 65, 68-71
  T4 51, 53, 59-63, 68, 71, 393
  T6r$^+$ 58
  T7 68
Balbiani's rings 167
Bar mutation 328
Base pairs 3-4
Bee embryo 136
Beta-rays 148
Biological clocks 335-337, 406
Blastomere 303
Blastula 233, 304-307
Bond
  covalent 44
  hydrogen 3
  intermonomer 3

Bond (Continued)
  internucleotide 3
  peptide 11
Bonner's hypothesis 336-372
5-Bromouracil 66, 111

Carp 229
  see Loach
Cell 2, 5, 184, 346
Centromere 170
Chicken 194, 199-200, 221, 229, 309, 316
Chinese hamster *see* Hamster
Chironomus dorsalis 158-160
Chironomus pallidivittatus 163
Chironomus tentans 155-157, 160-161, 168-169
Chloramphenicol 53, 56, 165-166, 335-336
Chloroplast 9
Cholinesterase 226
Chordomesoblast 304
Chromatid 130, 136-137, 154, 169, 178
Chromatin 9, 134, 137, 176, 274-279
Chromomere 170-172
Chromoneme 131
Chromosome 2, 4, 7-8, 42, 48, 81, 104-121, 129-139, 259-290, 295-296, 327, 330-331, 396
  bacterial 5, 81, 104-121, 295, 327
  giant 154-169
  growth point 111
  heteropycnotic 151
  lampbrush type 169-178, 298
  loops 169-178
  metacentric 130
  polytene 154-170
  puffs 153-169, 347
  replication 139
  X 150-152, 192, 248-249
  XX 150-151
  XY 150
  Y 150-152, 178, 248-249
Circadian rhythm 335-337
Cistron 30-37, 41-49, 54-56, 59-60, 63, 79-80, 83, 88, 127-129, 349, 352

INDEX

Clocks, biological 335-337
Cockroach 199
Code *see* Genetic code
Codon 4-8, 12, 18-19, 42, 61, 352
Colinearity 83, 105
Collagen 196-197, 316
Collagenase 226
Configuration 6
Conjugation, interrupted 97, 105
Controller genes 329-332
Corn 156, 329-332
Cortisone 323, 324, 406
Creatine kinase 230
Crepis 146
Cybernetics 373-376
Cycloheximide 166
Cysteine 99
Cytoplasm 6, 10, 14, 31-32, 36, 48, 80, 174, 296-303
Cytosine 60

Darwin's theory 345
Dehydrofolate reductase 53
Desoxycytidine pyrophosphatase 53
Desoxycytidine triphosphatase 58
Desoxycytidyl deaminase 53
Desoxycytidylate hydroxymethylase 53
Desoxycytosine 53
Desoxy-5-hydroxymethylcytidine-5-phosphate 53
Desoxyribonuclease 53, 136, 173, 221, 247
Desoxyribonucleic acid (DNA) 1-5, 79, 245-249, 261-290, 308, 396-400
  asynchrony 146-153
  breaks 68-69
  circular 45, 49
  denaturation 65-66
  double helix 3-5, 8-9, 49, 52, 65-68, 71, 105, 108, 116
  inducing factor 295-296
  labeled 66-68
  melting 55
  molecular weight 3-4
  nonsense strand 115

Desoxyribonucleic acid (Continued)
  Pauling-Corey model 3
  polymerase 3, 5, 60
  primer 5
  recombinase 67
  reference strand 115
  renaturation 65-66
  RNA hybrids 45, 48, 55, 70, 118, 399
  RNA interaction 115
  replication 4, 108, 140-144
  spiralization 139
  structure 3
  synthesis 3-5, 308, 396-400
  thermal conjugation 55
  viral 23
  Watson-Crick model 3-4, 108-109, 176
Differentiation, programmed 365-389
DOPA oxidase 192
Drosophila 146-148, 154-155, 161-166, 177-178, 192

Ecdysone 164-166
Ectoderm 312-313
Elastin 197
Embryogenesis 219, 233-241, 296-337, 351, 404-406
Embryonic induction 309-324
Endomitosis 154
Energy 10-11
Enzymes 2, 10, 82-83, 87, 218-232, 324, 346
  in Eskimos *see* Eskimo
  isozymes *see* Isozymes
Epimerase 87
Episome 108-111
Erythrocyte 185, 191, 203-207, 216-217, 347
Erythropoietin 216-217
Escherichia coli 9, 13, 16, 26, 28, 37-40, 45, 58-59, 70-71, 80, 83-87, 91-97, 110-121, 277, 346, 393, 395
  K-12, 98, 103
Eskimo 296
Estrogen 322, 406
Euglena gracilis 41

Euplotes edrystomus 144-145
Evolution 245
Eye
   amphibian 318
   anlage (optic cup) 313, 318-320
   primitive 306, 313-314

Feedback 296-298
Feedback inhibition 101
Feulgen reaction 170, 173
Fibroblast 318
5-Fluorouracil 60
Frog 195, 198-199, 207, 231, 244, 247, 303-308
Fructose-1,6-diphosphatase 223

Galactokinase 87
Galactose operon 87, 103
β-Galactosidase 83-87, 96, 103
Galactosidase permease 84
Galactosidase transacetylase 85
Gastrula 304, 312
Genes 2-3, 185, 347, 348,
   see Cistron
   controllers 329-332
   induction of activity 295-337
   see Operon
   regulator 83, 208-215, 297
   repressor 83-84
   structural 6, 88
Genetic clock 334-337
   code 4, 12, 18-20, 59, 391-392
   code degeneracy 7, 11, 18-19, 27, 59, 96
   control 185, 363-365
   loci 6-7, 63, 184, 295
   map 62-65, 69
   memory 2
   see Transcription
   see Translation
Genome 8, 48, 53, 60, 345
Giant chromosome see Chromosome
γ-Globulin 197-198
Glucagon 406

Glucokinase 224
Glucose-6-phosphatase 222
Glucose-6-phosphate dehydrogenase 151, 191, 222
Glucosidase 119
Glucosyltransferase 53
β-Glucuronidase 192, 332
Glutamate-alanine transaminase 224, 324
Glutamate dehydrogenase 226
Gonyaulax polyedra 335
Growth 169
Guinea pig 222

Hamster, Chinese 146, 150-152
Haptoglobulin 192
He-La cells 32, 34, 48, 146, 309
Hemoglobin 17-18, 33, 96, 184-189, 200-217, 332, 343-344
   abnormal 185-188, 208-214
   fetal 201-206, 332
   fractions 186-187
Hemolymph 165
Heteropycnosis 151
High HBF syndrome 208-217
Histidine 58, 88, 95, 101
   biosynthesis 89-91
   operon 88-91, 94, 100
Histidinol dehydrogenase 96
Histone 139, 142, 177, 260-290, 325, 403-404
Hormone 164, 231, 320-326, 406
   see respective hormone
Hydrocortisone 223, 324-325, 406
Hydroxyamino acid 136
Hydroxylamine 135-136
p-Hydroxyphenylpyruvic acid oxidase 223, 225

Immunofluorescence 203-205
Immunology 232
Induction
   of activity 295-337
   chemical 315-317
   embryonic 309-320

INDEX

Induction (Continued)
  heterotypic 320
  homotypic 320
Information
  carriers 6
  genetic 2-4, 6, 42
Informosomes 238, 327
Insulin 325
Intercistron punctuation 41
Invertase 119
Irradiation 234, 236-238
Isoleucine operon 92
Isozymes 189-190, 218, 226-232

Jacob-Monod model 82, 100

Keratin 197
Keratinization 197
Kinase 53

Labeling
  continuous 149
  pulse 149, 175
Lactate dehydrogenase 189-190, 226-232
Lactate isozymes 189-190, 226-232
Lactose operon 83-96, 214, 395
Lampbrush chromosome see Chromosome
Lamprey 199
Larva 164
Leucine operon 92
Leucylaminopeptidase 226
Life 1, 218
Liver enzymes 324
Loach (carp) 234, 236, 239
Lopashov's hypothesis 373-376
Luminescence 335-336
Lysozyme 40, 54, 58, 60

Macromolecules 1-3, 19-20, 42
Magnesium ions 8, 10
Malate dehydrogenase 230
Manganese ions 8

Medvedev's hypothesis 379-389
Meiosis 148
Meromyosin 193
Meselson-Stahl experiment 109
Mesosome 112
Metamorphosis 164, 231-232
Metaphase 154
Microautoradiography 149, 167
Mitochondria 9, 14, 194
Mitomycin C 38, 165-166
Mitosis 5, 130-132, 139-140, 148-149
Modulator 35
Modulator tripler 96
Monochromosome 154
Morphogenesis 1-2, 5, 11, 365-389
  see Bacteria
Mouse 228, 232
Mosaic oocytes see Oocyte
Multicellular organism see Organisms, multicellular
Muscle 318
Mutation 18, 60-64, 88, 163-164, 332
Mycoplasma gallisepticum 79
Mycoplasma laidlawii 79
Myoblast 318
Myofibril 193-194
Myogen 193
Myosin 193-195

Neural plate 312
  tube 315
Neuroepithelium 312
Neurula 306
Nitrate reductase 220
Nucleic acids see DNA, RNA
Nucleohistone 265-266
Nucleolus 6-7, 14, 36, 133
Nucleoprotamine 262
Nucleoside triphosphate 5
Nucleotides 3-4, 11-12, 80, 128
Nucleus 6, 14, 30-31, 48, 127-133, 149, 184, 233-235, 296-298
  juice 6
  membrane 133
  transplantation 303-309

Ontogenesis 24
Oocytes 169-178, 193, 233-241, 298-303
Oogenesis 176, 233-241, 298-303, 327
Ooplasmic differentiation 298-303
Ooplasmic segregation 298-303
Operon 42, 83-103, 210-212, 296, 349-350, 394-395
Operator 85-87, 100
Organisms, multicellular 2, 7, 127-178
Ornithine transcarbamylase 120
Ovulation 170, 174

Pactamycin 166
Pea seedling 9
Periodicity *see* Rhythm
Phages *see* Bacteriophage
Phosphophenylalanine 175
Phosphatases (acid and alkaline) 102, 119-120, 190-191, 219, 226, 230, 232
Phosphoenolpyruvate carboxylase 223
Phosphoglucose isomerase 222
Phosphoglucomutase 222
Phosphoribosyl pyrophosphate 88
Phosphoserine 135
Physarium polycephalum 107
Pituitary hormone 326
Placenta 191
Plant viruses *see* Viruses
Plastid 14
Plaut-Nash hypothesis 148
Polycistron 93-95
Polydesoxyribonucleotides 3
Polyserine 177
Polymerase 3, 5, 8, 31, 111, 166
Polynucleotides 3-8, 12-13, 18
Polyoperon 349
Polypeptide 4-8, 16-18, 187
Polyribonucleotide 18
Polyribosomes 2, 14-18, 32-33, 48
Polysomes *see* Polyribosomes
Polytene chromosome *see* Chromosome
Polyteny 154-155, 170
Proactinomyosin 195
Promoter 86
Prophase 154

Prosomyosin 194
Protamine 260-290
Protease 136, 239
Protein 1-4, 184-185, 192, 349
  chromosomal 259-290
  connective tissue 196-197
  conveyor principle 17-18
  inhibition 224
  plasma 197-200
  synthesis 8-20, 400-401
Protoplast 44
Protozoa 79
Pyrophosphorylase 96
Pulse labeling 149
Pupa 164-165
Puromycin 166, 335-336

Rabbit 195, 229
Rat 229, 323
Rays *see* Beta-rays
Receptor 297, 310, 315, 326
Recombinase 67
Regression 98-99
Releaser 99
Replication, semiconservative 3
Replicator 111-112
Replicon 108, 112, 113, 334
Repressor 53-54, 59, 85-88, 99-103, 266-267, 273-287, 395
Reticulocytes 33, 184
Retina 318
Rhynochosciara angelae 157
Rhythm 334-337
Ribonuclease 136, 167, 169-172, 221
Ribonucleic acid
  acceptor RNA *see* transfer RNA
  aminoacyl RNA 59
  alanine sRNA 14-15
  complementary RNA 32, 36, 38
  double helical RNA 31
  high polymer RNA 14
  labeled 55
  messenger RNA 5-18, 32-33, 42, 48, 52-56, 59, 61, 86, 89, 92-94, 144, 167, 233-249, 327

INDEX

Ribonucleic acid (Continued)
   polymerase 8, 31-41, 45, 54, 59, 274-275, 325
   primer 8, 31, 38
   replicase 38, 392
   ribosomal RNA 6-7, 10, 15, 53
   soluble RNA *see* tRNA
   synthease *see* polymerase
   synthesis 5-10, 116, 233-249, 323, 401-402
   transfer RNA (tRNA) 2, 6-7, 10-14, 17, 34, 242
   valine sRNA 14
   virus RNA 23
Ribonucleoprotein 24-30
Ribonucleotide triphosphate 8, 38
Ribosome 2, 6, 10, 14-18, 31-36, 41, 48, 61, 194, 239, 346
   30S 15
   50S 15
   70S 15
   80S 15
   dimer 15-16
   monomer 15-16
   polyribosomes *see* Polyribosomes
   tetramer 16
Rodent 199

Saccharomyces 87, 119
Salamander 207, 229
Salivary gland 146-147, 154, 155, 162-163, 167, 169
Salmonella typhimurium 88-92, 395
Sarcoplasm 196
Sea urchin 239-240
Serine 135-136
Serum albumin 197-198
Sex determinant 150
Sex hormones 320-322
Sheep 227
Sickle cell anemia *see* Anemia
Silkworm 308-309
Snail 263
Specificity 6-8
Speed modulator 35
Sperm 135

Spermatocytes 178
Spermatozoon 127
Styela 299-300
Suboperon 92
Succinate dehydrogenase 226
Synchronized growth 149

Tadpole *see* Frog
Telomere 170
Template 3, 6-13, 30-38, 41-42, 45, 52, 56, 59
Testosterone 323, 406
Thalassemia 208-214
Thiogalactoside transacetylase
Thymidine 148-149
Thymidylate synthease 53
Thymine 53, 58, 167
Thymus 267-268, 281, 290
Thyroxin 207, 231-232, 325-326
Tissue 184
   specificity 270-273
Tradescantia 132, 146
Transacetylase 96
Transcription 5-10, 42, 60, 89, 92-93, 95, 115-117, 326
Transferrin 192
Translation 6, 10-11, 18, 41-42, 59-60, 87, 326
Triplet 4-8, 12, 18, 34, 52, 59
   see Codon
   nonsense 8
Trisomy 151, 215
Tritium 148
Triton *see Triturus*
Triturus 131-132, 169-170, 174-177, 303
Trout 127, 207
Tryptophan operon 91-100, 395
   pyrollase 225, 325
   synthease 18, 83, 105
Tumor cell 306
Tyrosinase 226
Tyrosine 223
Tyrosine-α-ketoglutarate transaminase 325

Twins 191

UDP-glucose-pyrophosphorylase 223
UDR-glucose-glycogen transglucosylase 223
Uracil 58, 167
Uridine 167-168, 174-175
Uridyl transferase 87
Uterus 322

Valine operon 92
Valyl sRNA 102
Vicia faba 146
Viruses 2, 5, 7, 18, 23-77, 80, 351-356
   adeno 47-49
   barley striped mosaic 28
   broad bean mottle 28
   brome-grass 28
   DNA viruses 353, 393
   encephalomyocarditis 31
   evolution 41
   mengo 31
   morphogenesis 24-53
      of animal virus 30-36, 46-49
      of bacterial virus 23-24, 36-46
      of large virus 49-53
      of plant virus 23-30

Viruses (Continued)
   of small virus 37-38, 41-46
   mouse leukemia 31
   ontogenesis 24, 27-28
   ϕX174 42-46, 62, 70
   polio 31-34
   polyoma 49-50
   pseudorabies 47-49
   RNA viruses 392-394
   reo 31
   sendai 36
   smallpox 47
   tobacco mosaic 18, 24-28, 31
   tobacco necrosis 25-26
   turnip yellow mosaic 30
   vaccinia 47-48, 79
   wound tumor 28

Xenopus laevis 176, 305-306

Yeast 14

Zuckerkandl's hypothesis 210-215